International Max Planck Research School (IMPRS)
for Maritime Affairs
at the University of Hamburg

For further volumes:
http://www.springer.com/series/6888

Hamburg Studies on Maritime Affairs
Volume 29

Edited by

Jürgen Basedow
Monika Breuch-Moritz
Peter Ehlers
Hartmut Graßl
Tatiana Ilyina
Florian Jeßberger
Lars Kaleschke
Hans-Joachim Koch
Robert Koch
Doris König
Rainer Lagoni
Gerhard Lammel
Ulrich Magnus
Peter Mankowski
Stefan Oeter
Marian Paschke
Thomas Pohlmann
Uwe Schneider
Detlef Stammer
Jürgen Sündermann
Rüdiger Wolfrum
Wilfried Zahel

Jan Albers

Responsibility and Liability in the Context of Transboundary Movements of Hazardous Wastes by Sea

Existing Rules and the 1999 Liability Protocol to the Basel Convention

 Springer

Jan Albers
Hamburg
Germany

Dissertation zur Erlangung der Doktorwürde an der Fakultät für Rechtswissenschaft der Universität Hamburg
Vorgelegt von Jan Albers
Erstgutachter: Prof. Dr. Rainer Lagoni, LL.M. (Columbia)
Zweitgutachter: Prof. Dr. Dr. h.c. Peter Ehlers
Tag der mündlichen Prüfung: 29.01.2014

ISSN 1614-2462   ISSN 1867-9587   (electronic)
ISBN 978-3-662-43348-5   ISBN 978-3-662-43349-2   (eBook)
DOI 10.1007/978-3-662-43349-2

Library of Congress Control Number: 2014944723

Springer Heidelberg New York Dordrecht London

© Springer-Verlag Berlin Heidelberg 2015
This work is subject to copyright. All rights are reserved by the Publisher, whether the whole or part of the material is concerned, specifically the rights of translation, reprinting, reuse of illustrations, recitation, broadcasting, reproduction on microfilms or in any other physical way, and transmission or information storage and retrieval, electronic adaptation, computer software, or by similar or dissimilar methodology now known or hereafter developed. Exempted from this legal reservation are brief excerpts in connection with reviews or scholarly analysis or material supplied specifically for the purpose of being entered and executed on a computer system, for exclusive use by the purchaser of the work. Duplication of this publication or parts thereof is permitted only under the provisions of the Copyright Law of the Publisher's location, in its current version, and permission for use must always be obtained from Springer. Permissions for use may be obtained through RightsLink at the Copyright Clearance Center. Violations are liable to prosecution under the respective Copyright Law.
The use of general descriptive names, registered names, trademarks, service marks, etc. in this publication does not imply, even in the absence of a specific statement, that such names are exempt from the relevant protective laws and regulations and therefore free for general use.
While the advice and information in this book are believed to be true and accurate at the date of publication, neither the authors nor the editors nor the publisher can accept any legal responsibility for any errors or omissions that may be made. The publisher makes no warranty, express or implied, with respect to the material contained herein.

Printed on acid-free paper

Springer is part of Springer Science+Business Media (www.springer.com)

Realism should not imply resignation.[1]

---

[1] English translation of: "Realismus darf nicht zu Resignation führen." Kunig, *Reform der Charta der Vereinten Nationen*, in: Albrecht (ed.) (1998), at 156.

Preface

This study was accepted as a doctoral dissertation by the University of Hamburg in summer 2013. The topic originated from a suggestion of my doctoral advisor *Professor Dr. Rainer Lagoni*, former Director of the Institute of the Law of the Sea and of Maritime Law at the University of Hamburg. For his inspiring advice, critical feedback and constant support throughout the entire time of my research I am most grateful and deeply indebted to him. I would further like to thank *Professor Dr. Dr. h.c. Peter Ehlers* for his second opinion on my dissertation and his very valuable comments.

The research and writing of this study was conducted in large parts during my time as a scholar at the International Max Planck Research School for Maritime Affairs in Hamburg, which did not only grant a generous scholarship, but also provided me with a grant for the publication of this book. The interdisciplinary exchange of ideas and opinions among the directors, scholars, and friends of the Research School provided the fertile soil also for the completion of this study.

Moreover, this study would not have been possible without the invaluable advice and constant support of some of my best friends and colleagues. I feel obliged to all of them, but only a few can be mentioned here: I would like to thank *Thomas Wanckel* (partner at Segelken & Suchopar) for many inspiring discussions, *Barbara Schröder* for her coordinating efforts at the Research School, *Michael Friedman* for his invaluable efforts in decoding and proofreading the text as well as *Bärbel*, *Dan-Claas*, *Johannes*, *Julian*, *Mišo*, *Sara*, *Shermineh*, and *Verena*.

Special thanks are due to my mother *Helga* and my brother *Arne*, whose constant support and encouragement have made this book possible.

My deepest gratitude goes to *Nadja*. She supported me most and suffered most. Without her backing and untiring patience, I would have lacked the strength to make this book a reality.

This book is written in remembrance of my father *Dieter*.

Hamburg, March 2014   Jan Albers

Contents

Chapter 1: Introduction 1
  A. The Factual Perspective: Transboundary Movements
      of Hazardous Wastes by Sea 2
  B. The Legal Perspective: Existing Rules and the 1999 Liability
      Protocol to the Basel Convention 5
  C. The Structure of This Book 8

**Chapter 2: The International Trade in Hazardous Wastes
and Its Economic Background** 11
  A. Hazardous Wastes: Properties and Economic Importance 11
      I.   Sources and Composition of Hazardous Wastes 11
      II.   Volumes of Generated Hazardous Wastes 13
      III. Forms of Waste Treatment and Disposal 15
      IV. The Commercial Value of Hazardous Wastes 17
      V.   Summary 19
  B. The Transboundary Movement of Wastes 19
      I.   Reasons for the Emergence of Hazardous
          Waste Movements 19
      II.   Quantities and Typical Patterns of Hazardous
          Wastes Movements 21
      III. The Involvement of Waste Brokers and Waste Dealers 23
      IV. Illegal Traffic and Shipments Off the Official Path 24
      V.   Summary 25
  C. The Interests Involved 26
      I.   Private Parties 26
      II.   State Interests 28
      III. Hazardous Waste Trade as "Environmental Racism"? 31
      IV. Summary 32
  D. The Risk Potential of Hazardous Waste Movements 33

Chapter 3: The Present Legal Framework				35
A.	The Legal Concept of International Responsibility			35
	I.	The Principle of State Responsibility for Internationally Wrongful Acts		37
	II.	Is There a Need for Autonomous Rules on State Liability for Lawful but Injurious Activities?		41
		1. "False Cases of Liability Sine Delicto"		42
		(a)	The General Approach	42
			(aa) Obligations of Conduct and Obligations of Result	43
			(bb) Distinction Between Legal Acts and Factual Activities	44
			(cc) Implications for the Application of State Responsibility	44
		(b)	Further Enhancements of this Approach	46
			(aa) Equation of Activities and Its Consequences	46
			(bb) A Comprehensive Obligation to Prevent Damage	47
		(c)	Summary	49
		2. "True Cases of Liability Sine Delicto"		49
		(a)	Explicit Rules of State Liability	50
		(b)	Recognition as a General Principle of International Law?	51
		(c)	Implications of the Non-existence of a General Principle of State Liability	52
		3. The Relationship Between Both Concepts		54
	III.	The Importance of Civil Liability Conventions		55
B.	The Contribution of the International Law Commission			56
	I.	State Responsibility		57
	II.	State Liability		59
C.	State Responsibility in the Context of Transboundary Movements of Hazardous Wastes by Sea			63
	I.	Explicit Provisions of State Responsibility in International Treaty Law		64
		1. The Basel Convention		64
		(a)	Article 12	65
		(b)	Article 8	65
		(c)	Article 9	66
		(d)	Summary	67
		2. Other Conventions Relevant to the Trade in and Transport of Hazardous Wastes		68

	3.	The Law of the Sea Convention (UNCLOS)	70
	4.	Other Conventions Relevant to the Protection of the Marine Environment	72
	5.	Summary.	73
II.	The Customary Principle of State Responsibility		74
	1.	Act of the State	75

(a) State Organs and Persons Empowered by the State 76
(b) Persons in Fact Acting for the State 78
(c) Conduct Ultra Vires 78
(d) Interstate Attribution 79
(e) Persons Solely Acting in Private Capacity 80
(f) Summary 81

2. Breach of an International Obligation 81
   (a) Requirements in General 81
       (aa) Relevance of the Respective Primary Obligation 82
       (bb) A General Requirement of Damage? 85
       (cc) A General Requirement of Fault? 86
       (dd) Due Diligence as the Relevant Standard of Behaviour....................... 88
   (b) Obligations Arising from International Conventions............................ 90
       (aa) The Basel Convention 90
           (1) Background and Basic Legal Features of the Convention. 91
           (2) Obligations Imposed on States........ 95
               (i) General Obligation: Minimisation of Generation and Transportation of Hazardous Wastes 95
               (ii) First Tier: Absolute Trade Restrictions. 97
               (iii) Second Tier: Environmentally Sound Management 98
               (iv) Third Tier: Prior Informed Consent Principle 99
               (v) Further Obligations Imposed by the Basel Convention 103
           (3) The Application of the Basel Convention to End-of-Life Ships 104
               (i) The Dismantling of End-of-Life Ships 104

		(ii)	The Definition of "Waste"	105
		(iii)	Simultaneity of Ship and Waste?	109
	(4)	Summary		110
(bb)	Other Conventions and Regulations Relevant to the Trade in and Transport of Hazardous Wastes			111
	(1)	The Bamako Convention		111
		(i)	Background of the Convention	111
		(ii)	The Obligations Imposed by the Convention	113
		(iii)	Summary	115
	(2)	The Waigani Convention		116
	(3)	The Cotonou Agreement		117
	(4)	OECD Council Decisions		117
	(5)	European Union Legislation		119
(cc)	The Law of the Sea Convention			122
	(1)	Background and Basic Legal Features of the Convention		122
	(2)	Obligations Imposed on States		124
		(i)	The Obligation to Protect and Preserve the Marine Environment, Articles 192, 194(1)	124
		(ii)	The Obligation to Prevent Transboundary Pollution to the Marine Environment, Article 194(2)	127
		(iii)	Further Obligations	130
	(3)	Summary		131
(dd)	Further Conventions Relevant to the Protection of the Marine Environment			131
	(1)	MARPOL 73/78 Convention		131
	(2)	OPRC Convention		134
	(3)	London Dumping Convention		135
	(4)	The UNEP Regional Seas Programme		136
		(i)	Regional Seas Conventions	136
		(ii)	The 1996 Izmir Protocol to the Barcelona Convention	138
		(iii)	The 1998 Tehran Protocol to the Kuwait Convention	140

					(ee)	Conventions Relevant to the Trade and Transport of Hazardous Substances	141

- (ee) Conventions Relevant to the Trade and Transport of Hazardous Substances 141
  - (1) Rotterdam PIC-Convention 141
  - (2) Stockholm POPs Convention 142
  - (3) Hong Kong Convention 143
  - (4) Agreements and Conventions Concerning the Safe Transport and Handling of Dangerous Goods.... 144
- (c) Obligations Arising from General International Law 145
  - (aa) The Obligation not to Cause Significant Harm to the Environment of Another State's Territory 145
    - (1) Recognition in International Law 145
      - (i) The Origin of this Obligation ... 146
      - (ii) Exception from the Prevalence of Territorial Integrity in Case of Insignificant Harm 148
    - (2) The Content of this Obligation 149
      - (i) Geographical Scope 149
      - (ii) Prohibitive and Preventive Elements 150
      - (iii) Legal Character 151
    - (3) The Burden of Proof 151
    - (4) Application to Transboundary Movements of Hazardous Wastes by Sea 152
  - (bb) The Obligation to Protect and Preserve the Marine Environment 153
- (d) Circumstances Precluding Wrongfulness 154
- (e) Summary 155
3. Legal Consequences 157
  - (a) Continuation of the Primary Obligation 158
  - (b) Reparation 159
    - (aa) Causal Link 159
    - (bb) Forms of Reparation 162
  - (c) Invocation by the Injured State 164
  - (d) Serious Breaches of Peremptory Obligations 165
4. Jurisdictional Issues 166

III. Summary: State Responsibility 166

D.	Existing Civil Liability Conventions.		169
	I. 1999 Protocol to the Basel Convention.		169
		1. Evolution of the Protocol	169
		2. Legal Objectives and Main Content	172
	II. 1996/2010 HNS Convention		174
		1. The Evolution of the 1996 HNS Convention and Its Main Content	174
		2. Perceived Deficiencies and Obstacles to Ratification	176
		(a) Contribution of Packaged Cargo to the HNS Fund	177
		(b) Contributions of the LNG Account	178
		(c) Non-submission of Contributing Cargo Reports	179
		(d) Conclusion.	179
		3. The 2010 Protocol to the HNS Convention and Further Development	181
	III. Further Civil Liability Conventions		183
		1. Liability for Oil Pollution from Ships.	184
		2. Liability for Nuclear Damage	185
		3. Liability for the Carriage of Dangerous Goods by Land	185
		4. Other Civil Liability Conventions	186
	IV. Regional Civil Liability Regulations.		187
		1. 1993 Lugano Convention	187
		2. EU Environmental Liability Directive.	187
	V. The LLMC Convention.		188
	VI. Summary.		189
E.	Relationship Between Civil Liability Conventions and the Principle of State Responsibility.		190
F.	Summary: Responsibility and Liability *de lege lata*		191

Chapter 4: Attempting an Interim Conclusion: Preconditions for an Effective Legal Regime on Liability and Compensation 193

A.	Necessity of a Regime of Liability and Compensation	193
	I. Insufficiency of Non-financially Oriented Treaty Compliance Mechanisms.	193
	II. Liability Rules as a Remedy for Environmental Damage	196
B.	The Appropriate Form of Liability.	197
	I. National, Regional or Global Approach?.	198
	II. State Liability or Civil Liability?.	199
C.	Limitations Set by Other Areas of Law	201
D.	Summary	201

Chapter 5: The 1999 Basel Protocol on Liability and Compensation 203

A. The Regulatory Content of the Basel Protocol 203
    I. Scope of Application 203
        1. Transboundary Movements of Hazardous Wastes and Other Wastes 204
            (a) Wastes Subject to the Basel Protocol 204
                (aa) Hazardous Wastes 204
                    (1) Autonomous Definition 204
                    (2) National Definitions 205
                (bb) "Other Wastes" 206
                (cc) Wastes Excluded from the Scope of the Protocol 207
            (b) Transboundary Movements and Other Activities Covered by the Protocol 209
                (aa) Transboundary Movement 209
                (bb) Other Activities 210
        2. Incidents Covered by the Basel Protocol 211
            (a) Terminological Scope of "Incidents" 211
            (b) Geographical and Temporal Coverage of Incidents 212
        3. Damage Covered by the Protocol 214
            (a) Terminological Scope of "Damage" 214
                (aa) Personal Damage and Damage to Property 214
                (bb) Loss of Income 215
                (cc) Measures of Reinstatement and Preventive Measures 216
                    (1) Definitions 216
                    (2) Distinction from Salvage Remuneration 218
                (dd) Purely Ecological Damage 221
            (g) Geographical Scope of Covered Damage 222
        4. Summary 223
    II. Relationship to Other Civil Liability Regimes 225
        1. Requirements in General 225
            (a) Article 3(7) of the Protocol 225
                (aa) Legal Requirements 225
                (bb) Objections Raised Against This Provision ... 228
            (b) Article 11 of the Protocol 230
                (aa) Difficulties Related to the Dual Coverage of Movements 230

			(bb)	Meaning of "Portion of a Transboundary Movement"................................	231
			(cc)	Insufficiency of the Formal Criterion.......	232
	2.	Relationship to Single Civil Liability Instruments	234
		(a)	Relationship to the HNS Convention...........	234
		(b)	Relationship to the CRTD Convention	236
		(c)	Relationship to the Civil Liability Convention	237
		(d)	Relationship to Other Regimes of Liability and Compensation in the Field of Transboundary Movements of Hazardous Wastes..............	238
		(e)	Relationship to the OECD and EU Regulations ...	238
		(f)	Relationship to Regimes of Liability for Cargo Damage	239
		(g)	Relationship to the LLMC Convention	239
	3.	Summary..	241
III.	The Liability Regime of the Basel Protocol..............	243
	1.	The Basic Concept of Liability of the Basel Protocol...	244
		(a)	The Concept of Combining Strict and Fault-Based Liability....................	244
		(b)	No Subsidiary Liability of the State	245
		(c)	Common but Differentiated Civil Liabilities?.....	246
	2.	Strict Liability According to Article 4	248
		(a)	The Approach of Channelling Strict Liability According to Spheres of Responsibility.........	248
		(b)	Allocation of Strict Liability Under the Basel Protocol........................	251
			(aa)	The Regulation in Detail	251
			(bb)	The Temporal Break When Liability Shifts to the Disposer..................	254
			(cc)	Non-establishment of a General Secondary Liability of the Generator...............	255
			(dd)	Liability of the Generator for Aftercare Operations	257
		(c)	Exceptions to Strict Liability	258
			(aa)	Article 4(5)........................	258
			(bb)	Article 6(2)........................	261
	3.	Fault-Based Liability According to Article 5.........	261
	4.	The Requirement of a Causal Link................	263
		(a)	Differences Between Strict and Fault-Based Liability....................	263
		(b)	Combined Causes of Damage, Article 7	264
	5.	Contributory Fault, Article 9	265
	6.	Right of Recourse, Article 8	265
	7.	Summary......................................	267

	IV.	The Regime of Limitation of Liability	270
		1. Financial Limitation of Liability	270
		(a) Potential Conflicts with the LLMC Convention	271
		(b) The Legal Arrangement of Limitation of Liability	273
		(c) Practical Implications	275
		(d) Procedural Issues	277
		(e) Revision of the Financial Limits Established by Annex B	277
		(f) Summary	278
		2. Temporal Limitation of Liability	279
	V.	Further Financial Instruments Implemented by the Basel Protocol	281
		1. Compulsory Insurance or Similar Guarantees	281
		2. The Non-establishment of a Compensation Fund	282
		(a) The Need for an Additional Financial Instrument	282
		(b) The Utilisation of the Technical Cooperation Trust Fund	284
		(c) Interim Solution Insufficient Only upon Entry into Force of the Basel Protocol	286
	VI.	Rules of Procedures	290
		1. Rules on Competent Courts	290
		2. Rules on the Applicable Law	290
		3. Mutual Recognition and Enforcement of Judgments	291
	VII.	Overview of the Protocol's Final Clauses	292
B.	Excursus: The Cases of the M/V "Khian Sea" and the M/V "Probo Koala"		293
C.	An Assessment of the Basel Protocol		295
	I.	Summary of the Major Achievements and the Major Defects of the Basel Protocol	295
	II.	Reasons for the Protocol's not Entering into Force	299
		1. Lack of Political Incentive to Ratify the Basel Protocol	299
		2. Shortcomings of the Protocol and Obstacles for Implementation	301
		3. Summary	302
	III.	Consequences of the Protocol's not Entering into Force	303

Chapter 6: Concluding Summary 305

Appendix I: Text of the Basel Convention (Excerpts) 311

Appendix II: Text of the Basel Protocol. 319

Bibliography 335

Table of Cases. 357

**Table of International Conventions and Agreements, OECD,
EU and Other Legal Instruments.** 361

**About the International Max Planck Research School
for Maritime Affairs at the University of Hamburg** 369

Abbreviations

ADN	European Agreement Concerning the International Carriage of Dangerous Goods by Inland Waterways
ADR	European Agreement Concerning the International Carriage of Dangerous Goods by Road
AHWG	Ad Hoc Working Group of Legal and Technical Experts to Consider and Develop a Draft Protocol on Liability and Compensation for Damage Resulting from Transboundary Movements of Hazardous Wastes and their Disposal
ACP Group	African, Caribbean, and Pacific Group of States
Acta Jur.	*Acta Juridica*
add.	addendum
Afr. J. Int'l & Comp. L.	*African Journal of International and Comparative Law*
AJIL	*American Journal of International Law*
Am. U. J. Int'l L. & Pol'y	*American University Journal of International Law and Policy*
Ann. Dr. Mar. Océanique	*Annuaire de Droit Maritime et Oceanique*
ArchVR	*Archiv des Völkerrechts*
AU	African Union
BAN	Basel Action Network
BDGVR	*Berichte der Deutschen Gesellschaft für Völkerrecht*
Buff. L. Rev.	*Buffalo Law Review*
BYIL	*British Year Book of International Law*
Cath. U. L. Rev.	*Catholic University Law Review*
CILSA	*Comparative and International Law Journal of Southern Africa*

CLC	International Convention on Civil Liability for Oil Pollution Damage
Colo. J. Int'l Envtl. L. & Pol'y	Colorado Journal of International Environmental Law and Policy
Colum. J. Envtl. L.	Columbia Journal of Environmental Law
Colum. J. Transnat'l L.	Columbia Journal of Transnational Law
conf.	Conference
COP[no.]	Conference of the Parties to the Basel Convention, indicating its respective meeting
COTIF Convention	Convention Concerning International Carriage by Rail
CRTD Convention	Convention on Civil Liability for Damage Caused During Carriage of Dangerous Goods by Road, Rail and Inland Navigation Vessels
Delhi L. Rev.	Delhi Law Review
Denv. J. Int'l L. & Pol'y	Denver Journal of International Law and Policy
DGR	Dangerous Goods Code
Dick. J. Int'l L.	Dickinson Journal of International Law
Dir. Marit.	Il Diritto Marittimo
doc.	document
DVIS, Reihe A	Schriften des Deutschen Vereins für Internationales Seerecht : Reihe A: Berichte und Vorträge
e.g.	exempli gratia [for example]
EC	European Commission
Ecology L. Q.	Ecology Law Quarterly
ed. (eds.)	edition/editor (editors)
EEA	European Environment Agency
EEZ	Exclusive Economic Zone
EJIL	European Journal of International Law
Emory Int'l L. Rev.	Emory International Law Review
Env't Sci. & Tech.	Environmental Science & Technology
Envtl. L.	Environmental Law
Envtl. Pol'y & L.	Environmental Policy and Law
ESM	Environmentally Sound Management
et al.	et aliae [and others]
et seq.	et sequens [and the following one/ones]
etc.	et cetera
EU	European Union
EU15/25/27	European Union, at the time when it consists of the indicated number of Member States
EUR	Euro
Eur. Envtl. L. Rev.	European Environmental Law Review
FAZ	Frankfurter Allgemeine Zeitung
Fla. J. Int'l L.	Florida Journal of International Law

Abbreviations

Fund Convention	International Convention on the Establishment of an International Fund for Compensation for Oil Pollution Damage
Ga. J. Int'l .& Comp. L.	Georgia Journal of International and Comparative Law
GATT	General Agreement on Tariffs and Trade
GBP	Great Britain pound sterling
Geo. Int'l Envtl. L. Rev.	Georgetown International Environmental Law Review
GYIL	German Yearbook of International Law
Hague Y. B. Int'l L.	Hague Yearbook of International Law
Harv. Int'l L. J.	Harvard International Law Journal
HCR	Human Rights Council of the General Assembly to the United Nations
HKLJ	Hong Kong Law Journal
HNS	Hazardous and Noxious Substances
HNS Convention	International Convention on Liability and Compensation for Damage in Connection with the Carriage of Hazardous and Noxious Substances by Sea
i.e.	id est [that is]
IATA	International Air Transport Association
IBCs	Intermediate Bulk Containers
ibid.	ibidem [in the same place]
ICAO	International Civil Aviation Organization
ICC	International Chamber of Commerce
ICJ	International Court of Justice
ICJ Reports	International Court of Justice Reports of Judgments, Advisory Opinions and Orders
IGO	Intergovernmental Organisation
IJMCL	International Journal of Marine and Coastal Law
ILC	International Law Commission
ILM	International Legal Materials
ILO	International Labour Organization
IMCO	Inter-Governmental Maritime Consultative Organization
IMDG Code	International Maritime Dangerous Goods Code
IMO	International Maritime Organization
INCOTERMS	International Commercial Terms
Ind. Int'l & Comp. L. Rev.	Indiana International & Comparative Law Review
Ind. J. Global Legal Studies	Indiana Journal of Global Legal Studies
Int'l & Comp. L. Q.	International and Comparative Law Quarterly
Int'l Comm. L. Rev.	International Community Law Review

Int'l J. Hum. Rts.	*International Journal of Human Rights*
Int'l Law.	*International Lawyer*
Int'l Rev. L. & Econ.	*International Review of Law and Economics*
IOPC	International Oil Pollution Compensation
ISA	International Seabed Authority
ITLOS	International Tribunal for the Law of the Sea
ITLOS Reports	*International Tribunal for the Law of the Sea Reports of Judgments, Advisory Opinions and Orders*
J. Env. & Dev.	*Journal of Environment & Development*
J. Environ. Econ. Manage.	*Journal of Environmental Economics and Management*
J. Envtl. L.	*Journal of Environmental Law*
J. L. & Econ.	*Journal of Law and Economics*
J. Nat. Resources & Envtl. L.	*Journal of Natural Resources & Environmental Law*
J. W. T.	*Journal of World Trade*
JIML	*Journal of International Maritime Law*
ldt	Light displacement ton
LHD	*Legal Hukuk Dergisi*
LLMC Convention	Convention on Limitation of Liability for Maritime Claims
LMCLQ	*Lloyd's Maritime and Commercial Law Quarterly*
LNG	Liquefied Natural Gas
LOF 2000	Lloyd's Standard Form of Salvage Agreement of 2000
Loy. L. A. Int'l & Comp. L. J.	*Loyola of Los Angeles International and Comparative Law Journal*
LPG	Liquefied Petroleum Gas
LWG	Legal Working Group of the Basel Convention
M/V	Motor Vessel
Mar. Pol'y	*Marine Policy*
MARPOL 73/78 Convention	International Convention for the Prevention of Pollution from Ships
Max Planck YBUNL	*Max Planck Yearbook of United Nations Law*
MEAs	Multilateral Environmental Agreements
MEGCs	Multiple-Element Gas Containers
MOP	Meeting of the Parties to the Basel Protocol
Multinatl. Monit.	*Multinational Monitor*
N. Y. U. J. Int'l. L. & Pol.	*New York University Journal of International Law & Politics*
NGOs	Non-Governmental Organisations
no.	Number
Nord. J. Intl. L.	*Nordic Journal of International Law*

NUCLEAR Convention	Convention Relating to Civil Liability in the Field of Maritime Carriage of Nuclear Material
NuR	*Natur und Recht*
NYIL	*Netherlands Yearbook of International Law*
OAU	Organisation of African Unity
OBO-carrier	Cargo vessel carrying oil, bulk and ore cargos
OCHA	United Nations Office for the Coordination of Humanitarian Affairs
OECD	Organisation for Economic Co-operation and Development
OEWG	Open-ended Working Group of the Basel Convention on the Control of Transboundary Movements of Hazardous Wastes and Their Disposal
OILPOL Convention	International Convention for the Prevention of Pollution of the Sea by Oil
OPRC Convention	International Convention on Oil Pollution Preparedness, Response and Cooperation
OPRC-HNS Protocol	Protocol on Preparedness, Response and Co-operation to Pollution Incidents by Hazardous and Noxious Substances
OSPAR Convention	Convention for the Protection of the Marine Environment of the North-East Atlantic
Österr. Z. öffentl. Recht	*Österreichische Zeitschrift für öffentliches Recht und Völkerrecht*
p.	Page
p.a.	Per annum [per year]
para.	Paragraph/paragraphs
PCBs	Polychlorinated Biphenyls
PCIJ	Permanent Court of International Justice
PCIJ Series A	*Permanent Court of International Justice Series A: Judgments and Orders*
PCIJ Series A/B	*Permanent Court of International Justice Series A/B: Collection of Judgments, Orders and Advisory Opinions*
PCTs	Polychlorinated Terphenyls
PIC	Prior Informed Consent
PIF	Pacific Island Forum
Polish Y. B. Int'l L.	*Polish Yearbook of International Law*
POPs	Persistent Organic Pollutants
pp.	Pages
RabelsZ	*Rabel Journal of Comparative and International Private Law*
RdC	*Recueil des Cours*

RECIEL	*Review of European Community and International Environmental Law*
res.	Resolution
Resour. Conserv. Recycl.	*Resources, Conservation and Recycling*
Rev. Int. Econ.	*Review of International Economics*
RIAA	Reports of International Arbitral Awards
S.S.	Steam Ship
SCOPIC	Special Compensation P&I Clause
SDR	Special Drawing Rights
SIMPLY	*Scandinavian Institute of Maritime Law Yearbook*
SOLAS Convention	International Convention for the Safety of Life at Sea
Stan. J. Int'l L.	*Stanford Journal of International Law*
Suffolk Transnat'l L. Rev.	*Suffolk Transnational Law Review*
Temp. Envtl. L. & Tech. J.	*Temple Environmental Law & Technology Journal*
transl.	Translated
TOR	Terms of Reference of the Mechanism for Promoting Implementation and Compliance with the Basel Convention
Tul. Envtl. L. J.	*Tulane Environmental Law Journal*
Tul. J. Int'l & Comp. L.	*Tulane Journal of International and Comparative Law*
TWQ	*Third World Quarterly*
UCLA J. Envtl. L. & Pol'y	*UCLA Journal of Environmental Law & Policy*
UN	United Nations
UNCC	United Nations Compensation Commission
UNCLOS	United Nations Convention for the Law of the Sea
UN-ECE	United Nations Economic Commission for Europe
UNEP	United Nations Environment Programme
UNGA	General Assembly of the United Nations
Unif. L. Rev.	*Uniform Law Review*
UPR	Umwelt- und Planungsrecht
USA/US	United States of America
USD	United States dollar
USSR	Union of Soviet Socialist Republics
v.	Versus
V and. J. Transnat'l L.	*Vanderbilt Journal of Transnational Law*
VersR	Versicherungsrecht
VRÜ	Verfassung und Recht in Übersee
Vt. L. Rev.	*Vermont Law Review*

Wm. & Mary Envtl. L. & Pol'y Rev.	William and Mary Environmental Law and Policy Review
WTAM	World Trade and Arbitration Materials
WTO	World Trade Organization
WWF	World Wildlife Fund
Yale J. Int'l L.	Yale International Journal of International Law
Yb. Int'l Env. L.	Yearbook of International Environmental Law
YBICED	Yearbook of International Co-operation on Environment and Development
YBILC	Yearbook of the International Law Commission
ZaöRV	Zeitschrift für ausländisches öffentliches Recht und Völkerrecht
ZUR	Zeitschrift für Umweltrecht

Chapter 1
Introduction

In December 1999, in the wake of the 5th Conference of the Parties to the Basel Convention held in Basel, Switzerland, the then Executive Director of the United Nations Environmental Programme (UNEP), *Klaus Töpfer*, praised the recent adoption of the Basel Protocol on Liability and Compensation[1] as a "major breakthrough".[2] He claimed that "[f]or the first time, we have a mechanism for assigning responsibility for damage caused by accidental spills of hazardous waste during export or import". However, it did not take long before voices were being raised that cast a rather poor light on the Basel Protocol. The Protocol was criticised by legal scholars as being "far from perfect and in many respects […] unclear and confusing".[3] Others used even stronger language: "The Liability Protocol is […] a text with as many holes and exclusions as Swiss cheese" and "[it] is a dangerous precedent and is unlikely to ever, provide adequate relief for victims of toxic waste or serve as an incentive to avoid hazardous waste trafficking".[4] Similarly, it was charged that "the treaty offers very little that is positive and much that is highly negative", and, "[w]hat was adopted in Basel in 1999 […] represents a successful attack on the Basel Convention's own fundamental principles and a dangerous international precedent".[5] By means of this juxtaposition the nature of the major burden facing the Basel Protocol becomes plainly apparent: Diverging political, commercial and environmental interests put high requirements on a legal regime governing civil liability for damage resulting from the

---

[1] Basel Protocol on Liability and Compensation for Damage Resulting from Transboundary Movements of Hazardous Wastes and Their Disposal (to the Basel Convention) of 10 December 1999.
[2] Quoted from UNEP, press release of 14 December 1999, 'Compensation and Liability Protocol Adopted by Basel Convention on Hazardous Wastes'.
[3] Tsimplis, 'The 1999 Protocol to the Basel Convention', 16 *IJMCL* (2001), at 296.
[4] Kevin Stairs, political adviser with Greenpeace International, as quoted from Basel Action Network, press release of 10 December 1999, 'Hazardous Waste Agreement on Liability Protocol Reached at Basel Conference of Parties'.
[5] Sharma, 'The Basel Protocol', 26 *Delhi L. Rev.* (2004), at 196.

transboundary movement of hazardous wastes, and these can hardly be met by a compromise regulation as represented by the 1999 Basel Protocol.

The Basel Protocol on Liability and Compensation was adopted in 1999 to supplement the legal framework established by the 1989 Basel Convention on the Control of Transboundary Movements of Hazardous Wastes and their Disposal and to provide for rules imposing civil liability and making compensation available for the victims of pollution caused by hazardous wastes. The Basel Protocol, however, has yet not entered into force, and keeping in mind the harsh criticisms as outlined above, it is also questionable whether it will ever receive sufficient support from States to obtain the required number of ratifications in order to enter into force.

The present work starts exactly at this juncture. Its primary purpose is to outline the legal rules and regimes applicable for the imposition of responsibilities and liabilities for damage resulting from the transboundary movement of hazardous wastes by sea. Given the fact that the entry into force of the Basel Protocol is uncertain, it is necessary that this work begins with an analysis of responsibility and liability according to the rules of customary international law and according to the regulations of international conventions and regulations that are currently in force and applicable to the cases under consideration. The prospective regime of liability and compensation as proposed by the Basel Protocol can only be examined in detail in a second step, which may then illuminate the possible advantages and disadvantages of this potential solution *de lege ferenda*. As a result of this consideration it will be possible to make a recommendation whether it seems appropriate to agree with *Klaus Töpfer* and to further expedite and promote ratification of the Basel Protocol, or whether the criticisms voiced against the Protocol are actually true and efforts should rather be made to develop and strengthen other mechanisms to protect victims of pollution and the environment.

A. The Factual Perspective: Transboundary Movements of Hazardous Wastes by Sea

The transboundary movement of hazardous wastes represents a commercial activity promising huge returns for the persons engaged in the movement. On the downside, hazardous waste movements may pose a substantial threat for human health and the environment. Since, moreover, hazardous wastes are usually shipped in large amounts, incidents involving hazardous wastes are likely to affect large areas with adverse effects on a potentially large number of humans, animals and natural resources. The fact that incidents may occur during each stage of a transboundary movement of hazardous wastes by sea can be illustrated in the following examples[6]:

---

[6] It should be added, though, that the facts of the following cases have not been officially established. These cases are depicted as far as they are described in publicly assessable articles and contributions to newspapers and journals.

The M/V "Khian Sea" case[7]: In 1986 the city of Philadelphia in Pennsylvania (USA) instructed a private waste management company to export 14,000 tons of toxic ash derived from a municipal incineration plant. The bulk cargo was loaded on board the M/V "Khian Sea", which headed for the Caribbean to dump the ash on a man-made island in the Bahamas. The Bahamian government, however, refused to permit the discharge. The ash was then relabelled and the vessel tried to call at several other ports in the Caribbean, South America and Western Africa.[8] In spring 1988, the crew finally succeeded in discharging the ash onto a beach near Gonaives in Haiti by declaring it as topsoil fertilizer. After approximately 4,000 tons had been discharged, the Haitian government recognised the true nature of the cargo and ordered the vessel to reload, which, however, left Haitian waters without reassuming the ash.[9] Thereafter, the crew of the M/V "Khian Sea" was unsuccessful in its attempts to unload the remaining ash at several ports around the globe. The vessel was sold, reflagged and renamed, and when she arrived at Singapore in November 1988 her cargo was missing. It is assumed that the ash had been dumped into the Indian Ocean.[10] Whereas two executives of the vessel operating company were sentenced to imprisonment by US courts for the dumping into the high seas, no one could be held liable for the costs of re-importation and clean-up at the Haitian shore.[11] It was only in 2000 that a major part of the ash located in Haiti was re-shipped to the US and finally deposited at a landfill in Pennsylvania.[12]

The M/V "Khian Sea" case is only one striking example of major environmental incidents that have occurred in the context of transboundary movements of hazardous wastes by sea. Another example of wastes being sent around the globe is the M/V "Zanoobia" case from 1987. An Italian waste management company shipped approximately 2,200 tons of chemical waste to Djibouti where the drums were intended to be buried. After the local authorities refused to allow the

---

[7] For a detailed description of the M/V "Khian Sea" Incident see Gilmore, 'The Export of Nonhazardous Waste', 19 Envtl. L. (1988/1989), at 879–883; Pellow, Resisting Global Toxics (2007), at 107–123; see also Bruno, 'Philly Waste Go Home', 19 Multinatl. Monit. (1998).

[8] Gilmore, 'The Export of Nonhazardous Waste', 19 Envtl. L. (1988/1989), at 880; Pellow, Resisting Global Toxics (2007), at 108; Walsh, 'The Global Trade in Hazardous Wastes', 42 Cath. U. L. Rev. (1992/1993), at 106.

[9] Pellow, Resisting Global Toxics (2007), at 108; Rosenthal, 'Ratification of the Basel Convention', 11 Temp. Envtl. L. and Tech. J. (1992), at 62–63.

[10] Liu, 'The Koko Incident', 8 J. Nat. Resources and Envtl. L. (1992/1993), at 130; Pellow, Resisting Global Toxics (2007), at 108; Walsh, 'The Global Trade in Hazardous Wastes', 42 Cath. U. L. Rev. (1992/1993), at 106.

[11] Pellow, Resisting Global Toxics (2007), at 110; Tsimplis, 'The 1999 Protocol to the Basel Convention', 16 IJMCL (2001), at 298.

[12] Pellow, Resisting Global Toxics (2007), at 110–119. See also Greenpeace, press release of 29 October 1998, 'Philadelphia Incinerator Ash to Return from Haiti to the U.S.'; Haitian Government, press release of 22 April 2000, 'The Haitian People Achieve Environmental Justice for Earth Day'; 'Homeless for 16 Years, Barge of Garbage Returns to Pa.', Los Angeles Times of 11 August 2002.

discharge, the cargo was sent to Venezuela for interim storage, then to Syria and finally back to Italy. Several people that came into contact with the wastes during the movement fell ill.[13]

Further examples include, amongst others, the Kassa Island incident in Guinea,[14] the Koko Beach incident in Nigeria,[15] the *Thor Chemicals* incident in South Africa[16] and the *Formosa* incident in Cambodia.[17]

The *M/V "Probo Koala"* case[18]: The *M/V "Probo Koala"* was a Greek OBO-carrier chartered by an oil trading company with head offices in Amsterdam, Lucerne and London. In 2006, the vessel was used to temporarily store and process petrol blend stocks and naphtha while she anchored in the Mediterranean off the coast of Gibraltar. During this caustic washing process ("sweetening"), naphtha or petrol blends are mixed with caustic soda (liquid sodium hydroxide) to reduce the level of mercaptans in order to obtain tradable petrol for the African market. The highly toxic residues of this caustic washing were collected in the vessel's slop tanks.[19] In June 2006, the *M/V "Probo Koala"* called at the port of Amsterdam to refuel and to empty her slop tanks, whose content was declared as ordinary slops from oil tank washings. However, an unusual and pungent odour emanated from the samples taken by the port operator, so that the further discharging of the slop tank contents was prohibited.[20] The vessel then sailed to Paldiski in Estonia and loaded approximately 26,000 tons of petrol to be shipped to Nigeria. After delivery

---

[13] See Liu, 'The Koko Incident', 8 *J. Nat. Resources and Envtl. L.* (1992/1993), at 127–128; Wiedemann, 'Die schlimmste Fracht meines Lebens', *SPIEGEL* of 30th May 1988.

[14] See Gilmore, 'The Export of Nonhazardous Waste', 19 *Envtl. L.* (1988/1989), at 882; Liu, 'The Koko Incident', 8 *J. Nat. Resources and Envtl. L.* (1992/1993), at 130; Vir, 'Toxic Trade with Africa', 23 *Env't Sci. and Tech.* (1989), at 24.

[15] See on this case: Eguh, 'Regulations of Transboundary Movement of Hazardous Wastes', 9 *Afr. J. Int'l and Comp. L.* (1997), at 130–134; Liu, 'The Koko Incident', 8 *J. Nat. Resources and Envtl. L.* (1992/1993), at 131–134; Vir, 'Toxic Trade with Africa', 23 *Env't Sci. and Tech.* (1989), at 23–24.

[16] See Glazewski, 'Regulating Transboundary Movement of Hazardous Waste', 26 *CILSA* (1993), at 235; Lipman, 'Transboundary Movements of Hazardous Waste', *Acta Jur.* (1999), at 268; Poulakidas, 'Waste Trade and Disposal in the Americas', 21 *Vt. L. Rev.* (1996/1997), at 874.

[17] This incident is described by: Lohnes, 'Taiwanese Company Dumps 3000 Tons of Toxic Waste in Cambodia', 11 *Colo. J. Int'l Envtl. L. and Pol'y* (2000), at 264–270; Markus, 'Taiwanese Waste Sent to Europe', *BBC News* of 2nd March 2000.

[18] HRC Doc. A/HRC/12/26/Add. 2; COP8 Doc. UNEP/CHW.8/16, at 6–9. See also Fagbohun, 'The Regulation of Transboundary Shipments of Hazardous Waste', 37 *HKLJ* (2007), at 834–837; Knauer, et al., 'Profits for Europe, Industrial Slop for Africa', *SPIEGEL ONLINE* of 18th September 2006; Ognibene, 'Dumping of Toxic Waste in Côte d'Ivoire', 37 *Envtl. Pol'y and L.* (2007), at 31; Pratt, 'Decreasing Dirty Dumping?', 35 *Wm. and Mary Envtl. L.and Pol'y Rev.* (2011), at 582–584.

[19] HCR Doc. A/HRC/12/26/Add. 2, at 7–8; OCHA Doc. OCHA/GVA/2006/0190; Frenk, 'Was geschah an Bord der "Probo Koala"?', *FAZ* of 27th October 2006.

[20] The port operator instead suggested delivering the slop to a disposal facility based in Rotterdam possessing the required capability to incinerate the chemical residues. The costs, however, amounted to EUR 900 instead of EUR 20 per cubic metre.

of the cargo the vessel called at Abidjan in Ivory Coast on 19 August 2006 and emptied her slop tanks. Approximately 528 cubic metres of chemical wastes were delivered to a local waste management company[21] that had been founded only recently.[22] The liquid wastes were simply dumped at various sites in and around Abidjan lacking any kind of soil sealing. It is officially estimated that as a result of direct contact and indirect exposure by consumption of contaminated water, groundwater and food products, 15 residents died, 69 were hospitalised and more than 108,000 people sought medical attention because of intestinal and respiratory problems, nausea and vomiting.[23] When the international public became aware of this incident the oil trading company attempted to settle this matter by mutual agreement with the Ivorian government. According to this agreement the Ivorian government received a contribution of GBP 100 million towards the costs of restoration of environmental damage and towards compensation payments for the families of killed and injured residents.[24] Notwithstanding this step, a class action lawsuit aggregating 31,000 residents was instituted before the London High Court in 2009, this later being withdrawn after the oil trading company agreed to an out-of-court settlement paying GBP 1,000 to each victim.[25] The oil trading company was, furthermore, sentenced by an Amsterdam Court to pay a fine of one million Euros for concealing the hazardous character of the wastes when they were initially unloaded in Amsterdam.[26]

B. The Legal Perspective: Existing Rules and the 1999 Liability Protocol to the Basel Convention

International environmental law is a comparably young field of law. The vast majority of international conventions and agreements concerned with the protection of the environment and the conservation of natural resources have been

---

[21] The disposal fees amounted to USD 30–35 per cubic metre.

[22] OCHA Doc. OCHA/GVA 2006/0184; Ognibene, 'Dumping of Toxic Waste in Côte d'Ivoire', 37 *Envtl. Pol'y and L.* (2007), at 31.

[23] HCR Doc. A/HRC/12/26/Add. 2, at 8–9; COP8 Doc. UNEP/CHW.8/16, at 7.

[24] Fagbohun, 'The Regulation of Transboundary Shipments of Hazardous Waste', 37 *HKLJ* (2007), at 836; Pratt, 'Decreasing Dirty Dumping?', 35 *Wm. and Mary Envtl. L.and Pol'y Rev.* (2011), at 584.

[25] Dowell, 'Trafigura Settlement: A Drop in the Ocean?', *The Lawyer* of 28th September 2009; Leigh, 'Trafigura Offers £1,000 Each to Toxic Dumping Victims', *The Guardian* of 18th September 2009.

[26] 'Trafigura found guilty of exporting toxic waste', *BBC News* of 23rd July 2010; Corbett, 'Implications from 'Probo Koala' ruling', *Trade Winds* of 30th July 2010; Evans, 'Trafigura fined €1m for exporting toxic waste to Africa', *The Guardian*, of 23rd July 2010.

created subsequent to the 1972 Stockholm Conference.[27] Those multilateral environmental agreements (MEAs) are almost exclusively designed as sectoral or regional conventions dealing with the protection of specific environmental resources or with particular dangers to the environment.[28] A coherent conception or a structured development of international environmental law does not exist. International law rather emerges where States recognise a current need for binding rules in a specific sector. This, however, also means that as soon as a certain environmental issue vanishes from the public debate, it is questionable whether there will be sufficient political support to establish a new and far-reaching international legal regime.[29] Toxic waste exports represent such an environmental issue.

The question of liability and compensation for damage resulting from the transboundary movement of hazardous wastes by sea is an interdisciplinary matter that touches on several aspects and, hence, constitutes an area particularly ripe for disagreement. It not only involves different commercial interests on the waste exporting, handling and disposing side, but it also needs to take account of the shipping and insurance industries as well as the industrial sectors demanding a steady supply of raw materials. Moreover, diverging political interests and positions are involved and there is a strong need for the protection of human health and the environment as well as for the conservation of natural resources. All these different interests and positions require a careful balancing by a legal regime addressing this issue. This applies all the more since the exportation of toxic waste is a highly emotional issue that encompasses not only an "environmental component", but rather a conglomeration of human, social and ethical questions, the global distribution of responsibilities, the right to develop and to take part in the world trade market, and the burden of significant faults and failures from the past.[30]

The issue of liability and compensation for damage resulting from the transboundary movement of hazardous wastes by sea needs to be considered in the context of the surrounding branches of law. The most important legal framework of substantive rules, in which this topic is embedded, is provided by the widely accepted 1989 Basel Convention on the Control of Transboundary Movements of

---

[27] The United Nations Conference on the Human Environment held in Stockholm in June 1972 was concerned with the international protection of the environment and resulted in the foundation of the UNEP and the adoption of the 1972 Stockholm Declaration as well as the formulation of an Action Plan with 109 recommendations.

[28] A global convention dealing in general with the protection of the global environment does not exist. Non-binding rules are however contained in the 1972 Stockholm Declaration and in the 1992 Rio Declaration.

[29] Examples of crucial environmental issues that have vanished from the current focus of public awareness are: liability for nuclear damage, use of outer space, depletion of the ozone layer, acid rain and forest dieback, dumping of wastes and nuclear wastes, and pollution from ships.

[30] As to the difficulties generally faced by the international environmental law, see also Crawford, *Brownlie's Principles of Public International Law* (8th ed., 2013), at 352–355.

Hazardous Wastes as well as a number of regional conventions and agreements. Since waste movements mostly take part using the seas as a transport medium, the international law of the sea also provides for applicable substantive rules. While the general legal framework is defined by the UNCLOS, further sectoral conventions and agreements apply to specific aspects of marine transportation. This involves, for example, the MARPOL 73/78 Convention and the diverse Regional Seas Conventions, the latter also partially providing for specific rules dealing with the transboundary movement of hazardous wastes.[31] Overlaps in respect of this issue may arise from the existing and nascent legal regimes of civil liability, such as the 1996/2010 HNS Convention, the CRTD Convention or the regional or domestic liability regimes. Apart from all these connections to the existing and nascent treaty law, the issue of liability in the context of hazardous waste movements must also be related to the existing rules and principles of customary international law. The rules of State responsibility are subject to a steady process of development[32] and concern the question under which conditions particularly States may be held liable for damage due to internationally wrongful acts.

The particular difficulties related to the regulation of liability in the context of waste movements and their complex thematic classifications arise in the question of how to argumentatively approach this issue. The following steps of reasoning should be taken:

1. Which rules and provisions of the international law apply *de lege lata* as to the determination of responsibilities and liabilities for damage resulting in the context of transboundary movements of hazardous wastes by sea?
2. Does this existing law provide for a sufficient level of protection? Or is there a need for the establishment of a further legal regime? Are there general requirements on the prospective legal regime?
3. What solutions does the 1999 Basel Protocol provide for? Are these approaches *de lege ferenda* suitable and appropriate to meet the requirements?
4. In conclusion, is it appropriate to promote ratification of the Basel Protocol, or does it rather make sense to focus on the elaboration or further development of new or other legal instruments?

The objective of this book is to provide for an analysis and assessment of the existing rules and legal instruments relevant to the determination of responsibilities and liabilities for damage resulting from the transboundary movement of hazardous wastes by sea, as well as to give an estimation whether or not the regime of civil liability as envisaged by the 1999 Protocol to the Basel Convention

---

[31] See 1996 Izmir Protocol on the Prevention of Pollution of the Mediterranean Sea by Transboundary Movements of Hazardous Wastes and their Disposal; 1998 Tehran Protocol on the Control of Marine Transboundary Movements and Disposal of Hazardous Wastes.

[32] In this respect, particularly the 1938/1941 *Trail Smelter Arbitration Award*, 3 RIAA (1949), at 1905 *et seq.*, represents a milestone in the development of an international legal regime on the responsibility of States for transboundary harm, which finally led to the adoption of the ILC Draft Articles on State Responsibility in 2001.

constitutes a suitable and appropriate mechanism to compensate damage and provide a remedy to those who have been victimised by pollution. This work is intended to make a small contribution towards legal clarity in the area of liability in the context of hazardous waste movements. In the best case scenario, it should serve as a plea for the entry into force of the Basel Protocol, offering further stimulus for an arguably overdue ratification of the Protocol by the Contracting States of the Basel Convention.

C. The Structure of This Book

The structure of this book basically follows the steps of reasoning as outlined above. It is furthermore complemented by a description of the factual and economic background of hazardous waste movements.

As a first step, this book outlines the economic background of the international trade in hazardous wastes (Chap. 2). It describes common properties of hazardous wastes and explains why they are traded on a global scale. In this context, particular emphasis is put on a description of the respective commercial and political interests involved in hazardous waste movements. Subsequently, the applicable rules and provisions of current international law are outlined (Chap. 3). By means of this survey it is possible to obtain an overview of which branches of law currently provide for rules of liability applying to the cases under consideration. Conversely, it can be assessed which aspects of hazardous waste movements remain unregulated *de lege lata*. The analysis of current international law first outlines the legal conception of international responsibility and liability and its current state of development (Sects. "The Legal Concept of International Responsibility" and "The Contribution of the International Law Commission" in Chap. 3). It, then, addresses the customary principle of State responsibility and examines in which scenarios of a hazardous waste movement States may be held responsible for a failure to comply with the relevant international obligation. To this end, the relevant obligations deriving from international conventions, agreements and international customary law are analysed (Sects. "State Responsibility in the Context of Transboundary Movements of Hazardous Wastes by Sea" in Chap. 3). Following this, it is outlined which international civil liability conventions may apply and provide relevant rules regarding the movement of hazardous wastes (Sect. "Existing Civil Liability Conventions" in Chap. 3). The comprehensive analysis of current international law is, subsequently, followed by an intermediate examination regarding the necessity of the establishment of a regime of civil liability and compensation as a legal means to prevent and remedy pollution damage (Chap. 4). This intermediate examination also includes the determination of the necessary elements of an effective regime of liability and outlines the economic ramifications of environmental liability. In the final step, this book addresses in particular the provisions of the 1999 Basel Protocol and attempts to provide an assessment of whether this legal instrument is to be considered a reasonable and well-balanced legal regime of liability and compensation (Chap. 5). To this end, the individual provisions of the Protocol are examined and the overall concept and implications of the Basel Protocol are assessed. The work ends with a short summary and a final conclusion as regards the future steps that should be taken (Chap. 6).

Chapter 2
The International Trade in Hazardous Wastes and Its Economic Background

Before providing an examination of the existing legal framework governing the transboundary movement of hazardous wastes by sea and prior to an assessment of the particular provisions of the Protocol to the Basel Convention, it is necessary to first outline the practical circumstances of the underlying problem area and the typical constellations of hazardous waste shipments by sea. Consequently, at the outset of this work a whole series of questions arise: What is the practical sig nificance of hazardous waste movements and what amounts of wastes are shipped across the globe? What are the economic drivers and the interests of the parties involved? Which economic and political impulses influence States, economies and private players? How does "hazardous waste shipment" typically take place? And at what point might liability function as a trigger to correct possible misconduct by the parties involved? A brief introduction to these issues shall be given in this second chapter.

A. Hazardous Wastes: Properties and Economic Importance

The basic substance at issue throughout this book is described by the term "hazardous wastes". The meaning, properties and the economic importance of this substance shall thus be described at the outset.

I. Sources and Composition of Hazardous Wastes

Wastes may appear in diverse forms and originate from various activities. Consequently, many descriptions of this term can be found in common parlance. A rather pointed but also appealing version describes wastes as "the wrong material at the wrong time at the wrong place".[1] The most typical description defines wastes as unusable or unwanted substances or materials remaining from any production or consumption process.[2] This general understanding of wastes shows that a definition of this term cannot be made by means of purely objective criteria, but needs to take into account subjective elements of the person in possession of the wastes. Accordingly, the legal definitions of wastes in most international conventions combine objective and subjective elements and, as a general criterion, are based on the actual, intended or legally required disposal (including recovery operations) of the substances or materials in question.[3]

The term hazardous wastes (generally used synonymously with the terms toxic or dangerous wastes) is commonly understood as denoting wastes that are actually or potentially harmful to human health or the environment due to certain adverse characteristics or specific components of the wastes. The definitions of hazardous wastes used in international conventions follow this approach and provide for detailed lists of categories of waste streams and hazardous characteristics.[4] The hazardousness of wastes is usually defined by a combination of two elements, (1) the category of the waste stream, which is determined either by its origin or by certain compounds, and (2) certain hazardous characteristics or properties of these wastes. While some definitions cumulatively require both elements,[5] others only require that either the one or the other element applies to the wastes in question.[6] Again other definitions make the hazardousness of wastes merely conditional on whether the wastes display certain hazardous characteristics.[7] Household wastes and residues of the incineration of household wastes are either included in this definition or they are legally placed on the same level.[8]

It follows from this sophisticated legal definition that the term "hazardous wastes" basically covers a wide range of substances and materials. Individual waste streams, which represent the major part of hazardous wastes, include oil

---

[1] See the term "Abfall" (transl. "waste"), in the German Wikipedia as of June 2010.

[2] See e.g. "Abfall" in: *Brockhaus, Vol. 1*, (21st ed. 2006); "Abfall" in: *Meyers Enzyklopädisches Lexikon, Vol. 1*, (9th ed. 1971); "Waste" in: *The Oxford English Dictionary, Vol. XII*, (1978).

[3] Basel Convention, Article 2(1); Bamako Convention, Article 1(1); Waigani Convention, Article (1); OECD Council Decision C(2001)107/FINAL, A(1); EU Directive 2008/98/EC, Article 3(1); Izmir Protocol to the Barcelona Convention, Article 1(c); Tehran Protocol to the Kuwait Convention, Article 2(1).

[4] See also O'Neill, *Waste Trading Among Rich Nations* (2000), at 26–29.

[5] Basel Convention, Article 1(1)(a) in connection with Annexes I, III; Waigani Convention, Article 2(1)(a) in connection with Annexes I, II; OECD Council Decision C(2001)107/FINAL, A(2) in connection with Appendixes 1, 2; Tehran Protocol to the Kuwait Convention, Article 1(1)(a) in connection with Annexes I, III.

[6] Bamako Convention, Article 2(1)(a) and (c) in connection with Annexes I, III; Izmir Protocol to the Barcelona Convention, Article 3(1)(a) and (c) in connection with Annexes I, II.

[7] EU Directive 2008/98/EC, Article 3(2) in connection with Annex III.

[8] See Basel Convention, Article 1(2) in connection with Annex II; Tehran Protocol to the Kuwait Convention, Article 1(1)(b) in connection with Annex II.

mixtures, residues from industrial waste disposal, clinical waste, lead and lead compounds, tars, zinc compounds, paints and dyes, acids and asbestos.[9] But typical hazardous wastes are represented also by other remnants of chemical and industrial treatment processes associated with the production of, for example, pharmaceuticals, biocides, organic solvents, varnishes, resins, plasticisers and glues, as well as waste materials containing abstractly harmful components, such as certain metallic compounds, arsenic selenium and cadmium compounds, mercury, acidic and basic solutions, phenols, PCBs and PCTs.[10] Finally, hazardous wastes may display several characteristics and properties. They may, for example, be explosive, flammable, poisonous, infectious, toxic or ecotoxic, corrosive, oxidising or thermally unstable.[11]

II. Volumes of Generated Hazardous Wastes

The absolute amount of hazardous wastes generated annually around the world is unknown. It is, moreover, unlikely that reliable figures in this regard will ever be available.

This is to be explained by several reasons: On a global scale data regarding the generation of hazardous wastes are collected only by the Secretariat of the Basel Convention. However, the reporting requirement under the Basel Convention regarding such data is of a voluntary nature only.[12] Thus, only 37 States (about 20 % of all Contracting States) have reported data on the generation of hazardous wastes for 2009.[13] Further reasons for incomplete and unreliable data reports involve differences in the national definitions of hazardous wastes, deficiencies in national data collection, monitoring and enforcement capacities as well as missing explanations for large fluctuations in the reported data.[14] On a regional level, data and statistics regarding the generation of hazardous wastes are best available from the EU[15] and, to a lesser extent, also from the OECD. The data collected by the EU do not show the same deficiencies as those data collected by the Secretariat of the

---

[9] Basel Secretariat (ed.), *Global Trends 1993–2000*, at 15.

[10] See for example the enumerations in Annex I of the Basel Convention. See also Gwam, 'Adverse Effects of the Illicit Movement of Hazardous Wastes', 14 *Fla. J. Int'l L.* (2001/2002), at 431–432; Krueger, *International Trade and the Basel Convention* (1999), at 7–8 and 99–106.

[11] See for example the characteristic listed in Annex III of the Basel Convention.

[12] Basel Convention, Article 13(3)(b).

[13] See the "Reporting Database" on www.basel.int.

[14] See for a detailed analysis: COP10 Doc. UNEP/CHW.10/INF/4, at 6; Basel Secretariat (ed.), *Global Trends 1993–2000*, at 1, 8–11; Friedrich-Ebert-Stiftung (ed.), *Zehn Jahre Basler Übereinkommen* (1999), at 14–15; Walsh, 'The Global Trade in Hazardous Wastes', 42 *Cath. U. L. Rev.* (1992/1993), at 108–110.

[15] See Regulation (EC) No 2150/2002 of the European Parliament and of the Council of 25.11.2002 on waste statistics.

Basel Convention or the OECD, which is to be explained by the fact that by means of the EU regulations, reporting obligations are compulsory and the Member States, which mostly rank among industrial countries, also possess sufficient administrative capacities.[16] Nevertheless, the statistics provided by the EU only display a part of the worldwide production of hazardous wastes and only represent countries with a certain economic capability. These data, thus, cannot simply be projected on a global scale.

Data on hazardous waste generation are available as of 1990, as far as the level of the EU is concerned.[17] In EU15[18] the amount of generated hazardous wastes increased by 5 % *p.a.* from 36 million tons in 1997 up to 40 million tons in 2000.[19] Between 2000 and 2005 this amount increased by 4 % *p.a.* in EU15 and by 2 % *p.a.* in EU25.[20] Finally, in EU27 the amount of generated hazardous wastes increased from 88.5 million tons in 2004 to 100.6 million tons in 2006, followed by a slight decrease to 97.6 million tons in 2008, which corresponds to an average increase of 5.2 % *p.a.* between 2004 and 2008.[21] The generation of hazardous wastes accounted for a share of 3.7 % of the overall production of wastes in EU27 in 2008.[22] In contrast to these data provided by the EU, OECD statistics regarding hazardous waste generation are of little significance. The only conclusion that can be drawn from these data is that there has been, similarly, an increase in the overall production of hazardous wastes in OECD countries in the period between 1990 and 2005.[23] According to a rough estimate the share of the OECD countries in the overall production of hazardous wastes accounted for approximately 75 %.[24]

Although it is not possible to obtain reliable figures on the absolute amount of hazardous wastes generated worldwide, it is nevertheless possible to identify global trends in development, based on the fragmentary data available. For the

---

[16] As to the quality of the waste statistics provided by the EU see Report of the EC on waste statistics and their quality, EC Doc. COM(2011) 131 final.

[17] For an analysis of the data covering the period between 1990 and 1998 see European Environment Agency (ed.), *Hazardous Waste Generation in EEA Member Countries* (2002), at 14 *et seq.*

[18] The abbreviations EU15, EU25 and EU27 denote the number of Member States of the EU at that time.

[19] See Report of the EC, EC Doc. COM(2006) 430 final, at 6. See also the data provided in EC Doc. SEC(2006) 1053.

[20] Report of the EC, EC Doc. COM(2009) 282 final, at 5–6. See also the data provided in EC Doc. SEC(2009) 811 final, and European Environment Agency (ed.), *Transboundary Shipments in the EU* (2008), at 25–28.

[21] Data obtained from the database on hazardous waste generation on http://epp.eurostat.ec.europa.eu.

[22] European Union (ed.), *Eurostat Yearbook 2011*, at 481, 485.

[23] See OECD (ed.), *OECD Environmental Data, Compendium 2006–2008: Waste*, at 18–22; Basel Secretariat (ed.), *Global Trends 1993–2000*, at 11–12 and Appendix 3; Basel Secretariat (ed.), *Global Trends 2004–2006*, at 10 and Annex 2.

[24] COP10 Doc. UNEP/CHW.10/INF/4, at 6; Basel Secretariat (ed.), *Global Trends 2004–2006*, at 8–10.

period between 1950 and 1985 it has been estimated that the total amount of hazardous wastes generated annually increased 35-fold.[25] Rough estimates suggest that the absolute amount of hazardous wastes produced worldwide in 1990 lies between 250 and slightly over 600 million tons.[26] As of 1990 the data available for the EU and the OECD show that there is a constant increase in the average annual production of hazardous waste amounting to approximately 4–5 %. Since there is no obvious reason that in non-OECD or non-EU States the increase of hazardous waste generation would be less significant, it must be assumed that also on a global level the production of hazardous wastes has increased by not less than 4–5 %.[27]

A constant increase in the global generation of hazardous wastes, one which may be placed in the range of 4–5 % *p.a.*, is the only reasonable data that can be drawn from the available data. Although there are some attempts to estimate absolute amounts of global hazardous waste generation, such estimations must fail due to the large amount of missing and unreliable data. A further conclusion that can be drawn from the available data is that the entry into force of both the Basel Convention as well as the numerous regional conventions and regulations has obviously had no significant impact on the ever-increasing production of hazardous wastes.

III. Forms of Waste Treatment and Disposal

"Waste management" is used as an umbrella term to describe the entirety of all policies, research, tasks and measures related to the prevention, reduction, recovery and disposal of wastes. The particular tasks related to the handling and treatment of wastes include their collection and transport, the monitoring and collection of data, and the processing and disposal of waste materials. Waste disposal, in turn, comprises several methods.[28] The choice of the method used in a given case depends on various factors, like the amount of the particular waste stream, the availability of specialised disposal facilities, financial resources and the existing laws. Reliable figures on the respective disposal methods are available only at the level of the EU. According to this, landfilling and recovery are the most common disposal methods with a share of about 45 % each, while incineration

---

[25] This was calculated according to the increasing production of organic chemicals, see Kummer, *The Basel Convention* (1995), at 4. See also Walsh, 'The Global Trade in Hazardous Wastes', 42 *Cath. U. L. Rev.* (1992/1993), at 110–111.

[26] Kummer, *The Basel Convention* (1995), at 4; Poulakidas, 'Waste Trade and Disposal in the Americas', 21 *Vt. L. Rev.* (1996/1997), at 873; Rutinwa, 'Liability and Compensation', 6 *RECIEL* (1997), at 8; Williams, 'Trashing Developing Nations: The Global Hazardous Waste Trade', 39 *Buff. L. Rev.* (1991), at 276.

[27] A moderate increase is also assumed by COP10 Doc. UNEP/CHW.10/INF/4, at 6; Basel Secretariat (ed.), *Global Trends 1993–2000*, at 11–15; Basel Secretariat (ed.), *Global Trends 2004–2006*, at 8–10.

[28] A comprehensive list of disposal operations can also be found in Annex IV of the Basel Convention.

constitutes approximately 10 %.[29] But again here, one cannot simply assert that these figures claim global validity.

On a global scale, the most common method for hazardous waste disposal is geologic disposal at landfill sites. It is the least expensive option and, in the simplest case, it only requires a suitable storage ground. Therefore, it is the preferred disposal method especially in developing countries. On the downside, landfills that do not meet the technological minimum standards pose a significant risk of contamination of the ground and surface water. As can be seen in the case of the M/V "*Probo Koala*",[30] this may lead to significant environmental damage and present a serious health threat for a large number of people.[31]

Marine disposal or dumping at sea, including shipboard incineration, is also a cheap and easy method of hazardous waste disposal, particularly where no international regulation applies to the respective flag State or State of origin. Since marine dumping is not locally bound and since the dumping vessel in most cases cannot be identified, marine disposal often takes place outside any legal control. Moreover, the adverse long-term effects of hazardous substances dumped into the sea are unknown and cannot adequately be anticipated.

Incineration of hazardous wastes aims at the final destruction of the wastes and the neutralisation of the harmful substances. Due to air emissions and highly toxic residues, it generally poses a significant threat to human health and the environment. This risk is significantly minimised by modern incineration plants specialised for specific types of wastes. However, since the operation of such plants involves high financial expenditure and requires a steady supply of the target wastes, it is often not economically feasible for smaller economies.[32]

Recycling and the recovery of hazardous wastes are also widely prevalent. It should be considered the most environmentally sound disposal method since a large portion of the harmful substances can be recovered and is not yet released from the production circle. A major obstacle, however, is presented by the unstable market price for recycled products and, thus, the resulting disparity between supply and demand.[33]

Finally, processing or (pre)treatment of hazardous wastes should be mentioned. It may reduce the hazard level of wastes and may encompass physical, chemical and biological processes. Pre-treatment may take place at the site of generation of the wastes or it may be conducted in specialised facilities.[34]

---

[29] European Union (ed.), *Eurostat Yearbook 2011*, at 486.

[30] See *supra*, Sect. "The Factual Perspective: Transboundary Movements of Hazardous Wastes by Sea" in Chap. 1.

[31] Louka, *International Environmental Law* (2006), at 425.

[32] Louka, *International Environmental Law* (2006), at 427; see also Avery, 'Our Rubbish: Someone Else's Problem?', 2 *Int'l J. Hum. Rts.* (1998), at 24–25.

[33] Louka, *International Environmental Law* (2006), at 427.

[34] See for instance the required reception facilities for ship tank washing residues according to Annex II of MARPOL 73/78 Convention; *infra*, Sect. "MARPOL 73/78 Convention" in Chap. 3.

IV. The Commercial Value of Hazardous Wastes

The economic value of wastes is not intrinsic, but depends on the extent the wastes can be processed or recycled.[35] Non-hazardous wastes can be utilised to a large extent for recovery or recycling operations, or they can be sold to incineration plants. Thus, they often have a considerable positive economic value and—from an economic perspective—are to be considered as standard tradable goods. Hazardous wastes, by contrast, generally have to be disposed of through costly treatment methods and, therefore, rather represent a negative economic value for their owner, who has to bear the costs of disposal.[36] From an economic perspective, non-hazardous and hazardous wasted hence differ significantly.[37]

Regarding the trade in standard goods, the interests of the parties involved are balanced such that both the shipper and the consignee seek to ensure that the traded goods finally arrive at the consignee's place. The shipper only receives payment if delivery of the goods is actually performed to the consignee.[38] The consignee, in turn, having decided to purchase these goods, thus, has a genuine interest in taking delivery. Both parties, therefore, have an intrinsic economic interest in a sound and safe transport and delivery of the goods. Since this consideration applies for all goods with a positive economic value, it is also true for most non-hazardous wastes that are sold for a positive price on the world market.

This consideration, by contrast, does not apply with regard to the sale of hazardous wastes. Hazardous wastes mostly embody a negative economic value and are associated with negative externalities.[39] This means in practice that the seller

---

[35] van Daele/Vander Beken/Dorn, 'Waste Management and Crime', 37 *Envtl. Pol'y & L.* (2007), at 36.

[36] See also Abrams, 'Regulating the International Hazardous Waste Trade', 28 *Colum. J. Transnat'l L.* (1990), at 806.

[37] Moreover, one could think about defining wastes and hazardous wastes according to their economic value for their owner. So long as the wastes can be sold on the world market for a positive price, also taking into account transaction costs (e.g. costs for transportation or brokerage fees), this would reflect that the consignee has an economic interest in these wastes, e.g. for recovery or recycling operations. In this case the wastes could be considered non-hazardous. On the other hand, as soon as the seller has to pay additionally for selling the wastes, this is a strong indication that the wastes do not contain recyclable or recoverable substances or properties and, thus, are unwanted goods even from an economic perspective. Based on the assumption that costs for disposal increase with the hazardousness of the wastes, the latter type of wastes could be considered hazardous. This definition, to be sure, functions only in a very schematic way.

[38] This, of course, disregards the issue of determining when the risk of damage passes over to the consignee, which is most commonly specified by means of INCOTERMS.

[39] The economic perspective of establishing civil liability with regard to public goods (such as the "clean environment") is outlined by Feess, *Umweltökonomie* (3rd ed., 2007), at 37–57, 151–183; Reis, *Compensation for Environmental Damage under International Law* (2011), at 132–145; An economic analysis of the transboundary movement of hazardous wastes is provided by Cassing/Kuhn, 'Strategic Environmental Policies', 11 *Rev. Int. Econ.* (2003), at 495 *et seq.*; Copeland, 'International Trade in Waste Products', 20 *J. Environ. Econ. Manage.* (1991), at 143 *et seq.*; Hansen/Thomas, 'The Efficiency of Sharing Liability', 19 *Int'l Rev. L. & Econ.* (1999), at

has to pay for the readiness of the consignee to take delivery of the hazardous wastes. For this reason, the interests of the parties involved are different. Since the shipper is obliged to pay for the hazardous wastes only if they are actually delivered to the consignee, it is in his genuine economic interest that these wastes never arrive at the place of destination. The consignee, by contrast, has an interest in taking delivery of the hazardous wastes, because only in this case is he entitled to claim payment. This interest, however, is not an interest in receiving the hazardous wastes as such, but rather in receiving the additional payment. Therefore, none of the parties involved has a genuine economic interest in a safe and sound transport and delivery of the hazardous wastes.[40]

This, of course, only applies in a non-regulated world that disregards non-economic interests. An indirect economic interest, however, is created by imposing liabilities on the parties involved. Rules of liability[41] create an additional financial obligation that is contingent on a certain negligent behaviour of the person in charge of the wastes. Hence, liability rules may create an incentive ensuring that the wastes in question are safely and soundly delivered to the consignee in order to avoid the incurrence of financial liabilities. Therefore, the crucial effect of liability is not the imposition of financial burdens, but rather creating legal certainty, namely guaranteeing that, as long as certain behavioural requirements are fulfilled, the person in charge of the wastes is discharged from any liabilities, i.e. additional costs.[42] The creation of such behaviour-correcting incentives is one of the prime legal aims of civil liability conventions.[43]

However, this legal technique for shaping the economic incentives of the parties also entails disadvantages and certain risks. First, if the liability provisions are too rigorous and require too high a capital expenditure for loss prevention in order to enjoy relief from liability, then this rather provides an incentive for the shipper to elude the official method of waste shipment and to favour illegal shipments and dumping. Second, liability does not create a positive economic incentive in the shipper for the delivery of the wastes, but rather works as a negative incentive that imposes liability in the sense of an economic sanction. Hazardous wastes, thus, incur a double negative

---

(Footnote 39 continued)
135 *et seq.*; Kirstein, *Internationaler Müllhandel aus Sicht der ökonomischen Analyse des Rechts*, in: Eger/Bigus/Ott/von Wangenheim (ed.) (2008), at 443 *et seq.*; Rauscher, *International Trade, Factor Movements, and the Environment* (1997), at 91–121.

[40] See also Hackmann, 'International Trade in Waste Materials', 29 *Intereconomics* (1994), at 298–299; Marbury, 'Hazardous Waste Exportation', 28 *Vand. J. Transnat'l L.* (1995), at 259; O'Neill, *Waste Trading Among Rich Nations* (2000), at 34; Rauscher, *International Trade in Hazardous Waste*, in: Schulze/Ursprung (ed.) (2001), at 152.

[41] This basically applies to rules of fault-based liability, but may also apply to rules of strict liability, provided there are exclusions from liability in case of certain inevitable events.

[42] See Giampetro-Meyer, 'Captain Planet Takes on Hazard Transfer', 27 *UCLA J. Envtl. L. & Pol'y* (2009), at 77–84; Hackmann, 'International Trade in Waste Materials', 29 *Intereconomics* (1994), at 299–300.

[43] Lawrence, 'Negotiation of a Protocol on Liability and Compensation', 7 *RECIEL* (1998), at 250; Murphy, 'Prospective Liability Regimes', 88 *AJIL* (1994), at 62–63.

economic value that consists of not only the intrinsic negative value of the wastes as such, but also the attached risk of liability. The consequences are as follows: Since the shipper of the hazardous wastes cannot expect any positive financial return from delivering the wastes, but only bears the risk of additional liability, from a purely economic perspective there is no reason that the company performing the shipment remains operative. This creates a motivation to establish bogus firms that become insolvent in case of liability. A better solution, therefore, would likely be creating a positive economic interest in the delivery of the hazardous wastes, instead of imposing liabilities. However, due to the intrinsic negative value of most hazardous wastes, such a positive interest in hazardous wastes cannot be generated.

V. Summary

Wastes are commonly defined as substances or materials that are actually disposed of, intended to be disposed of, or are required by law to be disposed of. The hazardousness of wastes is usually defined by a combination of a particular waste stream and particular hazardous characteristics enumerated in comprehensive lists. The amount of hazardous wastes that are generated worldwide cannot be determined in absolute figures. Based on regional data it is, however, possible to identify trends in development. According to these figures, there are good reasons to assume that the volume of hazardous wastes generated annually is increasing by 4–5 %. The most common methods of hazardous waste treatment and disposal are geologic disposal at landfill sites as well as recovery and recycling. Incineration is also an important disposal method.

Since a positive economic value can be attached to non-hazardous wastes in most cases, the shipper has an intrinsic economic interest of the shipper to make sure that these wastes are shipped safely and soundly to the consignee's designated location. By contrast, such an incentive does not exist with regard to hazardous wastes. In such cases, it is rather in the shipper's economic interest that the wastes do not at all arrive at the final destination. Rules on liability may serve as an external corrective for this adverse economic incentive. However, imposing liabilities also poses risks. If liability is too stringent, it may have the opposite effect.

B. The Transboundary Movement of Wastes

I. Reasons for the Emergence of Hazardous Waste Movements

Transboundary movements of hazardous wastes emerged as a mass phenomenon in the 1960s and 1970s. The reason for the increasing importance of this kind of waste management was as banal as inevitable: Economic and political conditions stimulated such shipments.

In consequence of the increasing number of hazardous waste transports, the number of environmental catastrophes involving hazardous wastes, especially in the late 1970s and 1980s, increased as well.[44] These incidents attracted broad media coverage and gave rise to a heightened public awareness and an ecological sensitisation of the population primarily in industrial countries. Environmental organisations like *Greenpeace* and the *World Wildlife Fund (WWF)* rose in popularity, and at the political level a consciousness for the need of environmental regulations evolved.[45] In the following years, tougher environmental laws were enacted which established higher requirements for the sustainable treatment and disposal of hazardous wastes as well as higher requirements for employee protection.[46] This, in turn, caused a significant rise in the costs of domestic waste management.[47] In light of approximate disposal costs in industrial countries of up to USD 2,000 and, by contrast, disposal costs of about USD 2.50 to USD 50 in developing countries,[48] hazardous waste movements to developing countries for disposal became a quite economical solution even when taking into account additional transport costs.

A further consequence of the mentioned ecological sensitisation was the so-called NIMBY-syndrome.[49] Because of the media presence surrounding environmental incidents, broad segments of the population in industrialised countries became aware of the deleterious and harmful effects that were caused by hazardous wastes. Their concern was to keep treatment and disposal facilities for hazardous wastes away from their immediate vicinity. Due to the population's resistance, primarily in the USA, plans for the construction of waste incineration plans and disposal sites could not be realised. Disposal capacities decreased while the volume of produced hazardous wastes further increased.[50] Instead of pursuing a sustainable policy of waste avoiding, the solution for that capacity overload was

---

[44] Some examples are the *"Seweso Disaster"* in 1976, the case of the *M/V "Khian Sea"* in 1986 and the *"Koko Beach Incident"* in 1988.

[45] Friedrich-Ebert-Stiftung (ed.), *Zehn Jahre Basler Übereinkommen* (1999), at 9; Poulakidas, 'Waste Trade and Disposal in the Americas', 21 *Vt. L. Rev.* (1996/1997), at 877.

[46] Kummer, *The Basel Convention* (1995), at 6; Obstler, 'Toward a Working Solution to Global Pollution', 16 *Yale J. Int'l L.* (1991), at 78.

[47] The costs of hazardous waste in the USA increased from approximately USD 15 per ton in 1980 to approximately USD 250 regarding the disposal on landfill sites, or to approximately USD 2,000 for incineration, in 1988. See Friedrich-Ebert-Stiftung (ed.), *Zehn Jahre Basler Übereinkommen* (1999), at 9; Clapp, 'The Toxic Waste Trade', 15 *TWQ* (1994), at 506.

[48] Kummer, *The Basel Convention* (1995), at 6–7; Kitt, 'Waste Exports to the Developing World', 7 *Geo. Int'l Envtl. L. Rev.* (1994/1995), at 488.

[49] "NIMBY" is the acronym of "not in my backyard" and is also known as the "St. Florian's Principle".

[50] Friedrich-Ebert-Stiftung (ed.), *Zehn Jahre Basler Übereinkommen* (1999), at 9; Obstler, 'Toward a Working Solution to Global Pollution', 16 *Yale J. Int'l L.* (1991), at 76; Poulakidas, 'Waste Trade and Disposal in the Americas', 21 *Vt. L. Rev.* (1996/1997), at 877.

found by taking the "path of least resistance" and exporting those wastes to foreign countries with less stringent environmental rules.[51]

The fundamental economic and political conditions then prevailing in industrialised countries were accompanied by "ideal" framework conditions in the developing countries. At that time, the environmental and import regulations of most importing developing countries were weak and those countries often lacked sufficient enforcement capacities. Further relevant aspects included the prevalence of bribery, the need of foreign exchanges, and an undeveloped awareness of the population regarding adverse which hazardous wastes posed for human health and the environment.[52]

II. Quantities and Typical Patterns of Hazardous Wastes Movements

Data concerning transboundary movements of hazardous wastes, as far as the global scale is concerned, are best available from the Secretariat of the Basel Convention. Unlike the data collected by the Basel Secretariat on the generation of hazardous wastes, these data are quite reliable. This is to be explained by the fact that the reporting obligations under the Basel Convention are mandatory only in respect of transboundary movements of hazardous wastes.[53] Despite the mandatory nature of this obligation, there are, nevertheless, a large number of Member States which do not or only irregularly comply with the reporting duty.[54] In addition, differences in the national definitions of hazardous wastes and differences in the national reporting systems have a negative effect on the comparability of the collected data.[55] These deficiencies, however, can largely be overcome. This is due to the fact that the data submitted to the Basel Secretariat always contain information about the exporting and the importing country. By collecting and analysing the submitted data, missing data can in many cases by extrapolated.[56] In consequence, the statistics provided by the Secretariat of the Basel Convention can be

---

[51] Kummer, *The Basel Convention* (1995), at 6; Obstler, 'Toward a Working Solution to Global Pollution', 16 *Yale J. Int'l L.* (1991), at 75–80; Poulakidas, 'Waste Trade and Disposal in the Americas', 21 *Vt. L. Rev.* (1996/1997), at 910.

[52] Anand, *International Environmental Justice* (2004), at 64; Friedrich-Ebert-Stiftung (ed.), *Zehn Jahre Basler Übereinkommen* (1999), at 9; Kitt, 'Waste Exports to the Developing World', 7 *Geo. Int'l Envtl. L. Rev.* (1994/1995), at 488–490; Kummer, *The Basel Convention* (1995), at 7–8.

[53] Basel Convention, Article 13; see also Article 16.

[54] Only 58 Contracting States (ca. 33 %) have reported data on imports and exports of hazardous wastes for the year 2009; see the reporting database on www.basel.int.

[55] As to the reasons for incomplete or incomparable data see COP10 Doc. UNEP/CHW.10/INF/4, Annex 3, at 24 *et seq.*

[56] This methodology is explained in COP10 Doc. UNEP/CHW.10/INF/4, Annex 3, at 25–26.

considered a reliable approximation of the absolute amounts of hazardous wastes that are shipped across international boundaries. This applies at least to transports that are conducted legally and within the regulative framework of the Basel Convention.[57] The data collected at the regional level, such as by the EU or the OECD, will thus not be the basis of the further examination conducted in this present inquiry.

The majority of hazardous wastes generated worldwide is disposed of domestically. Older statistics suggest that approximately 10 % of all hazardous wastes are subject to transboundary movement,[58] whereas today a tentative estimate would be that only 1.5–3 % of all hazardous wastes are shipped across international boundaries.[59] The total volume of hazardous waste shipments was estimated to be about 7 million tons in 1999.[60] Between 2004 and 2006 the amount of hazardous wastes subject to transboundary movement then increased from 9.8 million tons to 11.2 million tons, which equals an increase of 15 % (7.5 % p.a.).[61] As of 2009, this amount has again increased in the same ratio.[62] From this it follows that despite the political efforts to reduce hazardous waste generation and movements, the absolute amount of hazardous wastes shipped across international boundaries is constantly increasing, and this at a significant order of roughly 7–7.5 % p.a.

Another issue concerns the question whether a typical global pattern of hazardous waste movements can be identified. A review of the relevant data reveals that the vast majority (approximately 90 %) of hazardous waste shipments takes place among so-called "Annex VII countries",[63] most of which are developed countries. Approximately 7 % of all hazardous waste shipments takes place among non-Annex VII countries, and shipments from Annex VII countries to non-Annex VII countries or vice versa represent only 1 and 2 %, respectively.[64] In general,

---

[57] As to the significance of illegal shipments see *infra*, Sect. "Illegal Traffic and Shipments off the Official Path".

[58] Kummer, *The Basel Convention* (1995), at 5; O'Neill, *Waste Trading Among Rich Nations* (2000), at 36; Valin, 'The Basel Convention', 6 *Ind. Int'l & Comp. L. Rev.* (1995), at 268.

[59] The 1980s estimation was based on an overall generation of 300–500 million tons. Given the increasing amount of generated hazardous wastes and the present estimations of transboundary movements as outlined in the following text, the 10 % estimation would only come to 112 million tons. See also Montgomery, 'Reassessing the Waste Trade Crisis', 4 *J. Env. & Dev.* (1995), at 4.

[60] Basel Secretariat (ed.), *Global Trends 1993–2000*, at 22.

[61] COP10 Doc. UNEP/CHW.10/INF/4, at 7, 26; Basel Secretariat (ed.), *Global Trends 2004–2006*, at 12.

[62] See the reporting database on www.basel.int, which, however, does not provide the evaluated and corrected data.

[63] Annex VII of the Basel Convention (which is not yet in force) comprises OECD Member States, EU Member States and Liechtenstein.

[64] These data are provided for 2006 by COP10 Doc. UNEP/CHW.10/INF/4, at 7–9; the same figures also apply for the following years; see the reporting database on www.basel.int.

more than 90 % of all hazardous waste shipments remain within the same region and take place between adjacent countries.[65] With regard to the intended form of disposal, it can be said that more than 80 % of the hazardous wastes subject to transboundary movements are destined for recovery or recycling operations.[66] Based on these figures it can be concluded that the main reason for transboundary movements is the different availability of specialised treatment capacities in the respective countries of export and of import. The level of wealth of the respective countries, in contrast, seems to be of no or only marginal significance.[67] This conclusion is furthermore confirmed by the fact that in comparison with the overall volume of movements, only negligible amounts of hazardous wastes are legally shipped from Annex VII countries to non-Annex VII countries.[68]

As a final remark, it should be stressed again that these figures and the resulting conclusions only take into account hazardous waste movements that are conducted in accordance with the applicable international law. Conclusions as to illegal traffic are not possible.

III. The Involvement of Waste Brokers and Waste Dealers

The central figures in the practice of transboundary hazardous waste shipments are waste brokers and waste dealers. Depending on the respective national regulations on waste management, the generator of hazardous wastes either enters into a contract directly with a waste broker or waste dealer, or he is obliged by law to deliver the wastes to a public waste management company. In the latter case this company may either perform the tasks of a waste broker or waste dealer itself, or it may sub-contract with a private waste broker or waste dealer.

In most cases, a waste broker is involved in the first instance. Waste brokers hold the function of an agent who is not buying the wastes itself, but who rather arranges on behalf of its principal for any task to be done with regards to the exportation of the hazardous wastes. Those tasks may involve, for instance, the packaging, collection, pre-treatment, separation, re-packing, marking or other preparatory operations related to the hazardous wastes. They also comprise the search for a buyer of the wastes or for a disposal or recycling plant, as well as the conclusion of the relevant contracts. Furthermore, the tasks of a waste broker encompass the organisation of the transport, including related tasks such as the organisation of the container packaging, customs clearance and insurance coverage. Waste brokers may

---

[65] Basel Secretariat (ed.), *Global Trends 2004–2006*, at 17.
[66] COP10 Doc. UNEP/CHW.10/INF/4, at 8; Basel Secretariat (ed.), *Global Trends 2004–2006*, at 15.
[67] COP10 Doc. UNEP/CHW.10/INF/4, at 8; Basel Secretariat (ed.), *Global Trends 2004-2006*, at 18.
[68] COP10 Doc. UNEP/CHW.10/INF/4, at 9.

also arrange for the fulfilment of the legal requirements for export, such as the notification procedures, documentation, and other national licensing requirements.

Waste dealers, by contrast, do not act as agents on behalf of their principals, but rather buy and re-sell the wastes on their own behalf.[69] It is rather unlikely that hazardous waste generators would directly enter into a contract with a waste dealer. This might, however, be different with regard to generators or public management companies having a large amount of production.

In a typical waste export and disposal scenario, several waste brokers and waste dealers are involved. They may collect, store and re-mix wastes on their own or on behalf of others and sell the particular wastes to further waste dealers specialised in the particular waste stream. Waste brokers and waste dealers often consist of small companies or even of one-man businesses. There is, in general, no legal requirement setting a minimum financial budget for such companies. Particularly with regard to waste brokers, there is also no actual need for their possessing either significant assets or a logistic infrastructure of their own. However, it should be noted that due to an advancing globalisation of the waste trading and waste broking markets as well as increasing competition, smaller companies are increasingly being driven out of the market and a market concentration towards some larger companies with a predominant market position is ongoing.[70]

IV. Illegal Traffic and Shipments Off the Official Path

An assessment of the data that have been reported by the Contracting States to the Secretariat of the Basel Convention certainly allows for some meaningful conclusions. The picture drawn by these data, however, is by far not complete and does not sufficiently reflect the actual volume of hazardous waste shipments and the global allocation of import and export roles. Shipments of hazardous wastes are often conducted illegally. A major incentive for illegal traffic is the low effort required and the huge profits that can be drawn in comparison with legal shipments.[71] Reliable data regarding the extent of illicit traffic are not available. However, some authors estimate that illegal shipments make up a large portion of hazardous waste shipments,[72] especially from developed to developing countries.[73]

---

[69] See e.g. Wynne, 'The Toxic Waste Trade', 11 *TWQ* (1989), at 130.

[70] van Daele/Vander Beken/Dorn, 'Waste Management and Crime', 37 *Envtl. Pol'y & L.* (2007), at 36; O'Neill, 'Out of the Backyard', 7 *J. Env. & Dev.* (1998), at 147.

[71] See e.g. van Daele/Vander Beken/Dorn, 'Waste Management and Crime', 37 *Envtl. Pol'y & L.* (2007), at 36.

[72] Kummer, *The Basel Convention* (1995), at 7; For a critical examination of the "tip of the iceberg theory" see Montgomery, 'Reassessing the Waste Trade Crisis', 4 *J. Env. & Dev.* (1995), at 13–20.

[73] An example of a recent instance of illegal dumping in the Czech Republic/Germany is given by Meßerschmidt, *Europäisches Umweltrecht* (2011), at 891.

Illegal traffic is deemed to be any transport that is conducted in contravention of the applicable rules governing the transboundary movement of hazardous wastes.[74] Such contravention may consist in circumventing the applicable rules, e.g. by shipping the wastes without any notification or approval, by mislabelling, mis-declaring or misreporting the wastes, or by breaching any other substantive requirement regarding the approval procedure or the actual shipment.

The typical procedure associated with an illegal shipment depends on the volume of hazardous wastes to be shipped. Where smaller amounts are shipped in packaged form, the cargo is often mislabelled and sent either to an intentionally incorrect or fictitious address where the cargo is unloaded. By the time the true content of the cargo is recognised, usually some time has passed and the original shipper will no longer be traceable. The larger the amounts of hazardous wastes to be shipped, the greater the efforts that need to be made. This may involve the payment of landlords to accept the storage of the wastes on their property, or the payment of bribes to local officials or other governmental decision-makers.[75]

In addition to the illicit traffic in hazardous wastes that takes place contrary to the relevant rules, there is also a certain "grey area". Relevant obligations are often circumvented by mixing hazardous wastes with non-hazardous wastes in order that the concentration of the entire waste remains below a predefined threshold that determines the hazardous character of that waste.[76] Another strategy is "creative labelling", which means the incorrect classification of wastes as products, recycling material or humanitarian aid, as well as "outright mislabelling", which denotes the process of changing the nature of wastes by mixing or processing it so that it may be treated as a raw material or energy source.[77]

V. Summary

The transboundary movement of hazardous wastes arose, ironically, as a mass phenomenon in the 1960s and 1970s as a consequence of the emergence of an ecologic awareness and a related political change in the political sentiment in the developed countries. In addition, increasing costs for domestic disposal creates the prospect of huge profits associated with the exportation of hazardous wastes to

---

[74] Illegal traffic is defined by the Basel Convention as any transboundary movement of hazardous wastes that does not conform with the requirements of the Basel Convention, Article 2(21), 9(1). Similar definitions can be found in Article 1(22), 9(1) of the Bamako Convention and in Article 1, 9(1) of the Waigani Convention.

[75] See Liu, 'The Koko Incident', 8 *J. Nat. Resources & Envtl. L.* (1992/1993), at 126.

[76] van Daele/Vander Beken/Dorn, 'Waste Management and Crime', 37 *Envtl. Pol'y & L.* (2007), at 35–36.

[77] van Daele/Vander Beken/Dorn, 'Waste Management and Crime', 37 *Envtl. Pol'y & L.* (2007), at 35–37; Giampetro-Meyer, 'Captain Planet Takes on Hazard Transfer', 27 *UCLA J. Envtl. L. & Pol'y* (2009), at 76.

countries with less stringent environmental regulations. The volume of hazardous wastes shipped abroad for disposal, thus, has constantly increased over the years and today continues to do so at the order of 4–5 % annually. The vast majority of legal hazardous waste shipments takes place among OECD countries (approximately 90 %). Only approximately 1 % is shipped from OECD countries to non-OECD countries. More than 80 % of the wastes are shipped for recovery and recycling purposes. These data, however, do not allow for conclusions regarding illegal traffic in hazardous wastes. The actual extent and the significance of illicit shipments remain unknown, although it is presumed to be a considerable amount. Illegal trafficking is fostered by the lack of sufficient governmental structures and enforcement capabilities especially in developing countries and, above all, by the huge profits that can be made by the persons involved. A further common way to circumvent legal requirements regarding the transboundary movement of hazardous wastes is the pre-treatment of such wastes. By mixing hazardous and non-hazardous wastes or by selling wastes as products or raw materials, attempts are made to avoid trade restrictions and requirements.

C. The Interests Involved

The transboundary movement of hazardous wastes is an exceedingly emotional and politicised topic. It does not comprise solely the economic activity of selling and shipping a tradable good from one place to another; rather it involves the fundamental issues of equal opportunity and equal treatment of humans around the world, a global allocation of responsibilities, the sharing of resources and economic benefits on a global scale, and the right to participate in global trade activities including the right to draw benefits.

I. Private Parties

Private parties are the main participants and factors in the field of hazardous waste shipments. The greatest share of transboundary movements are initiated and conducted by private parties; State owned companies play only a minor role. Private parties involved in hazardous waste shipments are the generators of the wastes, waste brokers and waste dealers, the carrier and the shipping company, and the disposer of the wastes. The terms notifier, shipper and consignee only denote specific functions of some of these involved parties.

The peculiarities of hazardous wastes as a tradable good have already been outlined above. The shipper has no genuine economic interest in the goods actually arriving at the place of final destination. The consignee is only interested in receiving the funds, rather than receiving the hazardous wastes themselves.[78] The interest of the private parties involved, however, is not only influenced by the value of the wastes themselves. There is rather a multitude of surrounding factors that affects the decision of private parties to participate in a transboundary movement. The primary drivers are surely economic incentives; however, legal and political factors as well as personal aspects also play a significant role.[79]

As regards economic factors that prompt hazardous waste movements, three issues are to be mentioned: First, the gap between the demand for raw materials and the amounts of raw materials that are available locally. Developing countries and countries with economies in transition need large amounts of the cheap raw materials that they can primary recover from recyclable wastes mixed with hazardous substances—such as electronic waste or metal scrap[80]—through the use of informal low technology installations or inadequately protected manual labour. Second, the gap between disposal costs in developed and in developing countries is still significant, mainly due to lower labour costs resulting from low environmental, health and labour standards. And third, the gap between the amounts of wastes generated in a certain area and the disposal capacities locally available. In order to be operated in a cost-effective manner, a waste treatment facility needs sufficient and consistent inputs. Since not every State can operate a specialised installation for any particular waste stream, waste movements are necessary to keep those specialised installations in continuous operation.[81]

Legal factors involve ineffective national legislation as well as a lack of legal clarity. The implementation of effective national legislation that sufficiently transforms the provisions of the Basel Convention into national law and, furthermore, makes use of the legal opportunities given by the Convention[82] represents a major challenge for all States Parties. It requires existing legal and administrative structures, considerable financial capacities and the involvement of well-qualified personnel. But above all, it requires an awareness concerning the importance of this issue. If the contracting State's focus is laid on another topic deemed of greater political profile, the lack of effective national legislation fosters waste trade and treatment in contravention of the existing international laws.[83] A similar meaning is attached to the availability of sufficient enforcement provisions and capacities. The best law is only as good as it can be enforced in practice. This

---

[78] *Supra*, Sect. "The Commercial Value of Hazardous Wastes".

[79] A comprehensive examination of the reasons for movements of hazardous wastes was prepared by the Indonesian-Swiss Country-Led Initiative to Improve the Effectiveness of the Basel Convention, as annexed to COP10 Doc. UNEP/CHW.10/INF/4.

[80] As regards the application of the Basel Convention to the export of end-of-life ships for dismantling see Lagoni/Albers, 'Schiffe als Abfall?', 30 *NuR* (2008), at 220 *et seq.*

[81] COP10 Doc. UNEP/CHW.10/INF/4, at 10–12.

[82] A Contracting State may, for instance, unilaterally ban any import of hazardous wastes with the legal consequence that the other Contracting States are obliged to ensure that no exports will be delivered to this particular State; see Basel Convention, Article 4(1)(b).

[83] COP10 Doc. UNEP/CHW.10/INF/4, at 12–14, 17.

includes not only the workforce and technical capacities and infrastructures for border controls and controls inside the country, but also the qualification of the competent authorities and the personnel in charge.[84]

All these factors lead to the economically driven incentive to not dispose of hazardous wastes at the place of generation, but to move them to other States for disposal. However, the increase in the amount of transboundary movements of hazardous wastes is not the only consequence. Due to the strict laws on waste shipments, particularly in developed countries, the recent development shows that in fact entire production processes, which are very waste-intensive, are moved to developing countries with less stringent laws. Although the settlement of production processes in developing countries has the advantage of creating additional employment in the respective regions, this does not compensate for the disadvantage of having generated a significant threat to human health and the environment by eluding the Basel Convention's requirement of environmentally sound management in the State of import.[85]

II. State Interests

Considering the adverse effects of hazardous wastes on human health and the environment, one could say that it is a harder question to determine why countries import wastes than to determine why they export them.[86] To give an answer to both questions, one should have a precise look at both varieties of States. Such an effort, however, is inevitably complicated by the common fallacy that waste exporting States are developed countries and waste importing States are developing countries, an assumption that fails to reflect the true picture of hazardous waste movements.[87] Rather, developed and developing countries exist on both the waste exporting and the waste importing side.

Based on the assumption that States predominantly act in the interest of their economies and pursue the concerns of their population,[88] State interests largely overlap with the private economic interests predominantly represented in that

---

[84] COP10 Doc. UNEP/CHW.10/INF/4, at 14–16; Alam, 'Trade Restrictions Pursuant to MEAs', 41 *J.W.T.* (2007), at 1003.

[85] Giampetro-Meyer, 'Captain Planet Takes on Hazard Transfer', 27 *UCLA J. Envtl. L. & Pol'y* (2009), at 76; Hackmann, 'International Trade in Waste Materials', 29 *Intereconomics* (1994), at 297; Suttles, 'Transmigration of Hazardous Industry', 16 *Tul. Envtl. L. J.* (2002/2003), at 1 *et seq.*

[86] See O'Neill, *Waste Trading Among Rich Nations* (2000), at 4.

[87] See *supra*, Sect. "Quantities and Typical Patterns of Hazardous Wastes Movements". For the diverging positions of the EU and the USA, both of which have highly advanced economies, see Dreher/Pulver, 'Environment as "High Politics"?', 17 *RECIEL* (2008), at 308 *et seq.*

[88] This, of course, excludes any political activity that is driven by corruption, the acceptance of advantages and other personal interests of the involved decision-makers.

State. Political and other interests may play an additional important role.[89] A State's incentive to export wastes may be based in its lacking sufficient domestic disposal capacities, which may be due to the already mentioned NIMBY-syndrome in developed countries or, as far as economies in transition are concerned, may be due to strong growth in the national economy which has not been accompanied by a corresponding growth of specialised disposal capacities. A further incentive of States to advocate for hazardous waste trades is the need to operate specialised waste disposal facilities in a cost-efficient manner, thus requiring a steady and sufficient supply of the particular waste stream. Since specialised installations for any particular waste stream cannot be operated by every State, it is necessary that particular waste streams are imported or exported in order to keep the available installations efficiently operating.[90] Most OECD countries, therefore, rejected the idea of a total ban on the trade of hazardous wastes. Moreover, a trade ban was expected to encourage illegal waste traffic and less environmentally sound means of disposal at the final destination. The OECD countries rather supported the adoption of an international regime establishing a requirement of notification and prior informed consent (PIC) and also imposing an obligation to ensure that the wastes are disposed of in an environmentally sound manner (ESM) in the State of import.[91] After the adoption of the Basel Convention laying down these principles, the EU took a unique approach and unilaterally banned any hazardous waste shipments from EU to non-OECD countries. This solution allows for transboundary movements among EU and OECD countries and thus takes account of the economic need to share specialised disposal capacities among adjacent countries. On the other hand, it prohibits environmentally risky exports of hazardous wastes to developing countries. A further basis for this political initiative of the EU may be seen in the political awareness that multilateral leadership is no less required in respect of environmental concerns than as regards economic or security issues.[92]

Regarding countries that are importing hazardous wastes, one incentive for accepting wastes is seen in the generation of jobs, income and technical advancement.[93] In the end, however, these advantages may turn out to be of less value since they are only short-term effects and, in the long run, the negative effects on human health and the environment will prevail.[94] An actual advantage, by contrast, is the steady and cheap supply of recyclable wastes (e.g. electronic waste and metal scrap which also contain rare earth elements) that allow the

---

[89] See also Rauscher, *International Trade in Hazardous Waste*, in: Schulze/Ursprung (ed.) (2001), at 157.
[90] See *supra*, Sect. "Private Parties".
[91] Johnstone, 'The Implications of the Basel Convention for Developing Countries', 23 *Resour. Conserv. Recycl.* (1998), at 4–5; Krueger, 'The Basel Convention', *YBICED* (2001/2002), at 45.
[92] Dreher/Pulver, 'Environment as "High Politics"?', 17 *RECIEL* (2008), at 311.
[93] Helfenstein, 'U.S. Controls on International Disposal of Hazardous Waste', 22 *Int'l Law.* (1988), at 788; Ovink, 'Transboundary Shipments of Toxic Waste', 13 *Dick. J. Int'l L.* (1994/1995), at 284.
[94] Poulakidas, 'Waste Trade and Disposal in the Americas', 21 *Vt. L. Rev.* (1996/1997), at 875.

recovery of raw materials, especially ferrous and non-ferrous metals. Depending on the respective demand for raw materials and the importance of the domestic waste recycling industry, countries with less-developed economies thus either advocate for less stringent rules on the transboundary movement of hazardous wastes, or they support national or regional import bans[95] and a corresponding export ban imposed on exporting countries.[96]

From all this it follows that exporting and importing States, irrespective of their economic strength, may have an interest in the hazardous wastes trade. Whether or not such an interest exist largely depends on their respective economies and either their dependence on an external supply of recyclable or recoverable waste or their possessing a surplus of such substances. The political reaction of adopting a unilateral or regional import ban or permitting hazardous waste imports is at the sole discretion of the respective State. From a legal perspective, hence, every State is sufficiently protected against unwanted hazardous waste imports.[97] It becomes obvious that the establishment of a total ban on hazardous waste shipments would exceed a reasonable and justifiable protection of vulnerable States and would only render impossible the necessary trade in hazardous wastes among countries having a comparable economic strength. A total trade ban would not eliminate illegal traffic and corruption, which is seen as the main cause of unwelcome damage to human health and the environment,[98] but it would rather intensify this phenomenon. Therefore, the proper answer can only be to reject any extreme form of interstate paternalism and to respect the sovereign interests of any particular State.[99] The PIC and ESM requirements of the Basel Convention in connection with regional political approaches, like the EU's ban on hazardous waste exports to non-OECD countries or the import bans of the Bamako and Waigani Conventions, thus represent a reasonable solution which serves to facilitate necessary trade among adjacent countries, on the one hand, and to prevent the abuse of economic imbalances between developed and developing countries, on the other hand.

---

[95] The Regional Workshop Aimed at Promoting Ratification of the Protocol held in Addis Ababa, Ethiopia in August/September 2004 under the aegis of the Basel Secretariat revealed that most African countries have established import bans on hazardous wastes, see COP7 Doc. UNEP/CHW.7/INF/11, at 3.

[96] Johnstone, 'The Implications of the Basel Convention for Developing Countries', 23 *Resour. Conserv. Recycl.* (1998), at 3; Krueger, 'The Basel Convention', *YBICED* (2001/2002), at 45.

[97] See also Montgomery, 'Reassessing the Waste Trade Crisis', 4 *J. Env. & Dev.* (1995), at 12.

[98] See also Fagbohun, 'The Regulation of Transboundary Shipments of Hazardous Waste', 37 *HKLJ* (2007), at 850–851.

[99] See also Hackett, 'Assessment of the Basel Convention', 5 *Am. U. J. Int'l L. & Pol'y* (1989/1990), at 298.

III. Hazardous Waste Trade as "Environmental Racism"?

The issue of transboundary movements of hazardous wastes has been criticised by some authors as a form of "environmental colonialism" or "environmental racism" comprising a part of the north-south-conflict.[100] It is not possible at this point to offer a detailed discussion of this emotionally charged theory which has involved a great deal of polemics and quarrelsomeness on both sides of the table. However, since this issue has consumed considerable space in public debate, its essential argument shall be outlined in brief.

The theory of "environmental racism" is understood by its supporters as an alternative or additional explanation for hazardous waste exports from developed to developing countries. It is based on the understanding that industrial countries are securing their superior economic position by consciously taking advantage of developing countries. Developing countries are used as supplier for raw materials and cheaply produced commodities. Once the lifetime of these products has expired, they are sent back to developing countries as "civil waste". The driving force behind this exploitation of developing countries is attributable to "a racist and classist culture and ideology within northern communities and institutions that view toxic dumping on poor communities of color as perfectly acceptable".[101] Developed countries, hence, are seen as being "willing to use developing countries as a dumping ground not because of cost or convenience but because of race and poverty".[102]

The "environmental racism" theory, however, begins with a false assumption; and even if this assumption were once true, at least since the adoption of the Basel and Bamako Conventions the fundamental conditions no longer apply. It has already been stressed above that both developed and developing countries are both exporters and importers of hazardous wastes. Furthermore, the portion of hazardous wastes that are moved from OECD countries to non-OECD countries only amounts to approximately 1 % of the overall transboundary movement. The assumption that industrialised countries are using developing countries as dumping grounds for their civil wastes, therefore, is simply not correct.[103] This is, furthermore, underscored by the fact that the EU has unilaterally banned hazardous waste exports to non-OECD countries. Finally, by adopting the Basel and Bamako Conventions, an international legal regime has been established that provides an

---

[100] Marbury, 'Hazardous Waste Exportation', 28 *Vand. J. Transnat'l L.* (1995), at 291–293; Pellow, *Resisting Global Toxics* (2007), at 9–10; O'Neill, 'Out of the Backyard', 7 *J. Env. & Dev.* (1998), at 142; see also Obstler, 'Toward a Working Solution to Global Pollution', 16 *Yale J. Int'l L.* (1991), at 80–81; Park, 'An Examination of International Environmental Racism', 5 *Ind. J. Global Legal Studies* (1997/1998), at 659 *et seq.*

[101] Pellow, *Resisting Global Toxics* (2007), at 9.

[102] See Park, 'An Examination of International Environmental Racism', 5 *Ind. J. Global Legal Studies* (1997/1998), at 660 with further references.

[103] Montgomery, 'Reassessing the Waste Trade Crisis', 4 *J. Env. & Dev.* (1995), at 3.

effective international instrument for the regulation of hazardous waste shipments and that allows for the recognition of a State's sovereign right to decide whether to ban or admit hazardous waste imports.

In summary, it must be stated therefore that, at least today, the theory of "environmental racism" has no validity as regards the transboundary movement of hazardous wastes.[104]

IV. Summary

The factors accounting for the emergence and tremendous increase of hazardous waste movements in the 1960s and 1970s are today no longer prevailing. The main initiators of hazardous waste shipments at present are private persons and companies located in those importing States requiring large amounts of cheap raw materials which can be recovered from hazardous wastes containing recyclable components. The main industrial sectors in this context are energy recovery and ferrous and non-ferrous metal extraction from electronic wastes and waste metals. A further reason for transboundary hazardous waste movements is the allocation of specialised incineration and disposal plants among adjacent countries or countries of comparable economic strength. The cost-efficient operation of specialised installations requires a steady supply of the respective waste stream at a sufficient volume. Nevertheless, considerably lower disposal costs in less-developed countries continue to encourage transboundary waste trade, such commerce being promoted also by insufficient national legislation and enforcement capacities. Today, economic interests are still the main drivers for hazardous waste movements, although political factors must not be underestimated. Those political factors may include an appreciation of a global political responsibility for environmental concerns and the exercise of multilateral leadership in this regard.

In addition to the adoption of the Basel Convention, regional political initiatives, like the EU ban on hazardous waste exports to non-OECD countries or the Bamako and Waigani Conventions, have created a comprehensive legal framework that allows for the consideration of the respective interests of the particular States. It is ensured that the sovereign right of each State to either participate in the hazardous waste trade or to be protected from such imports prevails. The most remarkable conclusion, however, is that the north-south current of the hazardous waste trade no longer applies.

---

[104] See also Park, 'An Examination of International Environmental Racism', 5 *Ind. J. Global Legal Studies* (1997/1998), at 700–702.

D. The Risk Potential of Hazardous Waste Movements

In consequence of the increasing number of hazardous waste shipments, the risk of incidents releasing hazardous wastes into the environment and the risk of pollution due to insufficient treatment methods has increased as well. Particularly lesser developed countries often lack the technology, infrastructure and know-how necessary for treating hazardous wastes in an appropriate manner. The transportation by sea poses further risks. Often significant volumes of hazardous wastes are concentrated on one means of carriage, and such shipments sail through areas where the cargo is beyond any legal or actual control. In respect of hazardous wastes that are shipped illegally, there is the additional risk that these wastes are mislabelled and essential safety instructions are omitted such that the actual hazard cannot be recognised by the persons handling and processing those wastes.

Human health and the environment may be affected in several ways by hazardous wastes. Whereas a given workforce may be directly exposed to hazardous wastes, it is also the case that a local community may be indirectly exposed, e.g. by burning or releasing toxic substances without any treatment on dumping grounds or into surface waters.[105] Hazardous wastes may pollute ground and surface waters or may lead to atmospheric pollution and soil contamination, potentially causing damage to ecologically important habitats or entire ecosystems.[106]

---

[105] A significant example of harm being caused to a great number of people by the release of hazardous wastes without any treatment into surface waters is offered by the case of the M/V "Probo Koala" in 1996; see *supra*, Sect. "The Factual Perspective: Transboundary Movements of Hazardous Wastes by Sea" in Chap. 1.

[106] COP10 Doc. UNEP/CHW.10/INF/4, at 30; Krueger, 'The Basel Convention', *YBICED* (2001/2002), at 43; Kummer, *The Basel Convention* (1995), at 12–16; Valin, 'The Basel Convention', 6 *Ind. Int'l & Comp. L. Rev.* (1995), at 270; Walsh, 'The Global Trade in Hazardous Wastes', 42 *Cath. U. L. Rev.* (1992/1993), at 105.

Chapter 3
The Present Legal Framework

In this chapter, the present state of the international legal framework governing liability and compensation for damage resulting from the transboundary movement of hazardous wastes by sea is outlined. To this end, this chapter not only examines the existing rules and provisions of international law, which apply *de lege lata* to transboundary movements of hazardous wastes by sea, but also describes and examines recent and current attempts at the international level to further develop this legal framework. In so doing, it is not only possible to give an overview of the rules and provisions currently applying to transboundary movements of hazardous wastes by sea, but also to take into account recent developments and emerging trends.

In the following sections it is examined, first, whether international responsibility and liability of States for damage resulting from the transboundary movement of hazardous wastes may arise from the legal concepts of either State responsibility or State liability. For this purpose, the general legal conception of international responsibility of States is outlined, and the attempts of the International Law Commission (ILC) to codify related rules are depicted. Subsequently, the principle of State responsibility is applied to the particular aspects of hazardous waste movements. In this context, also the applicable international and regional conventions and agreements providing for substantive rules relevant to the transboundary movement of hazardous wastes by sea are identified and examined in detail for their precise content and the impacts they hold for the constellations in question. Finally, the existing civil liability conventions and instruments shall be outlined and their legal ramifications shall be expounded upon.

A. The Legal Concept of International Responsibility

In order to examine the application of the existing rules and provisions of international law to the particular aspects and scenarios of transboundary movements of hazardous wastes by sea, it seems appropriate to first outline the legal concept of international responsibility and liability. In general, three legal constructs need to be taken into account: State responsibility, State liability and international uniform civil liability instruments.

First of all, however, some remarks concerning the terminology used in this context shall be made. The terms "responsibility" and "liability" are generally used in this book according to their common law meaning.[1] "Responsibility" means the State of being legally accountable for a certain situation and, thus, comprises any legal consequence that is attached to this state of accountability. The term "liability", by contrast, is narrower in meaning. It only denotes the state of being financially obligated as one of the possible legal consequences of being "responsible".[2]

Independent of this distinction, these meanings cannot simply be transferred to the terms "State responsibility" and "State liability". The issue of State responsibility and liability for wrongful and dangerous activities is the subject of a lively and controversial discussion, one which is compounded by inconsistent terminology.[3] There seems to be agreement, however, as to the use of the term "State responsibility". This term is generally understood to describe the state of being internationally obligated as consequence of an international wrongful act of State; hence, it requires a breach of an international obligation by the State in question. It becomes apparent that, different than the term "responsibility", the decisive characteristic for the legal concept of "State responsibility" is to be identified in the origin of liability (in this case "liability *ex delicto*") rather than in any other aspect, such as the particular legal consequence.[4] By contrast, no such consensus exists as to the meaning of the term "State liability". While some authors use this term to denote the legal consequence of being financially obligated (according to the meaning of the term "liability"), it is used in other contexts to denote the position of being internationally obligated in respect of damage that occurs due to injurious activities which are not deemed wrongful under international law

---

[1] In civil law systems the distinction between responsibility and liability is unknown. Both concepts are covered by the terms "responsabilité", "responsabilidad" or "Haftung", see Quentin-Baxter, ILC Doc. A/CN.4/334 and Add. 1 and 2 (YBILC 1980 II/1), at 250–251; Barboza, ILC Doc. A/CN.4/402 (YBILC 1986 II/1), at 145–146.

[2] See Barboza, 'International Liability for Lawful Acts', 247 *RdC* (1994), at 305–307; Boyle, 'State Responsibility and International Liability', 39 *Int'l & Comp. L. Q.* (1990), at 9; "Liability and Responsibility" in: *Black's Law Dictionary*, (9th ed. 2009).

[3] Reasons for this inconsistency may be seen in the mixture of common law and civil law influences as well as in the use of different denotations of the respective legal concepts in international conventions. Some examples in this respect are given by Lefeber, *Transboundary Environmental Interference* (1996), at 13–15.

[4] Bergkamp, *Liability and Environment* (2001), at 156; Crawford, *Brownlie's Principles of Public International Law* (8th ed., 2013), at 541–542; Evans (ed.), *International Law* (2nd ed., 2006), at 455, 463–467; see also Scovazzi, 'State Responsibility for Environmental Harm', 12 *Yb. Int'l Env. L.* (2001), at 43 *et seq.*

("liability *sine delicto*").[5] Pursuant to the conception of the term "State responsibility", it seems consistent to also classify the term "State liability" according to the respective cause of liability. The term "State liability", hence, will be used in this book to describe responsibility and liability arising from an injurious activity that is not deemed wrongful under international law.[6]

I. The Principle of State Responsibility for Internationally Wrongful Acts

The vast majority of rules in international law have been established in order to govern international relations among States and other subjects of international law. Irrespective of the respective source they derive from (conventions, custom, general principles of law, *etc.*), these rules constitute individual rights, stipulate standards of conduct and impose duties to act or, alternatively, to refrain from acting. They are commonly referred to as international "primary rules" or "primary obligations".[7]

Even though primary rules may establish a well-balanced substantive solution regarding a particular subject matter, they cannot prevent States from infringing these rules for a variety of reasons. In that case the question arises whether an infringement necessarily entails legal consequences and, if so, what the possible legal consequences might be. Explicit rules concerning this issue barely exist and are mostly annexed to the respective primary obligations.[8] An international treaty generally stipulating the legal consequences of any breach of an international obligation is not in force at present. In 2001, the International Law Commission (ILC) adopted its Draft Articles on Responsibility of States for Internationally Wrongful Acts, which have not yet entered into force.[9] These Draft Articles, albeit

---

[5] Ago, ILC Doc. A/CN.4/246 and Add.1-3 (YBILC 1971 II/1), at 203; Dahm/Delbrück/Wolfrum, *Völkerrecht, vol. I/3* (2nd ed., 2002), at 885–886; Kiss/Shelton, *Strict Liability in International Environmental Law*, in: Ndiaye/Wolfrum (ed.) (2007), at 1138–1140. See also Pinto, 'Reflections on International Liability', 16 *NYIL* (1985), at 24–28; Randelzhofer, 'Probleme der völkerrechtlichen Gefährdungshaftung', 24 *BDGVR* (1984), at 35 *et seq.*

[6] This terminology and conceptual distinction is in line with the work of the ILC. In 1997, the ILC decided to use the term "State responsibility" exclusively in connection with internationally wrongful acts and to use the term "State liability" to denote the state of being accountable for damage resulting from acts not prohibited by international law. See ILC Doc. A/9010/Rev.1 (YBILC 1973 II) at 169.

[7] ILC Doc. A/9010/Rev.1 (YBILC 1973 II) at 169; Ipsen, in: Ipsen (ed.), *Völkerrecht* (5th ed., 2004), at 619; see also Crawford, 'The ILC's Articles on State Responsibility', 96 *AJIL* (2002), at 876–879.

[8] See *infra*, Sect. "Explicit Provisions of State Responsibility in International Treaty Law".

[9] See in detail *infra*, Sect. "State Responsibility".

not legally binding as such, reproduce to a large extent existing rules of customary international law and are, therefore, used as a *de facto* binding instrument nevertheless.[10]

In order to understand the legal concept of State responsibility one must first take a step back. In respect of international law as such, scholars and jurists have long debated its fundamental claim to validity.[11] The international community lacks any superordinate political authority or central legislative body, and international law is characterised by the identity of its creator and the obligated subject(s).[12] Therefore, the binding force of international law, unlike law at the national level, cannot simply be derived from a sovereign instance but instead has to be explained on the basis of underlying principles of law. Several schools of thought have evolved attempting to derive the legal force of international law from different fundamental principles.[13] However, due to the rather philosophical nature of this issue, it might not be possible to determine one wholly accurate reason for the validity of international law. Therefore, one simply has to content oneself with the fact that international law is generally considered and treated as being valid

---

[10] See *infra*, Sect. "State Responsibility".

[11] For an overview of the historical development of this question see Dahm/Delbrück/Wolfrum, *Völkerrecht, vol. I/1* (2nd ed., 1989), at 34–44, and Doehring, *Völkerrecht* (2nd ed., 2004), at 3–11.

[12] Jellinek, *Allgemeine Staatslehre* (4th reprint of the 3rd ed., 1914), at 376; Scelle, *Droit International Public* (1943), at 21; Verdross/Simma, *Universelles Völkerrecht* (3rd ed., 1984), at 33–34; Graf Vitzthum, in: Graf Vitzthum (ed.), *Völkerrecht* (5th ed., 2010), at 22–24.

[13] Three main schools of thought have evolved. An early attempt to explain the validity of international law was provided by the theories of the State's will, according to which international law is deemed valid by virtue of the State's will either expressing a self-commitment of the State or aiming towards a common consent among States. See e.g. Hegel, *Grundlinien der Philosophie des Rechts* (edited by Georg Lasson, 3rd ed., 1930), at para. 330–340; Jellinek, *Allgemeine Staatslehre* (4th reprint of the 3rd ed., 1914), at 481, 484; Jennings/Watts (ed.), *Oppenheim's International Law* (9th ed., 1992), at para. 8; Triepel, *Völkerrecht und Landesrecht* (1899), at 88; see also *TheS.S. Lotus Case*, PCIJ Series A No. 10 (1927), at 19 *et seq*.

Subsequently, a normative school of thought emerged that tried to explain the binding force of international law by assuming a universally existing, constitutive norm of "*pacta sunt servanda*". This norm was considered to be an axiom or postulate that is *a priori* valid. This very positivistic approach was later extended with a view to non-treaty obligations. Accordingly, the binding effect of international law was derived from the basic principle to treat as law what is recognised and continuously applied as law by civilised nations, as well as from the principle of State sovereignty and integrity. See e.g. Dahm/Delbrück/Wolfrum, *Völkerrecht, vol. I/1* (2nd ed., 1989), at 43; Kelsen, *Reine Rechtslehre* (2nd ed., 1960), at 221–223; Anzilotti, *Lehrbuch des Völkerrechts* (translated 3rd ed., 1929), at 39; see also Kelsen, *General Theory of Law and State* (1949), at 349, who writes: "The States ought to behave as they have customarily behaved."

Further approaches were taken in accord with naturalistic and sociological positions which rely on fundamental values of mankind or on the social environment in which international law is embedded. See e.g. Verdross, *Die systematische Verknüpfung von Recht und Moral*, in: Sauer (ed.) (1950), at 12–13; de Visscher, *Théories et Réalités en Droit International Public* (4th ed., 1970), at 122.

and binding.[14] This conclusion is of importance also for the justification of State responsibility. If the binding force of international law is a matter of fact, then the necessary logical consequence is that any infringement of international law is to be considered as an injustice and may not remain without remedy. Otherwise, the legal force of international law would be devoid of meaning. It must, therefore, be seen as an intrinsic consequence of the legal force of international law that any breach of a primary obligation necessarily entails the state of being internationally responsible.[15] This outcome also corresponds to the legal concept of all domestic jurisdictions.[16] Accordingly, the correlation between a breach of international law and the resulting state of being internationally responsible represents a general principle of law.[17] Since this principle, furthermore, belongs to the customary international law established by international case law,[18] treaty practice[19] and legal doctrine,[20] it represents a valid part of the corpus of international law.[21]

The principle of State responsibility is to be seen as supplementing the primary obligations of international law. It functions as a trigger, which on the occasion of a breach of a primary obligation creates a new legal relationship between the acting State and the victim State or States. It furthermore invokes the application of an entirely different body of legal rules.[22] Since these rules are applied only in reaction to the breach of primary rules and solely within the new legal relationship,

---

[14] In the end, one has to be satisfied with the pragmatic reason that international law is binding because it is necessary; see Dahm/Delbrück/Wolfrum, *Völkerrecht, vol. I/1* (2nd ed., 1989), at 41; Ipsen, in: Ipsen (ed.), *Völkerrecht* (5th ed., 2004), at 16.

[15] *Factory at Chorzów Case*, PCIJ Series A No. 17 (1928), at 29; Ipsen, in: Ipsen (ed.), *Völkerrecht* (5th ed., 2004), at 618.

[16] Schweisfurth, *Völkerrecht* (2006), at 225.

[17] Ipsen, in: Ipsen (ed.), *Völkerrecht* (5th ed., 2004), at 618; von Münch, *Das völkerrechtliche Delikt* (1963), at 33.

[18] See e.g. *The S.S. Wimbledon Case*, PCIJ Series A No. 1 (1923), at 30; *Factory at Chorzów Case*, PCIJ Series A No. 17 (1928), at 29; *Phosphates in Morocco Case*, PCIJ Series A/B No. 74 (1938), at 28; *British Claims in the Spanish Zone of Morocco*, 2 RIAA (1949), at 641; *The Corfu Channel Case*, ICJ Reports 1949, at 23; *The Gabčíkovo-Nagymaros Project Case*, ICJ Reports 1997, at 56; *The M/V "Saiga" (No. 2) Case*, ITLOS Reports 1999, at para. 170–171.

[19] In respect of treaty obligations concerning State responsibility see *infra*, Sect. "Explicit Provisions of State Responsibility in International Treaty Law".

[20] Dahm/Delbrück/Wolfrum, *Völkerrecht, vol. I/3* (2nd ed., 2002), at 864; Jennings/Watts (ed.), *Oppenheim's International Law* (9th ed., 1992), at para. 145; Verdross/Simma, *Universelles Völkerrecht* (3rd ed., 1984), at para. 1262; see also the list of literature by Brownlie, *State Responsibility* (1983), at 7–8.

[21] This principle found its way into Article 1 of the ILC Draft Article on State Responsibility, which reads: "Every internationally wrongful act of a State entails the international responsibility of that State."

[22] Ago, ILC Doc. A/CN.4/246 and Add.1-3 (YBILC 1971 II/1), at 206; ILC Doc. A/9010/Rev.1 (YBILC 1973 II), at 174; de Aréchaga, *International Responsibility*, in: Sørensen (ed.) (1968), at 533; Dahm/Delbrück/Wolfrum, *Völkerrecht, vol. I/3* (2nd ed., 2002), at 867.

they are referred to as "secondary rules".[23] Secondary rules may cover two different functions: On the one hand, they may modify the legal conditions under which a State's conduct is to be considered a breach of the respective primary obligation; in other words, they may determine the "if" of State responsibility. On the other hand, secondary rules may define the legal consequences of a wrongful act; thus they may determine the "how" of State responsibility.[24]

Secondary rules may derive from any source of international law, such as from treaty provisions, custom or general principles of law.[25] Whereas secondary rules that determine the legal consequences of State responsibility are often laid down in express treaty provisions, secondary rules that modify the legal prerequisites of State responsibility mainly stem from customary international law. In case different obligations arise from customary law and express treaty provisions, the latter will prevail over customary rules by means of their specific nature (*lex specialis derogat legi generali*).[26]

It is not possible to describe the content of secondary rules in general terms. However, any secondary obligation serves three main objectives: The first is a preventive one. The threat of imposing secondary obligations on the offender serves as deterrence and creates the incentive to act in a legal way.[27] Secondary rules, therefore, help to strengthen legal certainty and predictability. The second function is a repressive or corrective one. After an infringement has taken place, secondary rules provide an instrument to safeguard legally protected interests of the impaired State. They therefore aim at the protection and enforcement of the international legal order and also embody a punitive character.[28] The third function, finally, covers the compensatory aspect of secondary obligations. By imposing duties of reparation and compensation on the violator of international law, secondary rules aim at shifting the injurious consequences from the impaired State to the creator or the source of harm and, thus, aim to restore the *status quo*

---

[23] ILC Doc. A/9010/Rev.1 (YBILC 1973 II), at 169; Lefeber, *Transboundary Environmental Interference* (1996), at 15. See also Combacau/Alland, "Primary" and "Secondary" Rules', 16 *NYIL* (1985), at 81 *et seq.*, Linderfalk, 'Primary-Secondary Rules Terminology', 78 *Nord. J. Intl. L.* (2009), at 56 *et seq.*

[24] See Cassese, *International Law* (2nd ed., 2005), at 244.

[25] The sources of international law are mentioned in Article 38(1)(a) to (c) of the Statute of the ICJ. Article 38(1)(d) does not describe a source of law, but rather mentions means for the determination and interpretation of international law. See Evans (ed.), *International Law* (2nd ed., 2006), at 129; Graf Vitzthum, in: Graf Vitzthum (ed.), *Völkerrecht* (5th ed., 2010), at 66.

[26] Dahm/Delbrück/Wolfrum, *Völkerrecht, vol. I/3* (2nd ed., 2002), at 868; Schweisfurth, *Völkerrecht* (2006), at 83. See also Article 55 of the ILC Draft Articles on State Responsibility.

[27] Gaines, 'International Principles for Transnational Environmental Liability', 30 *Harv. Int'l L. J.* (1989), at 326–328; Lefeber, *Transboundary Environmental Interference* (1996), at 1.

[28] Dahm/Delbrück/Wolfrum, *Völkerrecht, vol. I/3* (2nd ed., 2002), at 867; Lefeber, *Transboundary Environmental Interference* (1996), at 1; Wolfrum/Langenfeld/Minnerop, *Environmental Liability in International Law* (2005), at 455.

ante as far as possible.[29] Corresponding to these objectives the State whose particular rights were violated is entitled to certain remedies. Thus, as long as the internationally wrongful act is ongoing, the impaired State is entitled to demand the immediate cessation of the wrongful act and, if applicable, to require an appropriate guarantee of non-repetition.[30] Furthermore, the impaired State may demand the restoration of the previous *status quo* (*restitutio in integrum*), or in the event this is impossible or involves a highly disproportionate burden, to demand compensation for both material and immaterial damages.[31]

II. Is There a Need for Autonomous Rules on State Liability for Lawful but Injurious Activities?

The application of the principle of State responsibility presupposes the breach of an international primary obligation. Since, however, not every transboundary harm can be traced back to a breach of an international obligation, there seems to be a legal gap concerning liability for damage resulting from harmful activities which are not prohibited by international law. This issue of "State liability" has not developed in a systematic legal fashion, but rather arose within the scope of certain activities usually involving a high risk of transboundary damage. Especially since the 1960s, increased technological development has given rise to a number of commercial activities that are conducted on a global scale and that offer huge benefits for entire economies. On the downside, various aspects of these activities pose an increased risk for human health and the environment, e.g. the concentration of large amounts of harmful substances in a small area. Three typical activities warrant mention are: the peaceful use of nuclear energy, the peaceful use of outer space and the use of the sea as a resource and transport medium.[32] Correspondingly, a number of recent history's major environmental disasters can be attributed to these activities, including, for example, the large oil spills emanating from the *M/V "Torrey Canyon"* and the *Deepwater Horizon* accidents, the *Chernobyl Incident*, the *Cosmos 954 Crash* as well as incidents in connection with

---

[29] Dahm/Delbrück/Wolfrum, *Völkerrecht*, vol. I/3 (2nd ed., 2002), at 867; Lefeber, *Transboundary Environmental Interference* (1996), at 1; Gaines, 'International Principles for Transnational Environmental Liability', 30 *Harv. Int'l L. J.* (1989), at 324–326; Schachter, *International Law in Theory and Practice* (1991), at 376.

[30] This right actually arises from the respective primary obligation. See ILC Draft Articles on State Responsibility, Articles 29, 30; and in detail *infra*, Sect. "Continuation of the Primary Obligation".

[31] See ILC Draft Articles on State Responsibility, Articles 31, 34–38; and in detail *infra*, at Sect. "Reparation".

[32] Goldie, 'International Principles of Responsibility for Pollution', 9 *Colum. J. Transnat'l L.* (1970), at 283–298.

hazardous waste transports.[33] Nevertheless, these activities are not prohibited or even unwanted. In most cases they rather make good economic sense and have to be accepted as being economically indispensable.

Since it is not possible to fully eliminate the inherent risk related to these activities by the adoption of safety measures, a solution needs to be found by means of a legal regime that generally accepts the risk but provides for appropriate rules of allocating liabilities for damage. With this in mind, some authors advocate the recognition of an autonomous legal principle of State liability for lawful but injurious acts.[34] Others, in contrast, reject this concept of an autonomous legal principle of State liability and rather attempt to solve this issue by referring to the existing principle of State responsibility and by partially redefining the underlying substantive obligations of States. In the end, the existence and legal validity of a principle of State liability which impose liability *sine delicto* simply depends on whether such principle is generally recognised by international jurisdiction, State practice and legal literature.

This question will be addressed in the following sections. But before doing so, those cases shall be identified in which no actual need for an autonomous legal concept of State liability *sine delicto* exists. Only after having identified such "false cases of liability *since delicto*" is it necessary to ascertain whether, in respect of the remaining "true cases of liability *sine delicto*",[35] present international law supports a principle of State liability *sine delicto* for lawful but injurious acts.

1. "False Cases of Liability Sine Delicto"

The term "false cases of liability *sine delicto*" summarises all such cases in which transboundary environmental harm is caused by an activity that is not as such prohibited by international law, but in which the resulting damage is nevertheless considered to constitute a breach of an international obligation. This can be the case in the context of obligations of result, i.e. instances which do not require a certain conduct of the State but instead demand that a particular result is achieved regardless of the manner. In such cases, the failure of the State to achieve the prescribed result regardless of the means may be considered an internationally wrongful act. Hence, the principle of State responsibility would be applicable, although there is no particular conduct of the State that is in breach of an international obligation.

(a) The General Approach

At the outset of this inquiry, there must be an identification and precise construction of the respective international primary obligations governing at least

---

[33] See *supra*, Sect. "The Factual Perspective: Transboundary Movements of Hazardous Wastes by Sea" in Chap. 1.

[34] See *infra*, Sect. "Recognition as a General Principle of International Law?".

[35] This terminology is introduced by Akehurst, 'International Liability for Lawful Acts', 16 *NYIL* (1985), at 4 *et seq*.

partially the transboundary movement of hazardous wastes by sea. Of particular significance in this respect are obligations deriving from international environmental law, which mainly attempt to regulate and diminish certain adverse impacts on human health and the environment.

The following conception of the basic structure of international obligations underlies the approach of identifying "false cases of liability *sine delicto*":

(aa) Obligations of Conduct and Obligations of Result

International primary obligations can be divided, according to their regulatory content, into obligations of conduct and obligations of result. An obligation of conduct imposes on the State the duty to conduct itself or behave in a prescribed way or to refrain from a particular conduct or behaviour. Obligations of result, by contrast, stipulate a duty to ensure the occurrence or non-occurrence of a particular event or situation by whatever means.[36] An indication of an obligation of result can be seen in the use of the term "ensure", since this expression describes the final result of an activity rather than the actual activity leading to this result.

In addition, several variations and combinations of these types of obligations exist. This applies especially in respect of a more or less extensive margin of discretion that is left to the parties regarding the interpretation of terms and the final determination of particular standards or thresholds. For instance, international conventions may either specify fixed amounts or thresholds,[37] or they may leave the determination of such thresholds to the discretion of the respective States or an interstate committee.[38]

In respect of obligations of result, States are in general free to decide in which way to achieve the prescribed result.[39] However, the margin of discretion left to the States may be restricted in the individual case. Such restriction may be due in cases where it is established with almost absolute certainty that a particular activity will lead to the result prohibited by international law, without there being any opportunity for the State to avoid this result by taking any other measures apart from the complete prohibition of this activity. In this case, the margin of discretion

---

[36] Dahm/Delbrück/Wolfrum, *Völkerrecht, vol. I/3* (2nd ed., 2002), at 876; Wolfrum, *Obligation of Result Versus Obligation of Conduct*, in: Arsanjani/Cogan/Sloane/Wiessner (ed.) (2011), at 363 *et seq.*; The latter type of obligations of result is also denoted as "obligations of prevention".

See also Articles 20, 21 and 23 of the previous 1997 Draft Articles on State Responsibility (ILC Doc. A/32/10 (YBILC 1977 II/2), at 11–30; ILC Doc. A/33/10 (YBILC 1978 II/2), at 81–86). As to the reasons for the deletion of these Articles in the final 2001 ILC Draft Articles on State Responsibility, see Crawford, ILC Doc. A/CN.4/498 and Add.1-4 (YBILC 1999 II/2), at 20–29.

[37] Examples of fixed reduction rates and maximum levels of emissions can be found, for example, in Article 3, Annex B of the 1997 Kyoto Protocol and in Article IV, Annexes I, II of the 1972(1996) London Convention.

[38] Some examples are given by Wolfrum, 'Means of Ensuring Compliance and Enforcement', 272 *RdC* (1998), at 33–34.

[39] Dahm/Delbrück/Wolfrum, *Völkerrecht, vol. I/3* (2nd ed., 2002), at 876.

generally left to the State is narrowed to only one legal manner of conduct, i.e. the complete prohibition of this activity. In such constellations, where there is only one path of conduct left that may prevent the prohibited result, it seems appropriate to classify as unlawful not only the later occurrence of the prohibited result, but also any act of permission or connivance of the State which does not prevent the activity that would later cause the prohibited result.[40] Such obligations of result that exceptionally also comprise the activity causing this result may be denoted as "qualified obligations of result".

Finally, there is also a further type of international obligation that cannot be classified as either an obligation of conduct or an obligation of result. These are programmatic obligations which do not impose binding rules or obligations but which, rather, establish policy-oriented provisions or define legal goals.[41]

(bb) Distinction Between Legal Acts and Factual Activities

It has just been outlined that in respect of obligations of result, there is a margin of discretion left to the States regarding the manner of achieving the prescribed result. Possible legislative and administrative means for the achievement of such a result include a complete ban of the activity causing the adverse effects, the adoption of precautionary or preventive measures, or the regulation and restriction of such activities.

This consideration, moreover, shows that there is a basic distinction underlying the legal structure of internationally wrongful acts. On the one hand there is the legally relevant conduct of the State, which comprises the implementation of legislative and administrative acts. And on the other hand, there is the factual activity mainly conducted by private parties that actually leads to the unwanted result.[42] In line with this distinction, the term "activity" is mostly used to denote the factual conduct while the term "act" describes the conduct of the State in a legal sense.

(cc) Implications for the Application of State Responsibility

Given the distinctions between (i) obligations of conduct and obligations of result and (ii) legal acts and factual activities, it becomes apparent that in many of those

---

[40] This interpretation provides a higher level of protection for the victim State. It allows the victim State to demand the cessation not only of the ultimate impairment of its individual rights, but also of the activity leading to this later infringement. By this means, the victim State does not have to wait until the damage actually occurs but may undertake measures to prevent the occurrence of damage.

[41] See e.g. Wolfrum, 'Means of Ensuring Compliance and Enforcement', 272 *RdC* (1998), at 34; Wolfrum, *Obligation of Result Versus Obligation of Conduct*, in: Arsanjani/Cogan/Sloane/Wiessner (ed.) (2011), at 376–377.

[42] Akehurst, 'International Liability for Lawful Acts', 16 *NYIL* (1985), at 8, footnote 30; de la Fayette, 'The ILC and International Liability', 6 *RECIEL* (1997), at 327; Pisillo Mazzeschi, *Forms of International Responsibility*, in: Francioni/Scovazzi (ed.) (1991), at 26–27.

cases assuming a need for an autonomous legal concept of State liability for lawful but injurious activities, in fact the principle of State responsibility applies, which already provides for a relevant regime of liability and compensation for internationally wrongful acts.[43]

This may be the case because international law imposes an obligation of result on the State which does not require a certain conduct of the State but rather imposes a particular result to be achieved by whatever means. Hence, the State is not under the obligation to prohibit a certain activity that potentially causes the internationally prohibited result, provided the State is able to ensure the non-occurrence of this result by other legislative or administrative means.[44] This, however, also means that the permission of an activity that potentially causes an internationally prohibited result is not *per se* unlawful. For the same reason it is true that if the permission of an activity is in conformity with international law, this does not mean that the actual consequences of this activity are in conformity with the international law, too. It is rather possible that the actual consequences of such a lawful activity may constitute a breach of an international obligation of the State in that the State has failed to ensure the non-occurrence of this result by any other means.[45] Applied to the cases of lawful but injurious activities under consideration, this means, for instance, that the operation of an industrial plant close to the border to an adjacent country may represent a lawful activity under international law, whereas the causation of transboundary harm by this plant constitutes the breach of an international obligation of the State to ensure that damage to another State's territory is prevented.

From the distinction between legal acts and factual activities, however, it also follows that not any occurrence of an internationally prohibited result may be considered an internationally wrongful act of the State. The legally relevant acts of the State and the factual activities—which are mainly conducted by private parties—are not to be intermingled when assessing whether or not an international obligation has been breached. International obligations are principally addressed at States; private parties are in general not addressees of international norms.[46] Consequently, an international obligation may be breached only by an act of the

---

[43] Akehurst, 'International Liability for Lawful Acts', 16 *NYIL* (1985), at 8–9; Boyle, 'State Responsibility and International Liability', 39 *Int'l & Comp. L. Q.* (1990), at 8–16; Brownlie, *State Responsibility* (1983), at 50; Handl, 'Liability as an Obligation Established by a Primary Rule of International Law', 16 *NYIL* (1985), at 60; Pisillo Mazzeschi, *Forms of International Responsibility*, in: Francioni/Scovazzi (ed.) (1991), at 27–28; Scovazzi, 'State Responsibility for Environmental Harm', 12 *Yb. Int'l Env. L.* (2001), at 49.

[44] This is different only in case of qualified obligations of result, as outlined in the previous section.

[45] Akehurst, 'International Liability for Lawful Acts', 16 *NYIL* (1985), at 8; Boyle, 'State Responsibility and International Liability', 39 *Int'l & Comp. L. Q.* (1990), at 14.

[46] Exemptions from this principle exist for international crimes, such as slavery, piracy and war crimes, see Dahm/Delbrück/Wolfrum, *Völkerrecht, vol. I/2* (2nd ed., 2002), at 264–267; and in general Doehring, *Völkerrecht* (2nd ed., 2004), at 113–114; Evans (ed.), *International Law* (2nd ed., 2006), at 314–315.

State rather than by the factual activity itself.[47] Where an international convention attempts to regulate certain activities conducted by private persons by means of an obligation of result, it is not the activity itself which is judged against international law; the question is rather whether the entirety of the acts implemented by the State with a view to ensuring the prescribed result can be considered sufficient to comply with the international obligation. The standard of conduct to be complied with by the State is determined individually by the respective primary obligation. It is possible that a primary obligation imposes an absolute standard of conduct, according to which the State is absolutely responsible for the achievement of the prescribed result. In most cases, however, international obligations require a standard of due diligence, according to which the possibilities for the State to foresee the occurrence of damage and the capabilities of the State to regulate and control activities of private parties need to be taken into account.[48]

Notwithstanding this qualification, it becomes apparent that even if environmental damage was caused by an activity which is by itself considered to be in conformity with the international law, the consequences of this activity may nevertheless constitute a breach of an international obligation of the State. *Brownlie*, therefore, is right when he states that "[m]uch of State responsibility [...] is concerned with categories of lawful activities which have caused harm".[49]

(b) Further Enhancements of this Approach

Some authors even go beyond the basic approach to apply the principle of State responsibility to the unlawful consequences of a lawful activity. They advocate for the application of the principle of State responsibility *even to the lawful activities* that later cause an internationally prohibited result.

(aa) Equation of Activities and Its Consequences

According to one opinion, it may not be allowed to artificially distinguish between the conduct of an activity and the activity's factual consequences. Both components rather form parts of an indivisible whole that must be considered as single entity. From this it follows that in case the consequences of an activity constitute a breach of an international obligation, the conduct of this activity must be contrary to international law as well.[50] The main argument produced in favour of this view is the perception that a distinction between an activity's conduct and its

---

[47] The factual activity, however, may constitute a breach of an international obligation if it is conducted by the State itself.

[48] In respect of the standard of due diligence see *infra*, Sect. "Due Diligence as the Relevant Standard of Behaviour".

[49] Brownlie, *State Responsibility* (1983), at 50.

[50] Ipsen, in: Ipsen (ed.), *Völkerrecht* (5th ed., 2004), at 634.

consequences would be necessary only if culpability was a prerequisite for international responsibility. Only in this case would the wrongfulness of the conduct and the consequences of an activity have to be established independently of each other. Since, however, culpability is not seen as a general requirement of international responsibility,[51] there does not exist any reason to distinguish between the conduct of an activity and its factual consequences.[52]

This line of argumentation is not entirely convincing. It is true that today culpability is not considered to be a general requirement of international responsibility. However, from the distinction between primary rules and secondary rules of international law, it follows that the relevant primary rule remains decisive for the question of whether or not culpability is required to constitute a breach of that particular primary rule.[53] But of even greater importance is that the posited view suffers from a confusion concerning the terms "acts" and "activities". The relevant action to be judged against international law is the legal or administrative "act" of the State, which may consist in affirmative conduct or inactivity.[54] The statement that the conduct and the consequences of an actual "activity" need to be considered as an indivisible entirety does not change the fact that it is not the "activity", but rather the "act" of the State which is decisive in establishing an internationally wrongful act.[55] It may be true that in most cases where an activity causes an internationally prohibited result this constitutes the breach of an international obligation of result by a State due to the State's failure to ensure the non-occurrence of this result. However, since there are further legal prerequisites for the establishment of such an internationally wrongful act, the occurrence of the prohibited result and the State's act of non-prevention may not simply be equated. The argument that the unlawfulness of the consequences of an activity would at the same time establish the unlawfulness of the conduct of this activity may not be reconciled with the legal conception of international obligations and responsibility. Finally, since there seems to be agreement in international law that unlawfulness entails prohibition,[56] this consideration would lead to a *de facto* prohibition of all potentially harmful activities. This obviously cannot be an appropriate outcome.[57]

(bb) A Comprehensive Obligation to Prevent Damage

A second attempt to further enhance the application of the principle of State responsibility to lawful but injurious activities is provided by the view that favours

---

[51] See *infra*, Sect. "A General Requirement of Fault?".
[52] Ipsen, in: Ipsen (ed.), *Völkerrecht* (5th ed., 2004), at 634.
[53] See *infra*, Sect. "A General Requirement of Fault?".
[54] This is outlined *supra*, Sect. "Act of the State".
[55] This is also pointed out by de la Fayette, 'The ILC and International Liability', 6 *RECIEL* (1997), at 327.
[56] See ILC Draft Articles on State Responsibility, Article 30.
[57] See also de la Fayette, 'The ILC and International Liability', 6 *RECIEL* (1997), at 327.

a comprehensive coverage of any transboundary harm under the principle of State responsibility. The basis of this approach is an extensive construction and interpretation of the primary obligation of States to prevent transboundary harm, and, therefore, it concerns the scope of the substantive primary obligation rather than the principle of State responsibility. Due to its conceptual implications, this approach is nevertheless worth discussing at least in brief at this point.

According to this view, the underlying basic obligation of States to prevent the occurrence of transboundary harm[58] is construed and interpreted in such a way as to provide for a comprehensive scope.[59] This obligation is understood to prevent the occurrence of any transboundary harm, not being limited to significant damage.[60] It is construed as an obligation of result and, moreover, as an objective and absolute obligation. This means that a wrongful act of the State exists in any case where transboundary harm occurs, irrespective of which legislative and administrative measures have been taken by the State and irrespective of whether the State either is in a position to predict the later infringement or has the capability to regulate and control the activities conducted by private parties.[61] In consequence of this extensive interpretation, any causation of transboundary harm would be considered an internationally wrongful act, which would entail the application of the principle of State responsibility.

It must be admitted that this approach would indeed lead to a situation where there is no need for an autonomous legal concept of State liability for lawful but injurious activities, since any causation of transboundary harm would be deemed wrongful and correspondingly entail the principle of State responsibility. This, however, would also mean that an activity which involves only a potential threat of transboundary harm must be prohibited by the source State, even if the activity is deemed beneficial or necessary for the economy of that respective State. Particularly smaller countries would thus not be allowed to operate any pollutant-emitting plants, which obviously cannot be an appropriate solution. In addition, it must be objected that such an extensive construction and interpretation of the obligation to prevent transboundary harm is not recognised by international law. It goes far beyond the content of this obligation as accepted by international jurisdiction, State practice and legal literature[62] and, hence, may rather be understood merely as a well-meant but inapt political attempt at crafting a solution.

---

[58] This obligation is outlined in detail *infra*, Sect. "The Obligation not to Cause Significant Harm to the Environment of Another State's Territory".

[59] de la Fayette, 'The ILC and International Liability', 6 *RECIEL* (1997), at 324–327.

[60] de la Fayette, *ibid.*, at 324–325.

[61] de la Fayette, *ibid.*, at 326.

[62] See *infra*, Sect. "The Obligation not to Cause Significant Harm to the Environment of Another State's Territory".

(c) Summary

In summary, it can be concluded that the need for an autonomous legal concept of State liability does not exist with regard to cases in which transboundary environmental harm is caused by an activity that is not as such prohibited by international law, but in which the resulting damage is nevertheless considered to constitute a breach of an international obligation. Of importance for this consideration is the distinction between obligations of conduct and obligations of result as well as the distinction between a legally relevant "act" of the State and the factual "activity" conducted primarily by private persons. If an international obligation of result is in place that obliges the State to ensure the occurrence of a certain result, it is basically left to the discretion of the State in which way to achieve this result. A complete prohibition of the activity leading to the prohibited result is not required, provided the State is able to ensure the non-occurrence of the result by other preventive or precautionary measures. From this it follows that the lawfulness of a State's permission of a particular activity is to be distinguished from the lawfulness of the consequences of this activity. Consequently, even if a particular activity is deemed to be lawful under international law, the actual consequences of this activity may nevertheless constitute a breach of an international obligation of result.

By contrast, the further attempts to enlarge the coverage of the principle of State responsibility to lawful but injurious activities are not convincing and are not to be encountered in the body of valid international law.

2. "True Cases of Liability Sine Delicto"

The principle of State responsibility is not capable of either covering all transboundary impairments or providing for liability and compensation for every instance of environmental damage. Gaps of liability may exist especially with regard to damage that was caused notwithstanding the State's having complied with the required standard of due diligence.[63] Accordingly, it is being discussed whether and to what extent international law acknowledges a general legal concept of State liability within the meaning of strict liability *sine delicto*.

Rules of State liability *sine delicto* must be classified differently than rules of State responsibility *ex delicto*. Rules of State responsibility apply only in response to a breach of a substantive primary rule of international law and, thus, are referred to as secondary rules. They do not contain a legal valuation of their own, but merely adopt the legal valuation of the respective primary rule.[64] By contrast, rules

---

[63] Schröder, in: Graf Vitzthum (ed.), *Völkerrecht* (5th ed., 2010), at 591–592; Rudolf, *Haftung für rechtmäßiges Verhalten im Völkerrecht*, in: Damrau/Kraft/Fürst (ed.) (1981), at 541–542.

[64] See *supra*, Sect. "The Principle of State Responsibility for Internationally Wrongful Acts".

of State liability are not dependent on a breach or even on the existence of an underlying primary rule of international law. Rather, they predefine by themselves the substantive conditions under which liability shall incur and, thus, contain a legal valuation of their own. Rules of State liability, therefore, have to be classified among the body of primary rules of international law.[65]

(a) Explicit Rules of State Liability

Explicit rules of State liability can be found only sporadically in international law. For instance, in Article VII of the 1967 Outer Space Treaty and in Article II of the 1972 Convention on International Liability for Damage Caused by Space Objects, strict liability for transboundary damage caused by space objects is attached to the launching State. These rules are also referred to by several principles proclaimed by the UN General Assembly Resolution Relevant to the Use of Nuclear Power Sources in Outer Space.[66]

In UNCLOS, despite several explicit rules on State responsibility, no relevant rules on State liability can be found. Article 110(3) of UNCLOS establishes State liability, albeit not in the context of transboundary environmental damages.[67] Article 263(3) of UNCLOS is concerned with liability for pollution of the marine environment arising out of marine scientific research. It does not contain an independent basis for liability, instead referring to Article 235 of UNCLOS, which, in turn, refers to existing civil liability regimes[68] as well as to the general international law concerning State responsibility as a basis for liability.[69]

Finally, provisions in international conventions establishing a subsidiary liability of the State in case the otherwise liable individual or corporate entity is not available to provide compensation cannot be classified among the rules of State liability within the meaning of this consideration. This is because in such cases the State does not incur a genuine State liability, but rather assumes the liability of the involved private parties in the sense of a contingent liability. In this context, Article 9(2)(a) and (3) of the Basel Convention needs to be mentioned as an

---

[65] Quentin-Baxter, ILC Doc. A/CN.4/334 and Add.1 and 2 (YBILC 1980 II/1), at 253; Boyle, 'State Responsibility and International Liability', 39 *Int'l & Comp. L. Q.* (1990), at 10.

[66] UNGA Res. A/Res/47/68 of 23 February 1993.

[67] If a ship is stopped and searched on the basis of Article 110(1) of the UNCLOS due to a reasonable ground, e.g. a suspicion of piracy, this would represent a lawful act of the State even if this suspicion is later proven to be unfounded. The compensation the State has to pay in case of an unfounded suspicion according to Paragraph 3, thus, is based on liability *sine delicto*.

[68] According to Mensah, *Civil Liability in UNCLOS*, in: Basedow/Magnus/Wolfrum (ed.) (2010), at 4–5, UNCLOS only recognises the legal concepts of State responsibility and civil liability. This can be seen Article 229 of UNCLOS, which concerns the relationship of UNLCOS to civil liability regimes in general.

[69] Harndt, *Völkerrechtliche Haftung für die schädlichen Folgen nicht verbotenen Verhaltens* (1993), at 347; see also Nordquist/Rosenne/Yankov/Grandy (ed.), *UNCLOS—Commentary Vol. IV* (1991), at 412.

example which establishes a subsidiary liability of the State to ensure the re-importation or, alternatively, the environmentally sound disposal of hazardous wastes in the event of their illegal traffic.[70]

In summary, it must be stated, therefore, that, apart from space law, explicit rules on State liability are very rare in international law and remain an extraordinary appearance.

(b) Recognition as a General Principle of International Law?

In addition to the sporadic explicit rules on State liability established by sector-specific international conventions, it might also be possible that the concept of State liability for lawful but injurious activities is recognised as a general principle of international law.

This would be the case, first, if the concept of State liability represented a valid part of customary international law, as proven by State practice, international case law and legal literature. As set out above, only few instances of explicit rules on State liability can be found in the practice of States.[71] International case law acknowledging a general liability of States for lawful but injurious activities does not exist. And finally, contrary to a few voices that argue in favour of a general principle of liability *sine delicto* as derived from the principle "to use your own property so as not to injure another's" (*sic utere tuo ut alienum non laedas*),[72] international legal literature in general remains rather adverse to the recognition of a general principle of liability *sine delicto*.[73] As a result, it must therefore be stated that the concept of State liability for lawful but injurious activities does not represent a valid principle of customary international law.

Apart from this, the concept of State liability could form part of valid international law if it represented a general principle of law recognised by civilised nations.[74]

---

[70] Although such transports might be considered "illegal traffic", this does not mean that an internationally wrongful act of the State exists. See *infra*, Sect. "Article 9".

[71] A further situation in which State liability is recognised by customary international law concerns compensation for material losses caused by acts that are deemed lawful due to circumstances precluding wrongfulness, see ILC Draft Articles on State Responsibility, Article 27(b).

[72] Goldie, 'International Principles of Responsibility for Pollution', 9 *Colum. J. Transnat'l L.* (1970), at 306–309; see also Jenks, 'Liability for Ultra-Hazardous Activities in International Law', 117 *RdC* (1966), at 162–166.

[73] Beyerlin/Marauhn, *International Environmental Law* (2011), at 366–368; Birnie/Boyle/Redgwell, *International Law and the Environment* (3rd ed., 2009), at 219; Gründling, 'Verantwortlichkeit der Staaten für grenzüberschreitende Umweltbeeinträchtigungen', 45 *ZaöRV* (1985), at 280; Heintschel von Heinegg, in: Ipsen (ed.), *Völkerrecht* (5th ed., 2004), at 1062; Lefeber, *Transboundary Environmental Interference* (1996), at 187; Randelzhofer, 'Probleme der völkerrechtlichen Gefährdungshaftung', 24 *BDGVR* (1984), at 65–66.

[74] See Statute of the ICJ, Article 38(1)(c).

This is indeed assumed by some authors.[75] However, solid evidence for this assumption has not been produced. In consideration of the significant differences between the legal prerequisites and the legal consequences of the respective domestic concepts of liability *sine delicto*, it must rather be concluded that a clear and consistent standard does not exist. A general principle of law recognised by civilised nations comparable in content to the concept of State liability for lawful but injurious activities, therefore, cannot be ascertained.[76]

This conclusion also reflects the recent position of the ILC regarding the issue of State liability. The ILC, which has contributed with its work to the foundation of the concept of State liability and which has remained one of the strongest supporters of this concept for a considerable time, has recently abandoned its efforts to codify a regime of State liability and changed its strategic approach towards the codification of certain precautionary standards[77] as well as towards promoting the elaboration of international civil liability regimes.[78]

In summary, it can therefore be said, therefore, that a principle of State liability *sine delicto* for lawful but injurious activities does not find support *de lege lata* in international law.

(c) Implications of the Non-existence of a General Principle of State Liability

In the end, this result does not seem to be inappropriate.

Some authors assert that the existence of a principle of general State liability would involve certain advantages in comparison with the principle of State responsibility. These include that the concept of State liability would not entail the reproach of a violation of law upon the source State.[79] Furthermore, the violated State would not bear the burden of proving the entire range of a breach of an international obligation by the source State, but only the actual causal chain leading to the infringement.[80] And finally, the legal implication of prohibition would be avoided, this implication being associated with the wrongfulness of an

---

[75] Gaines, 'International Principles for Transnational Environmental Liability', 30 *Harv. Int'l L. J.* (1989), at 311 *et seq.*; Kelson, 'State Responsibility and the Abnormally Dangerous Activity', 13 *Harv. Int'l. L. J.* (1972), at 201.

[76] Birnie/Boyle/Redgwell, *International Law and the Environment* (3rd ed., 2009), at 219–220; Randelzhofer, 'Probleme der völkerrechtlichen Gefährdungshaftung', 24 *BDGVR* (1984), at 67; Schröder, in: Graf Vitzthum (ed.), *Völkerrecht* (5th ed., 2010), at 592.

[77] See in particular the 2001 Draft Articles on Prevention of Transboundary Harm from Hazardous Activities.

[78] The development of the work of the ILC is outlined in detail below, Sect. "State Liability".

[79] See e.g. Boyle, 'State Responsibility and International Liability', 39 *Int'l & Comp. L. Q.* (1990), at 5, 13; see also Quentin-Baxter, ILC Doc. A/CN.4/360 (YBILC 1982 II/1), at 60.

[80] See e.g. Boyle, 'State Responsibility and International Liability', 39 *Int'l & Comp. L. Q.* (1990), at 14.

act as resulting from the principle of State responsibility.[81] However, these arguments voiced in favour of the concept of State liability refer only to a comparison with the principle of State responsibility. Since the concepts are designated for very different scenarios, i.e. for lawful activities of private persons or, alternatively, for wrongful acts of the State, the correct question must not be whether the principle of State responsibility or the concept of State liability represents the more appropriate approach. Instead, it must be questioned whether there is an actual need for an additional concept of State liability in scenarios where the principle of State responsibility fails to apply, or whether in such constellations it seems more appropriate to accept a certain level of transboundary damage in order to facilitate necessary economic activities.

In this context, it should be kept in mind that the international relations among States are governed by the substantive rules and provisions of international law that impose obligations on the States to act or behave in a certain prescribed way. The entirety of the international primary rules, thus, does not only provide for a comprehensive legal valuation of what is considered to be lawful and what is deemed unlawful by the international community, it also attempts to reconcile the respective conflicting interests. Whereas rules on State responsibility simply adopt this legal valuation, rules on State liability rather establish a legal valuation of their own. In this context, the existence of a general principle of State liability applicable to any injurious activities would have a particular negative effect. By means of imposing a general standard of liability, the differentiated and more specialised legal valuation of the respective primary rules would be negated. It becomes apparent that such a "lawnmower-solution" may not be considered a suitable and appropriate solution regarding the considerable range of circumstances that need to be taken into account in respect of the possible commercial activities which pose a potential risk of causing transboundary harm. Therefore, a general principle of State liability *sine delicto* must be considered an improper, alien element in the overall structure of international primary rules and, consequently, should be rejected.

In conclusion, it can be summarised that the concept of State liability for lawful but injurious activities not only lacks recognition as a general principle of international law, but also that is fails to provide for a necessary or even appropriate approach which might contribute to a reconciliation of interests. Rather, it seems appropriate to restrict international liability to the general principle of State responsibility, through which it is ensured that the substantive valuation of the primary rules is adequately taken into account. This, of course, does not mean that it is unnecessary to specify existing primary rules or impose further ones on States in certain fields of law. Furthermore, the creation of explicit provisions on State liability in certain sector-specific conventions may be considered a suitable and appropriate legal instrument in the individual case.

---

[81] See *ibid.*, at 12–13.

3. The Relationship Between Both Concepts

Since international law does not embrace a general principle of State liability for lawful but injurious activities, the question regarding the relationship between the concepts of State liability and State responsibility only becomes relevant in respect of cases in which State liability is imposed by means of explicit convention provisions.

In this regard *Randelzhofer* proceeds on the assumption that the two concepts are mutually exclusive. As far as liability arising from one particular incident is concerned, he argues that the act of the State can either be lawful, or it can be wrongful, but by no means can it be both at the same time.[82] Although this statement is logically true, it nevertheless disregards the fact that both concepts concern very different premises of liability. Whereas State responsibility is concerned with internationally wrongful acts of the State, the concept of State liability focuses rather on the strict liability of the State for the injurious effects of lawful commercial activities conducted by private persons which fall under the jurisdiction of the liable State. Therefore, the assumption that a State incurs strict liability on account of an explicit provision of State liability allocating the adverse effects of a beneficial commercial activity to the source State does not preclude the finding that this State may at the same time be responsible for a breach of an international obligation, e.g. to ensure the non-occurrence of a particular type of damage or to implement certain supervisory instruments or precautionary measures.

It must be kept in mind that the concepts of State responsibility and State liability derive from different legal concepts. They have different legal prerequisites and impose different legal consequences; and in particular, they pursue different legal goals. Therefore, there is no reason to assume that one of these concepts encapsulates the other one, or that both concepts are mutually exclusive.[83] This conclusion is furthermore in line with the conception of the ILC, which in its earlier work did not restrict the concept of State liability to activities allowed under international law.[84] In summary, it can be concluded, therefore, that these concepts are not mutually exclusive; they rather complement each other and may principally be simultaneously applicable. Hence, the claimant State is given the opportunity to choose on which basis to bring its claim, taking account of the respective prerequisites and the legal content of the different claims.[85]

---

[82] Randelzhofer, 'Probleme der völkerrechtlichen Gefährdungshaftung', 24 *BDGVR* (1984), at 63.

[83] See also Boyle, 'State Responsibility and International Liability', 39 *Int'l & Comp. L. Q.* (1990), at 16, 22; Pinto-Dobernig, 'Liability for Transfrontier Pollution Not Prohibited by International Law', 38 *Österr. Z. öffentl. Recht* (1987), at 88–91; Randelzhofer, 'Probleme der völkerrechtlichen Gefährdungshaftung', 24 *BDGVR* (1984), at 61–62.

[84] This is why this concept is denoted by the ILC as liability for "acts not prohibited", rather than for "acts allowed". See Quentin-Baxter, ILC Doc. A/CN.4/360 (YBILC 1982 II/1), at 54, 59. See furthermore Boyle, 'State Responsibility and International Liability', 39 *Int'l & Comp. L. Q.* (1990), at 12.

[85] See also Lefeber, *Transboundary Environmental Interference* (1996), at 147.

III. The Importance of Civil Liability Conventions

The third component of the international legal framework governing responsibility and liability consists in civil liability conventions. By means of civil liability conventions States commit themselves to implement into their domestic laws an internationally uniform regime of civil liability, according to which certain minimum standards of liability and compensation are ensured among the contracting States. But civil liability conventions do not only stipulate a certain uniform standard of substantial liability; they also ensure a mutual recognition of court decisions that are based on this uniform law and lay down rules which in most cases create the opportunity for forum shopping.

Civil liability conventions today represent the most important legal instrument for addressing transboundary environmental damage.[86] This is mainly due to the numerous regulatory opportunities which may take account of the specific circumstances of the individual case. Thus, it is possible that by means of convention provisions, strict liability or fault-based liability or a combination of both forms is imposed on the responsible person. Moreover, a secondary liability of the State may be established, financial guarantees and insurance coverage may be required, and by means of exclusions of liability and maximum amounts of liability, the economic incentives of the private persons involved may be steered. The limitation of liability also facilitates the insurability of the financial risk attached to liability and, thus, ensures that compensation of the victim is available independently from the financial capacity of the liable party. Additional instruments, such as compensation funds, may ensure prompt emergency response and damage compensation. Finally, the codification of the preconditions for liability contributes to legal certainty and encourages economic activity as well as precautions being taken against damage. On the downside, it must be noted that a higher precautionary standard entails higher costs for the related preventive measures. If the precautionary standard and the standard of liability are set too high, this has rather the effect of encouraging the illegal conduct of commercial activities.

The high practical relevance of civil liability conventions is also in line with the recent approach of the ILC, which refrains from further attempts to codify general rules of State liability, but rather promotes the creation of sectoral regimes of civil liability.[87]

---

[86] In respect of civil liability conventions in the field of maritime law see Hafner, in: Graf Vitzthum (ed.), *Handbuch des Seerechts* (2006), at 395–398.

[87] The work of the ILC is outlined in the following section.

B. The Contribution of the International Law Commission

The allocation of responsibilities and the imposition of liability for transboundary environmental damage are governed by international law. Such rules cannot be dictated unilaterally but depend in their existence and scope on their being recognised by the community of States. Consequently, rules of international law are subject to a steady evolution which, due to the nature of affairs, proceeds without design. This applies particularly to customary international law. But also the international law found in conventions would remain fragmentary and incoherent in structure without any strategic guidance over the long term.

In order to determine the scope of existing rules in international law and to guide the codification of further rules, the installation of an interstate governmental organisation assigned with these tasks was envisaged as far back as the early 20th century. Thus, in 1924 the Committee of Experts for the Progressive Codification of International Law was established as a standing organ by the Assembly of the League of Nations. However, this Committee remained without practical significance as a result of the global turbulence witnessed in the 1930s and the subsequent outbreak of the Second World War.[88] After the Second World War, this purpose was taken up again during the foundation of the United Nations in 1945,[89] and in 1947 the United Nations General Assembly passed a resolution to set up the International Law Commission (ILC) as a permanent subsidiary organ.[90]

The ILC's goals are defined as promoting the progressive development and codification of international law. The Commission's focus lies primarily on public international law, but it is not precluded from considering private international law.[91] It consists today of 34 "recognized" international law experts,[92] and holds annual Sessions, with the 1st Session having been held in 1949.[93] Even though the statutory procedural rules of the ILC differ depending on whether the *progressive*

---

[88] As to the work of the Committee see: Dhokalia, *The Codification of Public International Law* (1970), at 112–133; Sinclair, *The ILC* (1987), at 3–5. See to the early development of State responsibility also Sucharitkul, 'State Responsibility and International Liability', 18 *Loy. L. A. Int'l & Comp. L. J.* (1995), at 823–828.

[89] Article 13(1) of the Charter of the United Nations defines as a task of the United Nations General Assembly to "initiate studies and make recommendations for the purpose of [...] encouraging the progressive development of international law and its codification."

[90] UNGA Res. A/Res/174 (II) of 21 November 1947, including the Statute of the future International Law Commission. As to the ILC's objectives, see Article 1 of this Statute.

[91] Statute of the ILC, Article 1. The ILC itself describes its *raison d'être* with the "idea of developing international law through the restatement of existing rules or through the formulation of new rules." See United Nations (ed.), *The Work of the ILC* (7th ed., 2007), at 1; see also Article 15 of the Statute of the ILC.

[92] Statute of the ILC, Article 2.

[93] As a subsidiary organ of the General Assembly, the Rules of Procedure of the General Assembly also apply to the procedure of committees; see Rule 161 of the Rules of Procedure of the General Assembly. See Rule 99(2) of the Rules of Procedure of the General Assembly in respect of the Sessions.

development or the *codification* of international law is concerned, this distinction has proven unworkable in practice, and the ILC rather follows a consolidated procedure that is divided into three stages.[94] Prior to the initiation of this procedure, the decision of the ILC to concern itself with a certain legal topic is precipitated either by a formal request of the General Assembly, by an ascertained need to progress or further develop previous works, or by recommendations of working groups.[95] Accordingly, in the first stage the ILC appoints one of its members as Special Rapporteur and collects information and documentation. The second stage comprises the first reading of the draft articles prepared by the Special Rapporteur and the submission of the approved drafts with commentaries to the General Assembly and to the governments. In the third and final stage, the Special Rapporteur reconsiders the draft articles after evaluating the respective government comments. A second reading of the revised draft takes place in plenary and the approved draft is again presented to the General Assembly along with recommendations regarding further actions.[96] Possible recommendations to the General Assembly include taking no action, taking note of or adopting the report, recommending the conclusion of a convention or convoking a conference to conclude the convention.[97]

I. State Responsibility

Already at its 1st Session in 1949, the ILC identified the topic of State responsibility as being suitable for codification.[98] Then, after a formal request of the General Assembly,[99] at its 7th Session in 1955 the ILC appointed *F.V. García-Amador* as Special Rapporteur, who presented six successive reports on this issue. He was succeeded in his position in 1963 by *Roberto Ago*, who similarly presented eight reports between 1969 and 1978. At its 21st Session in 1969, the ILC laid down criteria for further efforts on this issue. According to this, the work was to be confined to internationally wrongful acts; the question of liability for lawful acts was separated and deferred to a later stage.[100] This was confirmed at the 22nd Session when the ILC held the view that both issues could not be treated jointly and that the term "State responsibility" was to be understood as referring only to

---

[94] See ILC Doc. A/CN.4/325 (YBILC 1979 II/1), at 187–195; ILC Doc. A/51/10 (YBILC 1996 II/2), at 84, 86–87; United Nations (ed.), *The Work of the ILC* (7th ed., 2007), at Point 5.
[95] See ILC Doc. A/CN.4/325 (YBILC 1979 II/1), at 191; United Nations (ed.), *The Work of the ILC* (7th ed., 2007), at Point 5.
[96] See ILC Doc. A/CN.4/325 (YBILC 1979 II/1), at 195–200; United Nations (ed.), *The Work of the ILC* (7th ed., 2007), at Point 5. See also Articles 16, 17 and 18–23 of the Statute of the ILC.
[97] Statute of the ILC, Article 23.
[98] ILC Doc. "Report to the General Assembly" (YBILC 1949), at 280–281.
[99] UNGA Res. A/Res/799 (VIII) of 7 December 1953.
[100] ILC Doc. A/7610/Rev.1 (YBILC 1969 II), at 233.

internationally wrongful acts. A further innovation was the recognition of a distinction between primary and secondary rules of international law.[101] Pursuant to a recommendation of the General Assembly to prepare a first draft of articles on State responsibility,[102] the ILC adopted Part One of these draft articles concerning the origin of international responsibility, this occurring on first reading over the course of the 25th through 32nd Sessions held between 1973 and 1980.[103] Parts Two and Three of the draft articles concerning the content, forms and degrees of international responsibility as well as the legal consequences were prepared by the Special Rapporteurs *Willem Riphagen* and *Gaetano Arangio-Ruiz*, who were appointed in 1979 and 1987, respectively.[104] It was not until the 48th Session in 1996 that the ILC completed the first reading on Parts Two and Three and transmitted the provisionally adopted draft articles to the General Assembly and the governments for comments.[105] At the 49th Session, *James Crawford* was appointed as Special Rapporteur and the second reading of the draft articles took place. One of the major disputes in respect of these draft articles was the contemplated segregation of wrongful acts into crimes and delicts, which was addressed by Article 19. Since no consensus could be reached on this issue, it was agreed to defer this issue and to proceed without Article 19 as well as without the equally controversial provisions on dispute settlement.[106] The second reading was subsequently completed at the 53rd Session in 2001, where the ILC adopted the final draft articles on responsibility of States for internationally wrongful acts,[107] consisting in total of 59 Articles and the commentaries thereto.[108]

The General Assembly took note of the draft articles and repeatedly commended them to the attention of governments in 2001, 2004, 2007 and 2010.[109] However, no further significant steps have been taken so far. While some voices support an immediate or at least early convening of a diplomatic conference for the adoption of an international convention—pointing out that this would both be the next logical step and ensure legal certainty—other voices instead favour postponing the decision on

---

[101] ILC Doc. A/9010/Rev.1 (YBILC 1973 II), at 169–170. See also United Nations (ed.), *The Work of the ILC* (7th ed., 2007), at State Responsibility.

[102] UNGA Res. A/Res/3071 (XXVIII) of 30 November 1973.

[103] See United Nations (ed.), *The Work of the ILC* (7th ed., 2007), at State Responsibility.

[104] *Willem Riphagen* presented seven reports between 1980 and 1986 and *Gaetano Arangio-Ruiz* submitted between 1988 and 1996 eight reports on this topic to the ILC.

[105] Part One is published in ILC Doc. A/35/10 (YBILC 1980 II/2), at 30–34. Parts Two and Three are published in ILC Doc. A/51/10 (YBILC 1996 II/2), at 62–65.

[106] ILC Doc. A/53/10 (YBILC 1998 II/2), at 77. See on this issue also Crawford/Peel/Olleson, 'The ILC's Articles on State Responsibility', 12 *EJIL* (2001), at 976–979.

[107] See for a detailed analysis Caron, 'The ILC Articles on State Responsibility', 96 *AJIL* (2002), at 857 *et seq.*; Crawford, 'The ILC's Articles on State Responsibility', 96 *AJIL* (2002), at 874 *et seq.*; Crawford/Peel/Olleson, 'The ILC's Articles on State Responsibility', 12 *EJIL* (2001), at 963 *et seq.*

[108] ILC Doc. A/56/10 (YBILC 2001 II/2), at 25.

[109] UNGA Res. A/Res/56/83 of 12 December 2001; UNGA Res. A/Res/59/35 of 2 December 2004; UNGA Res. A/Res/62/61 of 6 December 2007; UNGA Res. A/Res/65/19 of 6 December 2010.

further actions. They emphasise that negotiating a convention would reopen controversial issues and jeopardise the delicate compromise represented by the present draft articles. Moreover, they argue that if only a small number of ratifications were to be recorded, this could hinder recognition of the draft articles as a part of accepted customary international law.[110] However, regardless of the future actions taken or not taken with a view to concluding a convention, the draft articles on State responsibility are already referred to extensively in the international practice of States, and they are already used as a *de facto* source of law by both the ICJ[111] and the ITLOS.[112,113]

II. State Liability

After the ILC decision to segregate the issue of State liability from the topic of State responsibility, between 1973 and 1977 the General Assembly urged to undertake a study on this issue.[114] At its 30th Session in 1978, the ILC commenced its work under the title "international liability for injurious consequences arising out of acts not prohibited by international law" and appointed *Robert Q. Quentin-Baxter* as Special Rapporteur on this issue. *Quentin-Baxter* presented five reports between 1980 and 1984 and, amongst them, also a schematic outline of the proposed codification.[115] This schematic outline is based on a "compound primary obligation" of States to prevent, minimise and remedy transboundary harm; it is, however, not conceived as an absolute obligation, so that failures to comply with this "negotiable duty"[116] would not entail the application of the principle of State

---

[110] See the "Summary of Work" of the 62nd and 65th Session of the UNGA concerning State responsibility, accessible at: www.un.org. See also Crawford/Peel/Olleson, 'The ILC's Articles on State Responsibility', 12 *EJIL* (2001), at 969–970; and Caron, 'The ILC Articles on State Responsibility', 96 *AJIL* (2002), at 861–866, who also stresses the "paradox that they could have more influence as an ILC text than as a multilateral treaty", *ibid.* at 857.

[111] See e.g. *The Gabčíkovo-Nagymaros Project Case*, ICJ Reports 1997, at 38–56; *Difference Relating to Immunity from Legal Process of a Special Rapporteur of the Commission on Human Rights*, ICJ Reports 1999, at 87, para. 62.

[112] *The M/V "Saiga" (No. 2) Case*, ITLOS Reports 1999, at 65, para. 170–171.

[113] This is due to the fact that these draft articles widely restate customary international law. See Caron, 'The ILC Articles on State Responsibility', 96 *AJIL* (2002), at 866; Wolfrum/Langenfeld/Minnerop, *Environmental Liability in International Law* (2005), at 459.

[114] UNGA Res. A/Res/3071 (XXVIII) of 30 November 1973; UNGA Res. A/Res/3315 (XXIX) of 14 December 1974; UNGA Res. A/Res/3495 (XXX) of 15 December 1975; UNGA Res. A/Res/31/97 of 15 December 1976; UNGA Res. A/Res/32/151 of 19 December 1977.

[115] Quentin-Baxter, ILC Doc. A/CN.4/360 (YBILC 1982 II/1), at 62–64; ILC Doc. A/37/10 (YBILC 1982 II/2), at 83–85.

[116] Quoted from Handl, 'Liability as an Obligation Established by a Primary Rule of International Law', 16 *NYIL* (1985), at 72. The schematic outline is also described as "a world in which nothing [is] either prohibited or made obligatory and everything [is] negotiable", see Boyle, 'State Responsibility and International Liability', 39 *Int'l & Comp. L. Q.* (1990), at 5.

responsibility.[117] *Quentin-Baxter* was succeeded in his position as Special Rapporteur by *Julio Barboza* in 1985. He, in turn, presented twelve reports between 1985 and 1996, amongst them the first ten draft articles in 1988 and, at the 42nd Session in 1990, a complete set of 33 draft articles on this issue.[118] These draft articles proceeded to conceive of State liability as an autonomous legal concept of primary international law that must be understood as being distinct from the principle of State responsibility.[119]

This very early position of the ILC to conceptualise State liability as an autonomous legal notion distinct from the principle of State responsibility has faced broad criticism since the beginning.[120] According to the ILC, this distinction was necessary since liability for environmental harm occurring in consequence of lawful but injurious activities cannot be traced back to a wrongful act of the State and, therefore, would have to be liability *sine delicto*.[121] This argumentation, however, has been seen as suffering from a "conceptual confusion"[122] and being "based on a fundamental misunderstanding".[123] As outlined above,[124] it seems instead preferable to accept the existence of general international obligations of States to prevent transboundary damage which—as far as lawful activities conducted by private persons are concerned—are transformed into obligations to control and prevent. In the event the source State has failed to comply with this obligation, it is internationally liable according to the State responsibility principle even though the activity remains lawful as such. From this it becomes apparent that: (i) relevant situations could, indeed, for the most part be addressed without an autonomous concept of State liability and (ii) the artificial distinction between both concepts as well as the ILC's approach to primarily apply the State liability concept to encounter lawful but injurious activities may be unfortunate and

---

[117] As to the schematic outline see Boyle, 'State Responsibility and International Liability', 39 *Int'l & Comp. L. Q.* (1990), at 4–6; Erichsen, 'Das Liability-Projekt der ILC', 51 *ZaöRV* (1991), at 101; Magraw, 'The International Law Commission's Study of "International Liability"', 80 *AJIL* (1986), at 309–314.

[118] Barboza, ILC Doc. A/CN.4/428 and Add.1 (YBILC 1990 II/1), at 105–109.

[119] The draft articles provide for reparation that has to be negotiated among the States involved as a means of a reconciliation of interests. Further innovations are a redefinition of risks which take account of both the degree of probability and seriousness of the potential damage. See on these draft articles Boyle, 'State Responsibility and International Liability', 39 *Int'l & Comp. L. Q.* (1990), at 6–8.

[120] See e.g. Akehurst, 'International Liability for Lawful Acts', 16 *NYIL* (1985), at 8; Boyle, 'State Responsibility and International Liability', 39 *Int'l & Comp. L. Q.* (1990), at 8 *et seq.*; Brownlie, *State Responsibility* (1983), at 50; de la Fayette, 'The ILC and International Liability', 6 *RECIEL* (1997), at 324–327.

[121] See e.g. ILC Doc. A/32/10 (YBILC 1977 II/2), at 87–88; Quentin-Baxter, ILC Doc. A/CN.4/334 and Add.1 and 2 (YBILC 1980 II/1), at 251; ILC Doc. A/51/10 (YBILC 1996 II/2), at 100.

[122] de la Fayette, 'The ILC and International Liability', 6 *RECIEL* (1997), at 323.

[123] Akehurst, 'International Liability for Lawful Acts', 16 *NYIL* (1985), at 8; Brownlie, *State Responsibility* (1983), at 50.

[124] See *supra*, Sect. "Implications for the Application of State Responsibility".

unnecessarily complicated. However, as far as the concept of State liability is concerned, it is true that it represents an autonomous concept distinct from the State responsibility principle and that is has to be classified as among primary international rules.[125] Due to the changes in the final work of the ILC, which will be outlined in the following paragraphs, this conceptual discord may, moreover, be of less significance in the end.

In consequence of the Special Rapporteur *Barboza*'s reports, at its 44th Session in 1992 the ILC acknowledged that the topic comprises issues of prevention as well as issues of remedial measures, and it decided that, initially, articles on preventive measures should be drafted and the drafting process should be expanded to remedial measures only after their completion.[126] This segregation was definitively accomplished at the 49th Session in 1997 when the ILC decided to split this topic into "prevention of transboundary damage from hazardous activities", on one side, and "international liability in case of loss from transboundary harm arising out of hazardous activities", on the other.[127]

The ILC proceeded first with the topic of "prevention of transboundary damage from hazardous activities" and at the same 49th Session appointed *Pemmaraju Sreenivasa Rao* as Special Rapporteur on this issue. During the following four Sessions occurring between 1998 and 2001, the ILC reconsidered and revised the draft articles previously received from Special Rapporteur *Barboza* at the 48th Session in 1996, and after a first reading transmitted to the governments a set of 17 draft articles for comments and observations. At its 53rd Session in 2001, the ILC adopted the final "Draft Articles on Prevention of Transboundary Harm from Hazardous Activities" (with commentaries thereto) and submitted them to the General Assembly along with the recommendation to elaborate a convention.[128] These final draft articles contain a basic obligation of States to prevent and minimise risks of transboundary harm and, to this end, establish certain procedural requirements such as notification and consultation rules.[129] The General Assembly took note of these final draft articles and requested the ILC to proceed with the second part of the topic.[130]

The ILC's conceptual approach for dealing with the issue of State liability has significantly changed under the guidance of Special Rapporteur *Rao* since the revision

---

[125] As to the relationship between both concepts see Magraw, 'The International Law Commission's Study of "International Liability"', 80 *AJIL* (1986), at 316–322.

[126] ILC Doc. A/47/10 (YBILC 1992 II/2), at 51.

[127] ILC Doc. A/52/10 (YBILC 1997 II/2), at 59. For a detailed discussion see also Fitzmaurice, 'International Liability', 24 *Polish Y. B. Int'l L.* (1999/2000), at 47 *et seq.*

[128] ILC Doc. A/56/10 (YBILC 2001 II/2), at 144–170.

[129] Articles 3 to 9 of the 2001 Draft Articles on Prevention of Transboundary Harm from Hazardous Activities. For a detailed analysis of these draft articles see Lammers, 'Prevention of Transboundary Harm from Hazardous Activities', 14 *Hague Y. B. Int'l L.* (2001), at 3 *et seq.*; Wolfrum/Langenfeld/Minnerop, *Environmental Liability in International Law* (2005), at 487–494.

[130] UNGA Res. A/Res/56/82 of 12 December 2001.

of the 1996 draft articles. It was recognised by the ILC that no agreement could be reached among States for the establishment of primary rules imposing strict liability on States for damages caused by lawful activities of private parties. The ILC, from that point forward, has changed its approach and has focused on the promotion of civil liability solutions modelled on the existing civil liability regimes concerning maritime transportation and nuclear activities.[131] According to this, the operator in command or control of the activity is seen and treated as the (strictly) liable party. The States are obligated insofar as they are required to implement into their domestic laws the rules and standards of liability set out in the ILC draft articles and, if necessary, to support financial remedies such as specially designated funds or a secondary liability of the State.[132] The advantage of this approach, despite the fact that the financial capacity of private parties is limited, is that it attracts a wide consensus among States and, moreover, creates a direct economic incentive in the acting parties to act in accordance with the procedural requirements in order not to lose insurance coverage.

In line with this reconceived approach and pursuant to a request by the General Assembly, the ILC, at its 54th Session in 2002, decided to take up its work on the second part of the State liability issue under the title "international liability in case of loss from transboundary harm arising out of hazardous activities", and to this end it appointed *Pemmaraju Sreenivasa Rao* as Special Rapporteur also for this issue. At its 56th Session in 2004 the ILC was intensively concerned with this issue and finally adopted on first reading a set of eight "Draft Principles on the Allocation of Loss in the Case of Transboundary Harm Arising out of Hazardous Activities", which it transmitted through the Secretary-General to governments for comments and observations.[133] The aim of the draft principles, which are formulated as soft law, is to ensure compensation for transboundary damage caused by hazardous activities by urging States to (i) implement into their domestic laws civil liability remedies imposed on the operator of the activity and (ii) adopt other supporting financial measures.[134] The second reading of these draft principles was completed at the ILC's 58th Session in 2006, at which the ILC recommended to the General Assembly that is endorse the draft principles by a resolution and urge governments to take national and international action to implement them.[135] At its

---

[131] Such as the CLC/FUND Convention, the HNS Convention, the Bunker Oil Convention, the Paris and Vienna and the NUCLEAR Conventions.

[132] Birnie/Boyle/Redgwell, *International Law and the Environment* (3rd ed., 2009), at 223–224, 319; Kiss/Shelton, *Strict Liability in International Environmental Law*, in: Ndiaye/Wolfrum (ed.) (2007), at 1139; Rao, 'International Liability for Transboundary Harm', 34 *Envtl. Pol'y & L.* (2004), at 225–226.

[133] ILC Doc. A/59/10, at para. 173–176.

[134] The content of these draft principles is thus in line with the ILC's recent aim of promoting the creation of civil liability conventions. See on the draft principles Birnie/Boyle/Redgwell, *International Law and the Environment* (3rd ed., 2009), at 319–321; and in particular Boyle, 'Globalising Environmental Liability', 17 *J. Envtl. L.* (2005), at 16–23; see also Lammers, 'New Developments Concerning International Responsibility', 19 *Hague Y. B. Int'l L.* (2006), at 91–93.

[135] ILC Doc. A/61/10, at para. 59–67.

Sessions in 2006 and 2007, the General Assembly commended these draft principles to the attention of governments.[136]

At its 65th Session in 2010 the General Assembly was, in respect of the State liability topic, mainly concerned with the question which form the draft articles and draft principles should take and in which way to proceed with this issue. The ILC had recommended that the draft articles on prevention should be adopted in an international convention, while the draft principles on allocation of loss should be endorsed in a resolution. However, no consensus in the General Assembly could be found in this regard. While some delegations expressed support for the idea of transforming the draft articles into a convention, emphasising the need for a uniform regime, the majority of the delegations opposed this idea and rather argued in favour of retaining the drafts in their current form and postponing the decision whether to adopt a convention.[137] Their main argument was that these drafts represent a progressive development of international law without yet claiming sufficient support in customary international law.[138]

For now, a conclusion of this codification process has not been reached and is not yet foreseeable. It remains an open issue whether and, if so, in which form the draft articles and draft principles will become binding law.

C. State Responsibility in the Context of Transboundary Movements of Hazardous Wastes by Sea

It has been outlined above that two different legal concepts of international law may be of practical relevance for the imposition of responsibilities and liabilities in the context of transboundary movements of hazardous wastes by sea. The first relates to explicit provisions and customary rules of State responsibility, according to which responsibilities and liabilities are imposed on a State accountable for an internationally wrongful act. The second relates to liability in accordance with civil liability conventions. In this section, the question is raised under which conditions the States may incur liabilities under explicit provisions of State responsibility as well as under the customary principle of State responsibility.

Since rules of State responsibility represent secondary rules of international law, the legal prerequisites for responsibility are dependent on the existence and the precise scope of the underlying primary obligations. Therefore, the primary

---

[136] UNGA Res. A/Res/61/36 of 4 December 2006; UNGA Res. A/Res/62/68 of 6 December 2007.

[137] See already Tomuschat, *International Liability for Lawful Acts*, in: Francioni/Scovazzi (ed.) (1991), at 65–67 who states at 65 that "[w]ith regard to the liability topic, ambitions cannot fly too high".

[138] See the "Summary of Work" of the 62nd and 65th Session of the UNGA concerning State liability, accessible at: www.un.org.

obligations of States relevant to the transboundary movement of hazardous wastes by sea are examined in detail in the context of the following investigation.

I. Explicit Provisions of State Responsibility in International Treaty Law

Explicit provisions concerning State responsibility that are established by international conventions represent *leges speciales* in relation to the customary principle of State responsibility.[139] Therefore, it shall be determined in a first step whether and to which extent explicit rules of State responsibility arise from international treaty law.

1. The Basel Convention

The 1989 Basel Convention on the Control of Transboundary Movements of Hazardous Wastes and Their Disposal (Basel Convention) represents the most important legal instrument on a global scale for the regulation of international hazardous wastes trades. It aims to achieve and strengthen the control and reduction of international hazardous waste streams by establishing a so-called "prior-informed-consent"-principle (PIC), which requires the previous written consent of all States involved in a hazardous waste transport before a transboundary shipment is allowed to take place.[140]

The Basel Convention does not contain an explicit regime of rules of State responsibility for the internationally wrongful acts of Contracting States. From the general conception of international law as well as from the explicit statement in Paragraph 15 of the Preamble to the Basel Convention, it rather follows that the provisions of the Basel Convention basically do not affect the application of the general principle of State responsibility. However, the application of this principle may be excluded by reason of specialty. This means that if the Basel Convention provides for a regime of rules conclusively governing the legal consequences of non-compliance by the State with regard to a particular subject matter—financial liability not necessarily having to be the specified consequence—such rules constitute a (sectoral) self-contained regime that overrules the general principle of State responsibility.[141] Hence, the question arises whether Article 12, 8 or 9 of the Basel Convention may be considered as constituting a sectoral self-contained regime.

---

[139] See ILC Draft Articles on State Responsibility, Article 55.

[140] The content of the Basel Convention is outlined in detail *infra*, Sect. "The Basel Convention".

[141] Dahm/Delbrück/Wolfrum, *Völkerrecht*, vol. I/3 (2nd ed., 2002), at 869; Pineschi, *Non-Compliance Procedures and the Law of State Responsibility*, in: Treves, *et al.* (ed.) (2009), at 483–484; Rauschning, 'Verantwortlichkeit der Staaten', 24 *BDGVR* (1984), at 20–22.

(a) Article 12

Article 12 of the Convention calls for the Contracting States to negotiate and adopt a supplementary protocol to the Convention that sets out in detail appropriate rules of liability and compensation (*"pactum de negotiando"*). Article 12, which is captioned as "Consultations on Liability", only provides for the mandate placed on the Contracting States to elaborate a liability protocol without setting out any substantial rules or standards to be met. It, particularly, does not determine whether the prospective protocol shall primarily be directed at private persons and entities, or at States. Therefore, it provides a margin of discretion to the Contracting States which entails not only the possible content of the prospective protocol, but also the basic decision of whether to elaborate an instrument of State responsibility and liability, or to codify a civil liability protocol.[142] However, it should be kept in mind that States are in general reluctant to enter into financial and political commitments, as would result in respect of a regime of State responsibility. Therefore, and in order to nevertheless gain the Contracting States' support for the adoption of a supplementary protocol on liability and compensation, the final text of the 1999 Protocol to the Basel Convention contains no rules imposing direct financial responsibilities or liabilities on States.[143]

It becomes apparent that neither Article 12 of the Basel Convention nor the provisions of the 1999 Liability Protocol are concerned with the international responsibility of States arising from the internationally wrongful acts of Contracting States themselves. These provisions rather govern the financial liabilities of individuals and corporate entities in consequence of their involvement in the harmful activity of shipping hazardous wastes. Therefore, a sectoral self-contained regime cannot be seen in Article 12 of the Basel Convention in connection with the 1999 Liability Protocol.

(b) Article 8

Article 8 of the Basel Convention provides that in case a transport of hazardous wastes to which the consent of the States concerned has been given cannot be completed in accordance with the terms of the contract, the State of export is under the obligation to ensure that the wastes are re-imported by the exporter into the State of export. By contrast, even a subsidiary obligation of the exporting State to re-import the wastes itself does not exist under the Basel Convention. In consequence, it must be stated that Article 8 is not concerned with illegal traffic under the Basel Convention and is even less concerned with the non-compliance of Contracting

---

[142] During the negotiations of the Convention the issue of liability was first and foremost discussed as applying at an interstate level. See Abrams, 'Regulating the International Hazardous Waste Trade', 28 *Colum. J. Transnat'l L.* (1990), at 835–836; Hackett, 'Assessment of the Basel Convention', 5 *Am. U. J. Int'l L. & Pol'y* (1989/1990), at 320–322.

[143] See *infra*, Sect. "Legal Objectives and Main Content".

States in respect of their obligations under the Basel Convention. Article 8, thus, does not contain a regime governing the legal consequences of internationally wrongful acts of Contracting States.[144] It rather deals with legal and permitted transports which entail an increased risk of harm due to the fact that these transports have not been completed. In this sense, Article 8 is comparable to the concept of State liability, albeit not being targeted towards financial compensation. It imposes a primary obligation on the Contracting States aiming at a reduction of the risk of further damage. Article 8 may not be considered a sectoral self-contained regime.

(c) Article 9

Article 9 of the Basel Convention is concerned with hazardous waste transports that are classified as "illegal traffic" due to non-compliance of the acting persons with the procedural requirements laid down by the Convention.[145] Article 9 establishes certain obligations in response to "illegal traffic". To this end, it distinguishes between cases in which the reason for "illegal traffic" lies with the exporter or generator on the outgoing side, and cases in which the reason lies with the importer or disposer on the incoming side. In the former case, the State of export is under the obligation to ensure that the illegally transported wastes are taken back to the State of export either by the exporter or generator, or, if necessary, by the State itself.[146] In the latter case, the same obligation is imposed *mutatis mutandis* to the State of import, which has to ensure that the wastes in question are disposed of in an environmentally sound manner by the importer or disposer or, if necessary, by the State itself.[147] Article 9, thus, determines the legal consequences of hazardous waste transports that are deemed illegal because of the conduct of individuals or corporate entities.

This raises the question whether the "illegal traffic" of private persons may at the same time constitute an unlawful act of the State and, therefore, a breach of an international obligation of that State. It is only in that case that Article 9 may have to be seen as a conclusive regulation of State responsibility that overrides the application of the general principle of State responsibility. However, the answer to this question cannot be given solely on the basis of Article 9, which only describes the consequences of "illegal traffic". Rather, it needs to be taken into account which particular obligations rest with the States in view of the conduct of private persons constituting the "illegal traffic". First of all, States are under the general obligation to implement the procedural rules stipulated by the Basel Convention into their national laws[148] as well as to comply with the Convention rules and

---

[144] See also Kummer, *The Basel Convention* (1995), at 221–222.
[145] Basel Convention, Article 9(1).
[146] *Ibid.*, Article 9(2)(a).
[147] *Ibid.*, Article 9(3).
[148] *Ibid.*, Article 4(4).

provisions which are addressed at the States itself. This involves, first, the obligation to install sufficient instruments and mechanisms of monitoring and control as well as the obligation to establish competent authorities.[149] Furthermore, the Basel Convention imposes on the Contracting States the obligation to prevent the conduct of any single transboundary movement of hazardous wastes, unless all notification and PIC-requirements have been fulfilled by the exporter or generator.[150] This obligation to not allow the commencement of non-consented transports is an obligation of conduct. The States are responsible to undertake a certain conduct, namely enforcing the non-permissibility of non-consented transports and the monitoring and control of relevant activities of private persons. In contrast, this obligation does not require the States to ensure that any kind of "illegal traffic" is prevented, within the sense of a State guarantee for the non-occurrence of such illegal transports. Therefore, it is possible that a particular hazardous waste movement which is deemed to be illegal within the meaning of Article 9 does not at the same time constitute a wrongful act of the State itself.[151] This, in turn, means that Article 9 is not necessarily concerned with internationally wrongful acts of States and, thus, cannot be classed among the body of rules of State responsibility. Article 9, therefore, cannot be considered a self-contained regime overriding the application of the general principle of State responsibility. It rather establishes a primary obligation of States to ensure the re-importation of illegally shipped wastes.

(d) Summary

In summary, it can be concluded that neither Article 12 in connection with the provisions of the 1999 Liability Protocol, nor Articles 8 and 9 contain rules of State responsibility. They rather establish the primary obligation of States to ensure the re-importation or proper disposal of hazardous wastes in case of illegal traffic or in case a transport cannot be completed according to the terms of the contract. These provisions, therefore, cannot be regarded as self-contained regimes of secondary rules conclusively governing the legal consequences of a breach of a primary obligation under the Basel Convention by the Contracting States. Hence, the general principle of State responsibility is not excluded by virtue of Article 12 in connection with the 1999 Liability Protocol or by virtue of Articles 8 or 9 of the Basel Convention.[152]

---

[149] *Ibid.*, Articles 4 and 5.

[150] *Ibid.*, Article 6(3) and (4).

[151] Thus, Kummer, *The Basel Convention* (1995), at 220, is not right when she states that in case of illegal traffic according to Article 9 of the Convention a wrongful act of the State simultaneously exists due to a failure of the State authorities to comply with obligations of control. See also Shibata, *Ensuring Compliance with the Basel Convention*, in: Beyerlin/Stoll/Wolfrum (ed.) (2006), at 75.

[152] See also Kummer, *The Basel Convention* (1995), at 221.

2. Other Conventions Relevant to the Trade in and Transport of Hazardous Wastes

Besides the Basel Convention, there are also regional conventions relevant to the trade in and transport of hazardous wastes which have a practical effect also on the global scale.[153] This includes in particular two conventions that have been elaborated in consequence of the Basel Convention's approach to implement a PIC-system instead of providing for a complete prohibition of hazardous waste trade. These conventions, therefore, establish a trade or import ban of hazardous wastes in their respective regional convention areas. The 1991 Bamako Convention[154] was developed under the auspices of the former Organisation of African Unity (OAU) and imposes a general import ban on hazardous wastes from non-Contracting Parties to African countries.[155] Likewise, the Contracting States of the 1995 Waigani Convention[156] agreed on a general prohibition of all imports of hazardous waste into the South Pacific Region.[157]

Except for the general PIC-approach, both Conventions are shaped according to the model of the Basel Convention. They do not contain their own regime of liability and compensation for damage arising from hazardous waste transports, but call for the Contracting States to elaborate a protocol setting out relevant rules of liability and compensation.[158] As of yet, such a protocol exists in respect of neither the Bamako nor the Waigani Convention. As with the Basel Convention, the mandate to negotiate in principle also allows for the elaboration of rules of State responsibility and State liability. Such obligations, of course, could only be imposed on the Contracting States of the respective convention, so that no responsibilities and liabilities could be imposed on non-Parties that are involved in hazardous waste exports into the respective convention area. The Bamako Convention, in addition, requires the Contracting States to impose strict and unlimited liability as well as joint and several liability on hazardous waste generators.[159]

Besides this *pactum de negotiando*, both Conventions stipulate in their Articles 8 and 9 obligations of the Contracting States to ensure the re-importation or proper

---

[153] A more detailed description of regional conventions is outlined *infra*, Sect. "The UNEP Regional Seas Programme".

[154] Bamako Convention on the Ban on the Import into Africa and the Control of Transboundary Movement and Management of Hazardous Wastes within Africa, see also *infra*, Sect. "The Bamako Convention".

[155] Bamako Convention, Article 4(1).

[156] 1995 Waigani Convention to ban the Importation into Forum Island Countries of Hazardous and Radioactive Wastes and to Control the Transboundary Movement and Management of Hazardous Wastes within the South Pacific Region, *infra*, Sect. "The Waigani Convention".

[157] Waigani Convention, Article 4(1)(a); the South Pacific Region as the "Convention Area" is defined in Article 1 of the Convention.

[158] Bamako Convention, Article 12; Waigani Convention, Article 12.

[159] *Ibid*, Article 4(3)(b).

disposal of hazardous wastes in case of "illegal traffic" or in case the transport could not be finished according to the terms of the contract. These rules are largely similar to Articles 8 and 9 of the Basel Convention and, thus, have to be construed in the same way. Since these obligations do not necessarily require an internationally wrongful act of the State, they cannot be classified among the rules of State responsibility. In summary, therefore, neither the Bamako nor the Waigani Convention contains explicit rules of State responsibility.

The same conclusion applies to the European Union (EU) Waste Shipment Regulation of 2006,[160] through which the requirements of the Basel Convention and Decision C(2001)107/FINAL of the OECD Council[161] are implemented in the EU area.[162] In its Articles 22–25, the EU Waste Shipment Regulation imposes on exporting States and in certain cases also on importing States the duty of either re-importation or proper disposal in case of illegal transports or in case the transport could not be finished according to the terms of the contract. As with Articles 8 and 9 of the Basel, Bamako and Waigani Conventions, Articles 22–25 of the EU Waste Shipment Regulation must be understood to contain substantive primary rules that apply irrespective of any international wrongful act of the State. Rules of State responsibility, therefore, cannot be found in the EU law relevant to the trade in hazardous wastes.

Finally, depending on the chemical composition of the hazardous wastes in question, also the 2001 Stockholm POP-Convention[163] may provide for applicable rules relevant in the context of transboundary movement of hazardous wastes. The Convention applies to an enumerated list of persistent organic pollutants (POPs). It restricts the production of POPs and imposes trade restrictions as well as other procedural requirements for their use and handling.[164] Concerning non-compliance mechanisms and rules of State responsibility, the Convention in Article 17 calls for the Contracting States to develop and approve procedures and institutional mechanisms for determining both the existence of non-compliance and the legal consequences of the same. In contrast to Article 12 of the Basel, Bamako and Waigani Conventions, Article 17 of the POP-Convention is more clearly directed towards the elaboration of rules concerning in particular the responsibility of States. However, besides this mandate to elaborate rules and procedures, no positive rules addressing the issue of State responsibility are included in the Convention. This is also acknowledged by the Convention itself, which in Paragraph 10 of its Preamble refers to the customary obligation of States not to cause

---

[160] Regulation (EC) No 1013/2006 of the European Parliament and of the Council of 14 June 2006 on shipments of waste.

[161] See *infra*, Sect. "OECD Council Decisions".

[162] Meßerschmidt, *Europäisches Umweltrecht* (2011), at 894.

[163] 2001 Stockholm Convention on Persistent Organic Pollutants (POPs), *infra*, Sect. "Stockholm POPs Convention".

[164] See *ibid*.

damage to the environment beyond the State's own territory. In summary, it can be stated that explicit rules of State responsibility cannot be found in the 2001 Stockholm POP-Convention.

3. The Law of the Sea Convention (UNCLOS)

In 1994 the United Nations Convention for the Law of the Sea (UNCLOS) entered into force, to which until now 165 States and the European Union have thus far become Contracting Parties.[165] Because of both the Convention's widespread acceptance as well as its universal application to marine legal issues and activities, the UNCLOS represents the major global instrument in the field of the law of the sea and, therefore, is often referred to as the "constitution of the oceans".[166] The UNCLOS, however, also represents a framework convention that in many aspects only provides for general rules combined with a mandate directed at the States to elaborate sectoral or regional conventions in the relevant fields.[167]

This structure particularly applies to Part XII of the UNCLOS which is concerned with the protection and preservation of the marine environment and which in its Article 197 mandates the Member States to elaborate further sectoral legal regimes. Part XII comprises 11 Sections, of which the single article of Section 9 is concerned with State responsibility and consists of only one article. Article 235 Paragraph 1 restates the general principle of international law according to which States are responsible for the fulfilment of their international obligations and, in case of non-compliance, are liable in accordance with international law.[168] This provision is a typical example of the framework character of the UNCLOS. Article 235 solely refers to the general principle of State responsibility recognised by customary international law and does not contain any legal rationale of its own. It is, therefore, to be seen as a simple declaratory or reference rule that cannot function as an autonomous basis for claims of compensation.[169] Furthermore, being "liable" within the meaning of Sentence 2 is neither to be understood in a way that this provision constitutes a legal basis for State liability *sine delicto*, nor

---

[165] See *infra*, Sect. "Background and Basic Legal Features of the Convention".

[166] This description was used already by the President of the Third UN Conference on the Law of the Sea, *Tommy Koh*, in his remarks under the title 'A Constitution for the Oceans'. See also Beyerlin/Marauhn, *International Environmental Law* (2011), at 120.

[167] See in respect of the structure and regulatory content of UNCLOS *infra*, Sect. "Background and Basic Legal Features of the Convention".

[168] UNCLOS, Article 235, Paragraph 1 reads: "States are responsible for the fulfilment of their international obligations concerning the protection and preservation of the marine environment. They shall be liable in accordance with international law."

[169] Harndt, *Völkerrechtliche Haftung für die schädlichen Folgen nicht verbotenen Verhaltens* (1993), at 347; see also Boyle, 'Marine Pollution under the UNCLOS', 79 *AJIL* (1985), at 367–368; Schneider, 'Environmental Aspects of the UNCLOS', 20 *Colum. J. Transnat'l L.* (1981), at 262; Hafner, in: Graf Vitzthum (ed.), *Handbuch des Seerechts* (2006), at 393.

in way that it refers to an existing principle of State liability under customary international law. Sentence 2 is rather to be seen in connection with Sentence 1 of the same paragraph, namely referring to the general legal consequences of an internationally wrongful act as being determined by the customary principle of State responsibility and as potentially including financial obligations.[170] Finally, also Paragraphs 2 and 3 of Article 235 have no particular significance as regards State responsibility. Paragraph 2 in connection with Article 229 incorporates a so-called "civil liability approach", according to which the States are obliged "to ensure that recourse is available" in accordance with either civil liability conventions or the domestic law in cases where damage was caused by pollution of the marine environment.[171] Paragraph 3 specifies the obligation of States to implement and further develop the international law relating to State responsibility and liability and establishes the obligation to implement further measures and instruments to safeguard the payment of adequate compensation, e.g. by compulsory insurance or compensation funds. In summary, Article 235 as a whole is of rather a declaratory and programmatic nature and thus does not provide any substantive rules of State responsibility.

In the UNCLOS no further rules of State responsibility can be found that are applicable to scenarios relevant in the context of transboundary movements of hazardous wastes. Article 139 Paragraph 2 contains an explicit rule establishing the international responsibility of States, this however being in connection with activities in the Area,[172] which is not relevant for the present consideration.[173] Furthermore, Article 232 concerns liability of States arising from unlawful enforcement measures with regard to the protection of the marine environment; and Article 262 concerns responsibility and liability of States in connection with marine scientific research. These provisions also lack a substantive rationale of their own, instead simply referring to the general rules of State responsibility being recognised as a general principle of customary international law. Finally, Article 304 clarifies in general terms that the provisions laid down by the UNLCOS regarding responsibility and liability for damage do not affect existing or future rules on State responsibility or State liability.

In summary, it can be said, therefore, that in the UNCLOS no explicit rules of State responsibility can be found that are relevant for the determination of responsibilities and liabilities of States with regard to damages occurring in the context of transboundary movements of hazardous wastes by sea.

---

[170] Nordquist/Rosenne/Yankov/Grandy (ed.), *UNCLOS—Commentary Vol. IV* (1991), at 412.

[171] Mensah, *Civil Liability in UNCLOS*, in: Basedow/Magnus/Wolfrum (ed.) (2010), at 4–5; Hafner, in: Graf Vitzthum (ed.), *Handbuch des Seerechts* (2006), at 395.

[172] "Area" is defined in Article 1 UNCLOS as "the seabed and ocean floor and subsoil thereof, beyond the limits of national jurisdiction".

[173] See also Harndt, *Völkerrechtliche Haftung für die schädlichen Folgen nicht verbotenen Verhaltens* (1993), at 340–341.

4. Other Conventions Relevant to the Protection of the Marine Environment

Besides the UNCLOS, further sectoral or regional conventions relevant to the protection of the marine environment need to be taken into account when assessing international responsibilities and liabilities of States for damage resulting from the transboundary movement of hazardous waste by sea. In this context, the MARPOL 73/78 Convention and a number of regional conventions elaborated within the framework of the UNEP Regional Seas Programme are of particular significance.

The MARPOL 73/78 Convention[174] is a global framework convention that only contains rules of control and enforcement with regard to the substantial provisions laid down in six Annexes to the Convention. For the transport of hazardous wastes by sea, Annexes II and III are of particular relevance. Annex II contains regulations for the control of pollution by noxious liquid substances transported in bulk; Annex III deals with the prevention of pollution of harmful substances transported in packaged form. Both Annexes also apply to wastes composed of noxious or harmful substances. Non-compliance rules are not contained in these Annexes, but Article 7 of MARPOL 73/78 establishes an explicit rule of State responsibility, according to which the State is responsible and liable to pay compensation if a ship has been unduly detained or delayed under Articles 4, 5 or 6 of the Convention (in connection with its Annexes). Although this provision concerns internationally wrongful acts of the State, it only encompasses the specific case that a ship has been unduly detailed or delayed. It does not apply to any other breach of a substantive obligation with regard to the protection of the marine environment as laid down in the Annexes to the Convention. Hence, the explicit rules regarding State responsibility established by MARPOL 73/78 are of no relevance for damages resulting from the transboundary movement of hazardous wastes by sea.

Relevant explicit rules on State responsibility may also arise from regional conventions aiming at the protection of the marine environment which have been elaborated in the framework of the UNEP Regional Seas Programme. This Programme was launched by the United Nations Environment Programme (UNEP) in 1974 in order to promote the creation of regional action plans and conventions for the protection of coastal and marine environments, taking account of the special requirements of the respective regions. Within this Programme, 13 Regional Seas Programmes have been established under the auspices of UNEP. Five further associated Programmes have been set up independently, outside the aegis of UNEP.[175] In respect of 14 of these 18 Programmes, legally binding conventions have been adopted aiming at the protection of the marine environment at a regional

---

[174] 73/78 MARPOL Convention for the Prevention of Pollution from Ships, see also *infra*, Sect. "MARPOL 73/78 Convention".

[175] See also *infra*, Sect. "The UNEP Regional Seas Programme". For a very instructive introduction see the UNEP Regional Seas Programme Website at http://hqweb.unep.org/regionalseas/about/default.asp.

level.[176] These conventions pursue the "framework convention" model consisting of a legally binding general agreement that is supplemented by associated, but legally independent protocols, each concerning a specific issue. However, neither these conventions[177] nor the respective Protocols concerning the transboundary movement of hazardous wastes[178] contain explicit rules through which responsibilities or liabilities are imposed on the Contracting States in response to an internationally wrongful act of the State.

5. Summary

In summary, it can be concluded that explicit rules of State responsibility for damage which result in the context of transboundary movements of hazardous waste by sea cannot be found in the relevant international conventions. Even though the general issue of State responsibility is addressed in most of the relevant conventions, this only concerns the inclusion of a declaratory rule or reference rule that makes recourse to the general principle of State responsibility or, alternatively, to a rule that mandates the respective Contracting States to elaborate and adopt further legal instruments on liability and compensation. If such a "*pactum de negotiando*" is included, this rule is mostly worded in a form that allows for the elaboration of civil liability conventions as well as rules of State responsibility and liability. The latter case is expressly mentioned in Article 13 of the 1982 Jeddah Convention. However, apart from the 1999 Liability Protocol to the Basel

---

[176] Regional Seas Conventions developed under the auspices of UNEP: 1976 Barcelona Convention amended in 1995 (Mediterranean Region); 1978 Kuwait Convention (Kuwait Region); 1981 Abidjan Convention (West and Central Africa Region); 1981 Lima Convention (South-East Pacific Region); 1982 Jeddah Convention (Red Sea and Gulf of Aden); 1983 Cartagena Convention (Wider Caribbean Region); 1985 Nairobi Convention amended in 2010 (Eastern African Region); 1986 Noumea Convention (South Pacific Region); 1992 Bucharest Convention (Black Sea Region); 2002 Antigua Convention (North-East Pacific Region).

Associated partner programs: 1991 Madrid Protocol on Environmental Protection to the 1959 Antarctic Treaty (Antarctic Region); 1992 Helsinki Convention on the Protection of the Marine Environment of the Baltic Sea Area, superseding its 1974 predecessor (Baltic Sea Region); 1992 OSPAR Convention (North-East Atlantic Region); 2003 Tehran Framework Convention (Caspian Sea Region).

[177] The following Conventions contain a mandate to elaborate and adopt supplementary rules of responsibility and liability of civil persons and/or of States themselves, "*pactum de negotiando*": Cartagena Convention, Article 14; Nairobi Convention, Article 16; Barcelona Convention, Article 16; Abidjan Convention, Article 15; Bucharest Convention, Article XVI; Jeddah Convention, Article XIII; Kuwait Convention, Article VIII; Lima Convention, Article 11; Noumea Convention, Article 20; Antigua Convention, Article 13; Madrid Protocol to the Antarctic Treaty, Article 16; Helsinki Convention, Article 25; Tehran Convention, Article 29.

[178] 1996 Izmir Protocol to the Barcelona Convention on the Prevention of Pollution of the Mediterranean Sea by Transboundary Movements of Hazardous Wastes and Their Disposal; 1998 Tehran Protocol to the Kuwait Convention on the Control of Marine Transboundary Movements and Disposal of Hazardous Wastes and other Wastes.

Convention, no such supplementary instruments have been adopted so far; and the 1999 Basel Protocol does not contain rules on State responsibility.

Declaratory or reference rules which turn to the customary principle of State responsibility can be found, for example, in Articles 139, 232, 263 and 304 of the UNCLOS. The Basel Convention also addresses the issue of State responsibility only by means of a reference to the customary principle of State responsibility in Paragraph 15 of the Preamble. The Basel Convention as well as the Bamako and Waigani Conventions impose on their Contracting States the obligation to reimport or, alternatively, to dispose of hazardous waste in case of illegal traffic and in case the transport of which could not be finished according to the provisions of the contract (Articles 8 and 9). Since, however, these provisions do not depend on a prior breach of an international obligation by the State, they cannot be attributed to the body of secondary rules concerning State responsibility.

II. The Customary Principle of State Responsibility

Having a particular practical importance, the customary principle of State responsibility is a key supplement to the few provisions of international treaty law explicitly imposing State responsibility on the respective Contracting States. In the following sections it is examined whether and under which conditions States that are involved in a transboundary movement of hazardous wastes by sea may refer to the customary principle of State responsibility as a legal basis for claims for compensation.

The legal prerequisites for bringing a claim under the principle of State responsibility can be derived from the 2001 ILC Draft Articles on State Responsibility, which, albeit not legally binding as such, are already used as a *de facto* source of law by international courts like the ICJ and the ITLOS.[179] According to these Articles the application of the principle of State responsibility presupposes an internationally wrongful act of the State.[180] An internationally wrongful act consists of two constituent elements, i.e. (1) a legally relevant act that is imputable to the State and (2) a breach of an international obligation of the State as resulting from this act.[181] After examining the particular scenarios in which an internationally wrongful act of the State may be assumed in the context of transboundary movements of hazardous wastes by sea, this section shall also outline the legal consequences of State responsibility.

---

[179] *Supra*, Sect. "State Responsibility".
[180] ILC Draft Articles on State Responsibility, Article 1.
[181] *Ibid.*, Article 2(a) and (b).

1. Act of the State

The application of the principle of State responsibility requires, first, an act of the State which consists of either an action or an omission that is attributable to the State.

In its external relations a State is considered as a uniform subject of international law with full legal authority to act under international law. As a legal entity, however, a State is not able to act by itself; its actions instead being understood as conduct of its organs and representatives which is attributable to the State.[182] This attribution is not a mere matter of causality, but rather a normative operation determined by international law. In this context, it should first be pointed out that the structure of the State and the functions that are performed by its organs and representatives are governed solely by the internal law of each State. Therefore, the domestic law remains applicable as to the determination of what constitutes a State organ and as to the designation of other persons empowered to exercise elements of governmental authority.[183] In contrast, the legal prerequisites for the international responsibility of States can be specified solely by international law itself. The conditions under which conduct is deemed to be conduct of the State for the purpose of State responsibility are, thus, independent from the respective domestic laws. This means that the conduct of certain institutions performing public functions (such as the police) may be imputed to the State by virtue of international law, even if such institutions are not deemed to be State organs and are not otherwise empowered to exercise elements of governmental authority according to the respective domestic law.[184] Relevant rules on the attribution of conduct to the State are laid down in Chapter II, Articles 4 to 11 of the ILC Draft Articles on State Responsibility. As a general rule, it can be said that the conduct of organs, agents and representatives of a State is attributable to that State, whereas the conduct of individuals and corporate entities acting solely in private capacity cannot be imputed to the State.[185]

It should be stressed, furthermore, that in general it makes no difference whether an action or an omission forms the legally relevant act entailing State responsibility. In certain situations, however, it may be difficult to identify the

---

[182] Commentaries to the ILC Draft Articles on State Responsibility, Article 2, at para. 5, where reference is made to the advisory opinion *German Settlers in Poland*, PCIJ Series B No. 6 (1923), at 22; Schweisfurth, *Völkerrecht* (2006), at 234. As to the historic development of this issue see Hessbruegge, 'The Development of the Doctrines of Attribution and Due Diligence', 36 *N. Y. U. J. Int'l. L. & Pol.* (2003/2004), at 265 *et seq.*

[183] Commentaries to the ILC Draft Articles on State Responsibility, Chapter II, at para. 6. This is also explicitly stated in Articles 4(2) and 5 of the ILC Draft Articles on State Responsibility.

[184] Commentaries to the ILC Draft Articles on State Responsibility, Chapter II, at para. 6–7; Ago, ILC Doc. A/CN.4/246 and Add.1-3 (YBILC 1971 II/1), at 238, states that the internal law is solely used as "a simple description of facts". See also the explicit statement in Article 3 of the ILC Draft Articles on State Responsibility.

[185] Commentaries to the ILC Draft Articles on State Responsibility, Chapter II, para. 2–3.

concrete conduct that ultimately constitutes the breach of international law. Particularly, if an omission is involved, it is necessary to determine first the concrete action that is required by law in order to be able to decide in a second step whether the related omission is of legal significance. Moreover, it is also possible that a combination of act and omission is to be seen as the relevant conduct.[186] Therefore, not too much emphasis should be put on the distinction between acts and omissions.

(a) State Organs and Persons Empowered by the State

First of all, the conduct of any State organ is to be considered an act of the State under international law.[187] The explicit emphasis on "any" State organ shows that this reference is to be understood in the most general sense. It does not only include State organs of the central government, but rather comprises any organ of the State and any governmental institution considered at all hierarchical levels exercising whatever functions. The capacity of acting as a State organ does, furthermore, not depend on whether legislative, executive or judicial functions are exercised.[188] Moreover, in federal States the conduct of the organs of individual States is in general attributed to the federal State.[189]

It has been shown that the status of State organ is determined by the internal law of each State.[190] This reference to the respective national laws and the possibly divergent definitions of State organs may lead to inconsistencies among States as to whether certain institutions are deemed State organs or not. However, in the end such divergences in the domestic concept of State organs have no practical impact. This is due to the fact that not only the conduct of State organs, but also the conduct of persons or entities that are otherwise empowered by the State to exercise elements of governmental authority is imputed to the State. Such attribution certainly requires that the person or entity is acting in that capacity in the particular instance.[191] This form of attribution comprises particularly parastatal entities and persons in charge of a semi-official office, both entities and persons in

---

[186] *Ibid.*, Article 2, para. 4.

[187] ILC Draft Articles on State Responsibility, Article 4, Paragraph 1.

[188] See in particular Commentaries to the ILC Draft Articles on State Responsibility, Article 4, para. 6. See further Ipsen, in: Ipsen (ed.), *Völkerrecht* (5th ed., 2004), at 637–639.

[189] Ago, ILC Doc. A/CN.4/246 and Add.1-3 (YBILC 1971 II/1), at 257–258; Dahm/Delbrück/Wolfrum, *Völkerrecht, vol. I/3* (2nd ed., 2002), at 892–893; Ipsen, in: Ipsen (ed.), *Völkerrecht* (5th ed., 2004), at 638–639; Wolfrum/Langenfeld/Minnerop, *Environmental Liability in International Law* (2005), at 470.

[190] ILC Draft Articles on State Responsibility, Article 4(2).

[191] *Ibid.*, Article 5.

charge of a semi-official office which have been empowered by the State to execute certain governmental functions. Thereby, it is of no relevance whether an entity is incorporated as a public, a semi-public or as a private entity.[192] It also makes no difference whether or not the State owns assets or shares of that company or exercises corporate or regulatory control, provided the entity is in each case empowered to exercise public functions normally exercised by State organs.[193]

With regard to the transboundary movement of hazardous waste by sea it needs to be taken into account that the regulatory structures of domestic waste management differ considerably among the individual States. The waste treatment and disposal process may be regulated, governed and supervised at a central governmental level or at a local level, and those tasks may be assigned to a specific central authority or they may be spread over several competent authorities. However, considering the main procedural rules imposed by the Basel Convention, it becomes apparent that usually two functional authorities are involved in hazardous waste movements, namely the notification authorities and the customs authorities. In the individual case further sectoral authorities may be involved, which include, for example, the port authorities, the offices for foreign trade and also environmental or veterinary authorities. As a general rule, the conduct and the knowledge of such authorities and offices is to be attributed to the State because either they are deemed State organs by virtue of the domestic law or they are empowered by the respective State to exercise elements of governmental authority.

However, the transboundary movement of hazardous wastes is regulated and supervised not only by the competent authorities of a State; it is in fact also carried out by waste management companies that are to a different degree empowered by the State to perform public tasks. The actual disposal and shipment of hazardous wastes may be carried out by a State company, by a private company on which the State has a corporate, regulatory or factual influence, or by an autonomous private company. If such a company is empowered by the State to exercise certain tasks of public authority while carrying out the actual measures related to the treatment and shipment of hazardous wastes, this conduct of the company is attributed to the State irrespective of the particular corporate structure of the company. Relevant cases may involve waste management companies that undertake, at the same time, to perform certain functions on behalf of the State, such as tasks related to the notification and customs procedures. As a further example, one can mention companies that are designated by the State to have the exclusive right to perform hazardous waste shipments and, thus, hold a *de facto* monopoly.

If, in contrast, the company in question is not empowered by law to exercise governmental authority, its conduct may only be imputed to the State by means of the further rules of attribution as outlined in the following sections.

---

[192] For example, in some countries private security firms exercise public tasks. In Germany "*Beliehene*" are empowered to charge truck tolls.

[193] Commentaries to the ILC Draft Articles on State Responsibility, Article 5, para. 1–3.

(b) Persons in Fact Acting for the State

In addition to the conduct of State organs and persons or entities otherwise empowered to exercise elements of governmental authority, also the conduct of private persons or entities may be imputed to the State, provided that this conduct is directed or controlled by a State. This particularly applies to cases in which a person or entity is in fact acting on the instructions of, or under the direction or control of a State in carrying out the conduct.[194] Such constellations, hence, concern cases of factual State influence rather than cases of delegated public authority. The required degree of direction or control to be executed by the State is relatively high. A mere general dependence or support is not sufficient. An attribution of this conduct to the State rather requires that there is an actual direction or control of the State related to the particular conduct in a given case.[195]

In the field of transboundary movements of hazardous wastes such constellations of attribution mainly concern private waste management companies which are not empowered with public authority. Irrespective of any corporate linkage or the devolution of governmental powers, the conduct of such companies is considered conduct of the State only on the condition that the State by means of instruction, regulation or control in the individual case actually influences the conduct of the private entity. The particular difficulty with this rule of attribution is not a legal, but a practical one. If an injured State claims compensation for damage suffered as a result of a wrongful act of another State under the principle of State responsibility, the burden of proof lies in principle with the claimant State that has to establish in full the legal prerequisites of this claim, particularly the degree of direction or control performed by the opponent State in the individual case.[196] From an external position, however, the claimant State has no insight into the internal structures and decision-making processes of the opponent State and the acting private company. Furthermore, it has no legal or actual possibility to investigate these internal affairs of the opponent State and, therefore, will typically not be able to gather sufficient evidence.

(c) Conduct Ultra Vires

In addition to the rules of attribution outlined in the previous sections, international law also acknowledges the attribution of conduct to the State in a further constellation relevant to typical scenarios in the context of transboundary movements of hazardous wastes by sea. According to this approach, the conduct of State organs and other persons or entities empowered with public authority is attributed

---

[194] ILC Draft Articles on State Responsibility, Article 8.
[195] Commentaries to the ILC Draft Articles on State Responsibility, Article 8, para. 1–6.
[196] Commentaries to the ILC Draft Articles on State Responsibility, Article 19, para. 8; Lefeber, *Transboundary Environmental Interference* (1996), at 107.

to the State even if the person or entity acting in official capacity is exceeding the authority or competence which has been delegated by the individual instructions or respective internal laws of the State.[197] This attribution of conduct *ultra vires* to the State is a necessary extension of the attribution of the regular conduct of State organs and other persons empowered with public authority. Since the internal law of the States remains applicable as to the determination of the notion of State organs and as to the assignment of public authority to other persons or entities, it would be possible for a State to avoid the attribution of conduct by simply referring to its internal laws, regulations or instructions. Consequently, there is agreement on the international level that a State must be responsible for all "official" acts of persons exercising governmental powers, irrespective of whether such conduct is within or outside the particular powers, e.g. criminal conduct in office.[198]

In the context of transboundary movements of hazardous wastes, this rule of attribution is of major significance in respect of cases of corruption and other types of official misconduct by the acting person in charge. It makes clear that the State is in principle also responsible if damage occurs which can be attributed to the conduct of an official person that has, for instance, arranged for or permitted an illicit export or import of hazardous wastes, or has accepted bribes securing the unlawful notification or approval of transboundary hazardous waste shipments.

(d) Interstate Attribution

A different kind of attribution concerns the interstate attribution of conduct which is relevant for the determination of the responsibility of a State in connection with the act of another State. According to this rule of attribution, in fact two different operations of attribution take place. In the first step, the conduct of State organs and persons or entities otherwise empowered with public authority is imputed to the State according to the general rules of attribution. Then, in a second step, the question arises whether such conduct of a State can be attributed to another State. International law provides for detailed rules in which such interstate attribution is endorsed.[199] The only constellation that is relevant in the context of transboundary shipments of hazardous wastes relates to the case where a State aids or assists another State in the conduct of an internationally wrongful act. Such aid or assistance is attributable to the State of origin provided three conditions are fulfilled: First, the State conducting the aid or assistance must be aware of the circumstances of this conduct; second, the aid or assistance is given with a view to

---

[197] ILC Draft Articles on State Responsibility, Article 7.

[198] A comprehensive examination of relevant case law can be found at Commentaries to the ILC Draft Articles on State Responsibility, Article 7, para. 2–6.

[199] These rules are laid down in Chapter IV, Articles 16–18, of the ILC Draft Articles on State Responsibility.

facilitating this conduct; and third, the act would be wrongful if committed by the State itself.[200] Such attribution is of course without prejudice to the international responsibility of the State ultimately committing the act in question.[201]

With regard to the transboundary movement of hazardous wastes, it should be kept in mind that the rules of procedure laid down by, for example, the Basel Convention require a constant cooperation of the States involved in a movement in the form of mutual notification and approval. Therefore, it is well possible that one authority comes to know that the authority of another State is not acting in conformity with international requirements. It is, furthermore, possible that this authority, or rather the official person in charge, has a certain personal interest in nevertheless granting permission for the particular transport in question. This might be due to, for example, the acceptance of bribes, unofficial orders, or simply due to reasons of convenience. Provided the claimant State (which might be a State of transit that suffered damage caused by an unapproved transport) is in the position to prove these circumstances, the State that is aware of and supports the wrongful act of another State is internationally responsible for the wrongful act of the other State in consequence of the interstate attribution of this wrongful act. This, of course, has no effect on the international responsibility of the State actually conducting the wrongful act.

(e) Persons Solely Acting in Private Capacity

As a general rule, the conduct of private persons and corporate entities which do not represent State organs and are not otherwise empowered to execute governmental powers, and which do not act under the State's factual direction or control, cannot be imputed to the State.[202] This, however, does not mean that States cannot at all be responsible with regard to the conduct of individuals or corporate entities under their jurisdiction. The responsibility of States for such conduct may be incurred in an indirect way, namely by taking into account the State's own obligations under international law with regard to the conduct of private parties. A State, thus, may be under an international obligation of result, according to which it is obligated to ensure that a particular activity is prevented or that certain procedural requirements are complied with. If in such case the State knows or ought to know about any wrongful conduct of private parties and does not prevent

---

[200] ILC Draft Articles on State Responsibility, Article 16. See Commentaries to the ILC Draft Articles on State Responsibility, Article 16, para. 1–6.

[201] This is made clear by Article 19 of the ILC Draft Articles on State Responsibility.

[202] Commentaries to the ILC Draft Articles on State Responsibility, Chapter II, para. 3, with reference to the *Case Concerning United States Diplomatic and Consular Staff in Tehran*, ICJ Reports 1980, at 29–30, para. 57–61; Ipsen, in: Ipsen (ed.), *Völkerrecht* (5th ed., 2004), at 644–645; Lefeber, *Transboundary Environmental Interference* (1996), at 56.

this conduct, although it is able to do so, the State incurs international responsibility for its failure to comply with its own obligations under international law.[203]

In the context of transboundary movements of hazardous wastes the States involved are under the obligation of result to implement the procedural rules and to establish effective control mechanisms as prescribed particularly by the Basel Convention. This may also include ensuring the strict observance of absolute trade restrictions.[204] Obligations of States may furthermore derive from the customary international law, such as from the obligation to control and prevent as far as possible harmful activities conducted within their jurisdiction.[205] The particular content of these obligations will be outlined in the following sections.

(f) Summary

In summary, it can be said that due to the harmonisation of procedural rules at the international level, State authorities are to a large extent involved in the administration of transboundary movements of hazardous wastes. The conduct of such authorities is generally to be considered an act of the State. The same conclusion basically applies to illegal transports that take place with the knowledge and/or support of the local administration. Under certain conditions, illegal conduct of State representatives can, moreover, be attributed to other States involved. Considering this, it becomes clear that in most cases of illegal transports of hazardous wastes, the conduct of the persons in charge can be attributed to the State. By contrast, the conduct of individuals or corporate entities acting in solely a private capacity cannot be attributed to the State. However, the State can be responsible for such conduct indirectly, namely if in such cases the State at the same time fails to comply with an international obligation to control and to ensure the non-occurrence of the transports in question.

2. Breach of an International Obligation

(a) Requirements in General

The second central element of State responsibility is the breach of an international obligation caused by the act of the State. The term "breach of an international obligation" is defined as an act of the State that is not in conformity with what is

---

[203] *Case Concerning United States Diplomatic and Consular Staff in Tehran*, ICJ Reports 1980, at 29–30; Dahm/Delbrück/Wolfrum, *Völkerrecht*, vol. I/3 (2nd ed., 2002), at 910.

[204] See in detail *infra*, Sect. "First Tier: Absolute Trade Restrictions".

[205] See in detail *infra*, Sect. "The Obligation not to Cause Significant Harm to the Environment of Another State's Territory".

required of it by an international obligation.²⁰⁶ However, on a closer view this definition is of no real assistance. It does not determine the conditions under which an act of a State is deemed to be "not in conformity with an international obligation". In other words, it is unclear what precisely is required to constitute a breach of an international obligation. Does an act of a State, in order to be deemed internationally wrongful, generally require a violation of a legal position of another State, or even physical damage? And is personal fault in terms of the culpability of the person acting on behalf of the State principally a necessary prerequisite? The answer to these questions can certainly not be conceived in general and abstract terms. It rather depends to a large extent on the particular content of the respective primary obligation. The primary obligations, therefore, represent the major yardstick for the determination of the precise requirements for an act being deemed in non-conformity with a respective obligation. Which basic conclusions can be drawn from the international primary obligations and whether or not there are, in addition, general requirements that must be fulfilled irrespective of the content of the particular obligation will be examined in the following sections.

(aa) Relevance of the Respective Primary Obligation

Whether or not the conduct of a State is in conformity with an international obligation can be established only on the basis of the respective primary obligation itself.²⁰⁷ From this obligation it can be ascertained, for example, what substantive conduct is required, which standard of conduct needs to be observed, what result needs to be achieved, or which further particular requirements are to be fulfilled.²⁰⁸

According to the ILC Draft Articles on State Responsibility, it is made clear that State responsibility may arise out of a breach of primary obligations "regardless of its origin or character".²⁰⁹ This means that there is no difference as to whether the obligation arises from international treaty law, customary international law or from other sources of international law (such as unilateral acts) or as to whether the obligation has an international, regional or bilateral scope.²¹⁰ By also referring to the "character" of the obligation, it is made clear that the character, meaning the particular properties of that obligation and its classification, is of no relevance. Thus, the principle of State responsibility applies to obligations of

---

[206] ILC Draft Articles on State Responsibility, Article 12.

[207] This decisive role of the particular primary obligation is expressly stated in Article 12 of the ILC Draft Articles of State Responsibility, which provides that a breach of an international obligation requires that an act of the State is not in conformity with what is required of it "by that obligation".

[208] Commentaries to the ILC Draft Articles on State Responsibility, Chapter III, para. 2.

[209] ILC Draft Articles on State Responsibility, Article 12.

[210] Commentaries to the ILC Draft Articles on State Responsibility, Article 2, para. 7, and Article 12, para. 3–10; Schweisfurth, *Völkerrecht* (2006), at 239.

result as well as to obligations of conduct.[211] It further applies irrespective of the standard of conduct to be complied with by the State. Also, no distinction is made between international crimes and delicts.[212] The only requirement is that the obligation is in force for the State in question at the time of the legally relevant act.[213]

Even though it is not possible to specify in abstract terms the requirements for a breach of an international obligation, it is nevertheless possible to identify some general rules that derive from the characteristics of the respective obligation. In this context, some conclusions can be drawn from the character of the obligation as one of conduct or one of result. The decisive factor for the distinction between these types of obligations is whether the obligation's essential purpose is to require the State to perform a particular act or to ensure the achievement of a certain situation.[214] The common wording "to take appropriate measures to ensure" can, as a general rule, be understood to indicate an obligation of result. As far as an obligation of conduct is concerned, an act of the State is in breach with this obligation if the actual conduct imputable to the State does not correspond with the conduct required by the obligation, irrespective of the factual consequences of this conduct.[215] By contrast, an obligation of result is breached on the condition of any divergence between the result prescribed by that obligation and the result that actually occurred. If an obligation of result prescribes that a certain situation is to be prevented, this obligation is breached if this situation occurs nevertheless. In any case entailing an obligation of result, the legal or administrative measures actually undertaken or omitted by the State to achieve this result are of no relevance.[216]

Some peculiarities must be observed in respect of international conventions that were created with a view to regulating the conduct of private persons and companies. This, particularly, also applies to the Basel Convention, bearing in mind that most transboundary movements of hazardous wastes are actually carried out by private parties. The Basel Convention as well as the relevant regional conventions, thus, are not exclusively concerned with the conduct of States, but rather

---

[211] Commentaries to the ILC Draft Articles on State Responsibility, Chapter III, para. 2, and Article 12, para. 11.

[212] See Köck, *Staatenverantwortlichkeit und Staatenhaftung im Völkerrecht*, in: Köck/Lengauer/Ress (ed.) (2004), at 204–208. Serious breaches of international law, which are denoted as international crimes, may, however, entail specific consequences, see *infra*, Sect. "Serious Breaches of Peremptory Obligations".

[213] ILC Draft Articles on State Responsibility, Article 13.

[214] See *supra*, Sect. "Obligations of Conduct and Obligations of Result".

[215] Wolfrum, *Obligation of Result Versus Obligation of Conduct*, in: Arsanjani/Cogan/Sloane/Wiessner (ed.) (2011), at 375–376.

[216] *Trail Smelter Arbitration Award*, 3 RIAA (1949), at 1905 *et seq.*; *Cases No. A15(IV) and A24 (Iran v. United States)*, 11 WTAM (1999), at para. 95; Dahm/Delbrück/Wolfrum, *Völkerrecht*, vol. I/3 (2nd ed., 2002), at 876; Wolfrum, *Obligation of Result Versus Obligation of Conduct*, in: Arsanjani/Cogan/Sloane/Wiessner (ed.) (2011), at 371–373.

ultimately aim to regulate the conduct of the private parties involved.[217] Since, however, international law is solely addressed at States and other subjects of international law, individuals and corporate entities are not directly bound by international law.[218] In order to apply the rules and provisions laid down by international conventions, such as the Basel Convention, it is necessary that these rules are "made valid" for private parties. This can be achieved either by means of an act of transformation, by which international rules are reworded as part of the domestic law, or by means of an act of adoption or incorporation, by which international rules as such are given domestic legal force.[219] Only after such act of transformation or incorporation are individuals or corporate entities (indirectly) bound by an international obligation. By contrast, the State remains responsible at the international level for the fulfilment of its international obligations arising from the respective convention. This involves, first, that the State complies with the obligations directly addressed at the States, such as the duty to establish competent authorities, to co-operate with a view to the exchange of information, or to fulfil a financial obligation. In addition, and with regard to rules and obligations ultimately addressed at private parties, the State is required to transform or incorporate these rules into the domestic law within a certain timeframe. And even after having undertaken this act of transformation or incorporation, the State is not discharged from this obligation. From the perspective of the State, an obligation to transform or incorporate substantive rules into the domestic law changes upon the act of transformation or incorporation into an obligation to further regulate, control and ensure the enforcement of such rules.[220] Because of the fact that the obligation of States to implement the substantive rules of an international convention into the domestic law and to further regulate, control and ensure the enforcement of such rules is not explicitly laid down in international conventions, but inherently results from the international obligation of the State, such derivative obligations are also denoted "relative obligations".[221]

---

[217] See already Nanda/Bailey, 'Export of Hazardous Waste and Hazardous Technology', 17 *Den. J. Int'l L. & Pol'y* (1988/1989), at 201.

[218] Exemptions from this principle are international crimes, such as slavery, piracy or war crimes, see Dahm/Delbrück/Wolfrum, *Völkerrecht, vol. I/2* (2nd ed., 2002), at 264–267; and in general Doehring, *Völkerrecht* (2nd ed., 2004), at 113–114; Evans (ed.), *International Law* (2nd ed., 2006), at 314–315.

[219] As to the different approaches of the doctrine of transformation and the doctrine of incorporation see Cassese, *International Law* (2nd ed., 2005), at 217–220; Schweisfurth, *Völkerrecht* (2006), at 197–199.

[220] *Case Concerning United States Diplomatic and Consular Staff in Tehran*, ICJ Reports 1980, at 29–30; Dahm/Delbrück/Wolfrum, *Völkerrecht, vol. I/3* (2nd ed., 2002), at 909–910; Evans (ed.), *International Law* (2nd ed., 2006), at 461–462; Ipsen, in: Ipsen (ed.), *Völkerrecht* (5th ed., 2004), at 645–646; Lefeber, *Transboundary Environmental Interference* (1996), at 34.

[221] Pisillo Mazzeschi, 'The Due Diligence Rule', 35 *GYIL* (1992), at 15. From the perspective of the conduct of individuals, this obligations—in case of a breach—results in "indirect responsibility", see Eagleton, *The Responsibility of States* (1928), at 214–215.

It follows from the foregoing, that the origin and character of the respective primary obligation may provide some general conclusions as to the particular conditions of a breach of this obligation. However, the general conclusions that can be drawn from the obligation's basic character cannot hide the fact that in the end the precise content of the individual primary obligation remains the only decisive criterion for determining whether a State's act is not in conformity with this obligation.

(bb) A General Requirement of Damage?

It has been the subject of debate in international scholarship whether the State claiming compensation under the principle of State responsibility must have necessarily sustained damage, irrespective of the content of the obligation that has been breached.[222] In order to give an answer one needs to draw a distinction: As far as damage is understood in terms of material damage, such as environmental damage or other damage to the territory of the affected State, such general requirement cannot be based on a corresponding practice of States or international jurisdiction. Moreover, such requirement would contradict the common practice of States, according to which, for example, the failure of a State to implement the substantive provisions of an international convention into the domestic law is recognised as an internationally wrongful act, even though in this case no material damage has been sustained. If, in contrast, damage is conceived in a wider meaning including the non-material violation of legal positions under international law, this general requirement would be superfluous since it would be fulfilled in any case. International obligations are always established with a view to other States and, therefore, a breach of an international obligation necessarily entails the violation of a corresponding right of the respectively affected State.[223]

In conclusion, it must be stated, therefore, that damage within the meaning of a non-material violation of a legal position of another State is not a general requirement for the establishment of State responsibility. The question of whether physical damage is required to constitute a breach of an international obligation, in contrast, depends on the respective primary obligation.[224] A general requirement of physical damage does not exist.[225]

---

[222] See Cassese, *International Law* (2nd ed., 2005), at 252–253; Evans (ed.), *International Law* (2nd ed., 2006), at 465–466; Ipsen, in: Ipsen (ed.), *Völkerrecht* (5th ed., 2004), at 630.

[223] Commentaries to the ILC Draft Articles on State Responsibility, Article 2, para. 8–9; Ago, ILC Doc. A/CN.4/246 and Add.1-3 (YBILC 1971 II/1), at 223; Cassese, *International Law* (2nd ed., 2005), at 252; Ipsen, in: Ipsen (ed.), *Völkerrecht* (5th ed., 2004), at 630; von Münch, *Das völkerrechtliche Delikt* (1963), at 140.

[224] For example, the customary obligation not to cause substantial damage to another State's territory requires as an obligation of result the actual interference, meaning physical damage, of the territory of the affected State. See *infra*, Sect. "The Obligation not to Cause Significant Harm to the Environment of Another State's Territory".

[225] Commentaries to the ILC Draft Articles on State Responsibility, Article 2, para. 9; Evans (ed.), *International Law* (2nd ed., 2006), at 465–466.

(cc) A General Requirement of Fault?

One of the most controversial issues in the law of State responsibility has been seen in the question of whether or not fault constitutes a necessary element of an internationally wrongful act that must be established irrespective of the content of the particular primary obligation.[226] This question involves the very fundamental dispute in international law of whether subjective responsibility (like in most national laws) or objective responsibility represents the more suitable approach to determine responsibilities of States in interstate constellations.

Fault in this context is described as the particular subjective and psychological attitude, the moral culpability of the individual actor that consists either in the intentional (malice, *dolus*) or at least negligent (recklessness, *culpa*) causation of the particular conduct or result.[227] The theory of fault responsibility goes back to *Grotius*, and in its initial strict form it required that the State itself be at fault in order to establish an internationally wrongful act.[228] Other supporters later modified this doctrine that henceforth has focused on the personal fault of the person acting on behalf of the State, and has imputed this personal fault to the State.[229] According to this doctrine, fault forms a necessary subjective element of a wrongful act and, thus, provides in the first place a criterion to distinguish between adverse effects that simply have to be accepted by the international community and conduct entailing State responsibility. The criterion of fault is supposed to be of decisive importance particularly in respect of State obligations with a view to the conduct of private parties. But also in respect of obligations directly addressed at States, the requirement of fault takes on an indirect effect. In the latter case the presence of fault is deemed to be legally presumed unless the lack of fault is clearly established, such as in case of fortuitous events or *force majeure*.[230] However, there are strong conceptual arguments articulated against this doctrine of fault responsibility. Specifically, it encounters quite fundamental complaints when the subjective concept of fault, which is based on the individual and psychological attitude of humans, is being applied, or even attributed, to the State as an abstract and artificial entity. Furthermore, the community of States is built on objective relations and obligations to which the attribution of subjective categories like

---

[226] As to the different legal schools of thought see Pisillo Mazzeschi, 'The Due Diligence Rule', 35 *GYIL* (1992), at 11–21. See also García Amador, ILC Doc. A/CN.4/125 (YBILC 1960 II), at 60–65; Brownlie, *State Responsibility* (1983), at 37–48; von Münch, *Das völkerrechtliche Delikt* (1963), at 152–169.

[227] Cassese, *International Law* (2nd ed., 2005), at 250–251; Pisillo Mazzeschi, 'The Due Diligence Rule', 35 *GYIL* (1992), at 9; Smith, *State Responsibility and the Marine Environment* (1988), at 13.

[228] See Grotius, *The Law of War and Peace, Book III*, in: Grotius (1925), at 597 *et seq.*

[229] See e.g. Ago, 'Le Délit International', 68 *RdC* (1939), at 450–498; Lauterpacht, *Private Law Sources and Analogies of International Law* (1927), at para. 58–62; Verdross, *Völkerrecht* (5th ed., 1964), at 376–369.

[230] Ago, 'Le Délit International', 68 *RdC* (1939), at 450–498; See also Pisillo Mazzeschi, 'The Due Diligence Rule', 35 *GYIL* (1992), at 11–13.

dolus, culpa and fault seems to be incongruous.[231] In response to these objections there has been an attempt to maintain the doctrine of subjective responsibility by conceiving the notion of fault to be a normative one that is established not by a subjective or psychological attitude of the acting person, but rather by any behaviour which is non-compliant with an international obligation.[232] However, such a view renounces the essential meaning of the notion of fault and would in the end amount to an objective conception of responsibility.[233]

In contrast, the doctrine of objective responsibility does not consider subjective fault as a necessary element of an internationally wrongful act. According to this theory, the responsibility of States can be determined by purely objective criteria. What is decisive is, first, whether there is a violation of the right of another State established by any non-conformance with an international obligation[234] and, second, the imputability of this conduct to the State.[235] Such pure objective consideration, however, would be tantamount to the State being absolutely responsible for complying with its international obligations. This makes it necessary to establish a normative adjustment of this outcome. In contrast to the subjective theory, this adjustment is not achieved by an additional requirement of fault, but by an interpretation of the content of international obligations in a way that the standard of behaviour, which must be complied with by the State, is not an absolute one, but a standard of due diligence.[236]

In international jurisdiction, references can be found both to the subjective theory[237] and to the objective theory.[238] This, however, also means that there is no consistent case law in support of an additional general requirement of fault. Of particular significance, furthermore, is the recent work of the ICJ which, when

---

[231] Combacau/Sur, *Droit International Public* (9th ed., 2010), at 544–545.

[232] See e.g. Accioly, 'Principes Généraux de la Responsibilité Internationale', 96 *RdC* (1959), at 369–370; Cheng, *General Principles of Law* (1953), at 218–232; Salvioli, 'Les Régles Générales de la Paix', 46 *RdC* (1933), at 97–100.

[233] Pisillo Mazzeschi, 'The Due Diligence Rule', 35 *GYIL* (1992), at 13.

[234] Exceptions to this rule are circumstances precluding wrongfulness like error, fortuitous events or *force majeure*, which function as objective defences.

[235] See already Eagleton, *The Responsibility of States* (1928), at 213; Lauterpacht, *Private Law Sources and Analogies of International Law* (1927), at para. 58–62; Triepel, *Völkerrecht und Landesrecht* (1899), at 334–335; see also de Aréchaga, 'International Law in the past Third of a Century', 159 *RdC* (1978), at 267–269; Brownlie, *State Responsibility* (1983), at 38–40; Ipsen, in: Ipsen (ed.), *Völkerrecht* (5th ed., 2004), at 628–629; Shaw, *International Law* (5th ed., 2003), at 698–700.

[236] As to the due diligence standard see the following section.

[237] *Davis Case, Merits*, 9 RIAA (1959), at 463; *Salas Case*, 10 RIAA (1960), at 720; *Affaire de Casablanca*, 11 RIAA (1961), at 119 *et seq.*; *Home Missionary Society Case*, 6 RIAA (1955), at 44.

[238] *Owners of the Jesse Case*, 6 RIAA (1955), at 59; *TheNeer Case*, 4 RIAA (1951), at 62 *et seq.*; *Estate of Jean-Baptiste Caire Case*, 5 RIAA (1952), at 529–531; *The Corfu Channel Case*, ICJ Reports 1949, at 14 *et seq.*; *Case Concerning United States Diplomatic and Consular Staff in Tehran*, ICJ Reports 1980, at 33 *et seq.*

adopting the Draft Articles on State Responsibility in 2001, did not include fault as a necessary element of State responsibility.[239] The international practice of States, therefore, strongly tends towards the theory of objective responsibility.

Today, the theory of objective responsibility in its strict form has been further developed. Under a recent approach there is still no general requirement of fault in order to establish a breach of an international obligation. However, it is stressed that the relevant primary obligation remains decisive for the determination of the particular requirements of an act being in non-compliance with this obligation. It is, therefore, possible that a particular obligation requires fault on the part of the person acting on behalf of the State in order to establish an internationally wrongful act.[240] A general requirement of fault, however, is not imposed by current international law.

(dd) Due Diligence as the Relevant Standard of Behaviour

It has been outlined that the particular prerequisites for an act of a State being in breach of an international obligation are determined predominantly by the content of the respective primary obligation. Furthermore, international law rather follows the approach of objective responsibility and does not in general require fault on the part of the person acting on behalf of the State. If, however, the international responsibility of States would only be dependent on the fact that a particular conduct or result is from a purely objective perspective not in conformity with what is required under an international obligation, a too far-reaching standard eventually amounting to an absolute responsibility of the State would result that. Therefore, a normative adjustment of this outcome is necessary, which is achieved by imposing a standard of due diligence on State conduct.[241]

The determination of the level of care required by the standard of due diligence is not an easy task. The particular level of care differs depending on the respective primary obligation and must be determined individually for each single obligation.[242] The determination of the applicable level of care is not to be confused with the question of whether the primary obligation is one of conduct or result. It rather concerns the question of whether the State is either absolutely responsible for the particular conduct or for the achievement of the particular result, or whether it is responsible for this outcome only within the realms of a certain degree of diligence taking account of the individual circumstances of the case and the possibilities for

---

[239] Commentaries to the ILC Draft Articles on State Responsibility, Article 2, para. 10.

[240] de Aréchaga, 'International Law in the past Third of a Century', 159 *RdC* (1978), at 270–271; Brownlie, *State Responsibility* (1983), at 37–49; Dahm/Delbrück/Wolfrum, *Völkerrecht, vol. I/3* (2nd ed., 2002), at 946–947; Evans (ed.), *International Law* (2nd ed., 2006), at 465.

[241] de Aréchaga, 'International Law in the past Third of a Century', 159 *RdC* (1978), at 270; Guggenheim, *Traité de Droit International Public* (1954), at 54–55; see also Pisillo Mazzeschi, 'The Due Diligence Rule', 35 *GYIL* (1992), at 17.

[242] For an overview of the position of arbitral and other international courts see Barnidge, 'The Due Diligence Principle under International Law', 8 *Int'l Comm. L. Rev.* (2006), at 81 *et seq.*

action. The issue of which level of care needs to be observed, furthermore, is to be distinguished from the question whether or not a State may invoke general defences precluding the wrongfulness of the act of the State, such as fortuitous events or force *majeure*.[243]

From the objective conception of interstate relations and obligations it follows that the applicable level of care must also be an objective one that is based on and conceived in objective terms. This standard thus forms part of the international obligation itself, rather than being attributable to an additional subjective criterion of fault.[244] This does not mean that the fault or negligence of the acting person is of no relevance, but such behaviour is considered to be a part of the State's conduct as a whole.[245] The particular level of care must be determined individually for each primary obligation. In this context, the respective obligation itself provides the major criterion for the determination of the applicable level.[246] The level of due diligence which is derived from the respective obligation, then, applies to all States in equal measure. This may be different only if the provision contains an individualised obligation depending on the affiliation of the State to a certain regional area or on the economic performance and industrial capacity of the State.[247] Finally, it should be noted that the "usual" level of care applied by States in their internal affairs does not have any implication on their international obligations.[248]

The standard of care required by an international obligation can be an absolute one. This would be the case, for example, if an obligation requires the States to ensure that a certain result is achieved in any event, notwithstanding any possible objections or defences such as the lack of actual control or foreseeability of the course of events. The typical consequence of a breach of such obligations is that responsibility of the State pursuant to the principle of State responsibility would be absolute (by contrast, responsibility of the State would be strict if the State is allowed to rely on certain predetermined objections or defences, as a result of certain circumstances deemed to break the chain of causation). The majority of international obligations, however, do not require an absolute level of care excluding such objections and defences. This raises the question of how to determine the exact level of care which is due in respect of obligations which do not demand an absolute level of care. For this purpose, three criteria have been developed which need to be taken into account and weighed according to their

---

[243] Dahm/Delbrück/Wolfrum, *Völkerrecht*, vol. I/3 (2nd ed., 2002), at 948; Pisillo Mazzeschi, 'The Due Diligence Rule', 35 *GYIL* (1992), at 45.

[244] *TheNeer Case*, 4 RIAA (1951), at 62; *TheVenable Case*, 4 RIAA (1951), at 229; *TheNoyes Case*, 6 RIAA (1955), at 311; Dahm/Delbrück/Wolfrum, *Völkerrecht*, vol. I/3 (2nd ed., 2002), at 948; Pisillo Mazzeschi, 'The Due Diligence Rule', 35 *GYIL* (1992), at 42–44.

[245] Pisillo Mazzeschi, 'The Due Diligence Rule', 35 *GYIL* (1992), at 42.

[246] Boyle, 'Nuclear Energy and International Law', *BYIL* (1989), at 272–273.

[247] This may be the case with regard to obligations imposing common but differentiated responsibilities.

[248] Pisillo Mazzeschi, 'The Due Diligence Rule', 35 *GYIL* (1992), at 41.

significance in the particular case. These criteria include (1) the ability to control and influence the actual activity that causes the threat of an infringement, (2) the degree of the predictability of harm, and (3) the importance of the right or legal interest to be protected. How due diligence is to be understood will be a function of how these three criteria are collectively weighted and assessed. Easily controlled activity which carries a readily foreseeable and significant risk of harm would thus be held to a more exacting expectation of care.[249]

In summary, it can be concluded, therefore, that the specific level of care compelled by the standard of due diligence must be determined on a case-by-case basis. The assessment depends on the content of the particular obligation and needs to take account of the particularities of the case, such as the possibility of the State to control the activity causing the breach of an obligation, the probability of damage and the significance of the impaired rights and interests.

(b) Obligations Arising from International Conventions

In this section, obligations of States arising from international conventions relevant to the transboundary movement of hazardous wastes by sea shall be outlined and their particular content as well as their significance for the transports in question shall be examined. This examination will allow for an assessment of which scenarios and under which particular circumstances a State is in contravention of its respective obligation and, consequently, will be internationally responsible under the principle of State responsibility.

Relevant provisions imposing international obligations on States may derive from global, regional and bilateral instruments. They may be established as a part of the international law concerning the transportation of hazardous wastes, from the body of the law of the sea, including several sectoral conventions, as well as from conventions established in other, related legal fields.

(aa) The Basel Convention

The central legal instrument on a global scale for the regulation and control of international hazardous waste streams is to be found in the 1989 Basel Convention on the Control of Transboundary Movements of Hazardous Wastes and Their Disposal (Basel Convention). The Basel Convention imposes on its States Parties the obligation to reduce as far as possible the generation and transport of hazardous wastes and, if a transboundary movement of such wastes cannot be avoided, requires the observance of strict procedural rules, including the prior informed consent of all States concerned and an environmentally sound management of hazardous wastes.

---

[249] Dahm/Delbrück/Wolfrum, *Völkerrecht*, vol. I/3 (2nd ed., 2002), at 948; Pisillo Mazzeschi, 'The Due Diligence Rule', 35 *GYIL* (1992), at 44.

(1) Background and Basic Legal Features of the Convention

First attempts to elaborate a legal regime governing the trade in and transport of hazardous wastes were made in the beginning of the 1980s.[250] In 1981, the United Nations Environment Programme (UNEP) convened the first Montevideo Programme for the Development and Periodic Review of Environmental Law and charged it with the task to developing a strategic guidance plan in the field of environmental law. This Programme recommended to UNEP that it work towards the conclusion of international agreements and the development of international principles, guidelines and standards relevant, amongst others, to the transport, handling and disposal of toxic and dangerous wastes.[251] UNEP thereupon established a working group on this issue, whose endeavours finally led to the approval of the UNEP Cairo Guidelines in 1987,[252] which laid down major principles related to the management of hazardous wastes. Since, however, these guidelines were conceived as a form of non-binding soft-law which sought to induce the States to implement these principles into their domestic law, it lacked real authority.[253] At the same time, UNEP convened a working group representing about 100 States, international agencies and NGOs to elaborate an international binding convention on the control of transboundary movements of hazardous wastes. On the working group's sixth and final meeting on 22 March 1989 in Basel, Switzerland, the final draft of the Basel Convention was adopted and subsequently signed by 53 States and the EU.[254] The Basel Convention entered into force on 5 May 1992. Since then, it has become the most important legal instrument governing international hazardous waste trades, being commonly accepted on a global scale. It has been ratified so far by 180 States and the European Union.[255]

The Basel Convention aims to achieve and strengthen the control and reduction of international hazardous waste trade streams. Its objectives are to minimise the amount and hazard level of wastes, to avoid the generation and transportation of hazardous wastes as far as possible and to encourage the environmentally sound management (ESM) of hazardous wastes by facilitating the transfer of technology and know-how. Furthermore, waste transports and their way from the source to the

---

[250] The legal situation regarding the transport and transboundary movement by sea of waste oil prior to the entry into force of the Basel Convention is outlined by Lagoni, *Altöl und Seeschiffahrt*, in: Becker/Bull/Seewald (ed.) (1993), at 1014–1021.

[251] See Conclusion (a)(iii) of the Programme for the Development and Periodic Review of Environmental Law, UNEP Decision 10/21 of 31 May 1982.

[252] Cairo-Guidelines and Principles of the Environmentally Sound Management of Hazardous Wastes, of 17 June 1987; available in printed form at (1986) 16 *Envtl. Pol'y & L.* 31–33.

[253] Choksi, '1999 Protocol on Liability and Compensation', 28 *Ecology L. Q.* (2001), at 516; Kummer, *The Basel Convention* (1995), at 38–40.

[254] Kummer, *The Basel Convention* (1995), at 40–41.

[255] A list of the States Parties is available at: www.basel.int/ratif/convention.com. The only developed country that has not yet ratified the Convention is the USA.

disposal site are to be traceable in order to avoid "disappearance" or improper handling as well as illegal shipments outside the legal frame of the Basel Convention.[256]

Concerning its regulatory content the Basel Convention first adopts general rules and principles related to the reduction of the generation and transboundary movement of hazardous wastes. This includes the obligation of States to reduce the generation of hazardous wastes to a minimum[257] as well as the proximity and self-sufficiency principles according to which each State must ensure the availability of sufficient domestic waste disposal facilities to enable the treatment and disposal of hazardous wastes as close as possible to the place of its generation.[258] Thus, with regard to the admissibility of transboundary hazardous waste movements, the Basel Convention provides a restrictive system under which such transports are permitted and, to this end, pursues a three-tier concept of requirements:[259] At the first tier, the Basel Convention prohibits any transport of hazardous wastes to non-Parties of the Convention and to the Antarctic.[260] Furthermore, it reproduces the general rule that every Party has the sovereign right to unilaterally ban the import of hazardous wastes, entailing a corresponding obligation of the other Parties to prohibit the export of hazardous wastes to that Party.[261] At the second tier—as far as an export to a particular State is not generally prohibited—hazardous waste transports are only allowed to take place on condition that it is ensured that the import State will manage the wastes in an environmentally sound manner[262] and, in addition, that either the State of export lacks appropriate disposal capacities or the wastes are needed as raw materials in the State of import.[263] At the third and final tier, a requirement is set that prior to and during the conduct of a transport all parties involved must fulfil the formal requirements of notification and approval. The State of export is obliged to notify the transport to both the State of import and transit States.[264] Those, in turn, have to give their written consent to the particular transport before it may commence (so-called prior informed consent—PIC).[265]

---

[256] See the Preamble and Article 4 of the Basel Convention. See also Gudofsky, 'Transboundary Shipments of Hazardous Waste', 34 *Stan. J. Int'l L.* (1998), at 226–228; Kummer, *The Basel Convention* (1995), at 55–60; Widawsky, 'In My Backyard', 38 *Envtl. L.* (2008), at 588–593.

[257] Basel Convention, Article 4(2)(a).

[258] *Ibid.*, Article 4(2)(b) and (d); Louka, *International Environmental Law* (2006), at 429.

[259] This is outlined e.g. by Beyerlin/Marauhn, *International Environmental Law* (2011), at 216–217; Louka, *International Environmental Law* (2006), at 429–430.

[260] Basel Convention, Article 4(5) and (6).

[261] *Ibid.*, Article 4(1)(a) and (b) and 4(2)(e). These provisions actually reiterate customary international law deriving from the principle of sovereign equality of all States and, therefore, are of declaratory nature only, see Preamble, Paragraph 6.

[262] *Ibid.*, Article 4(2)(e) and (g) and 6(3)(b).

[263] *Ibid.*, Article 4(9).

[264] *Ibid.*, Article 6(1). According to Article 7, the requirement of Article 6(1) applies *mutatis mutandis* to transit States which are not Parties to the Basel Convention.

[265] *Ibid.*, Article 6(2) and (3); in respect of States of transport Article 6(4) applies.

Any non-compliance with these procedural requirements of formal notification and approval as well as fraud, misrepresentation or non-conformance of the actual transport with the transport documents renders the transport "illegal traffic"[266] that has to be prohibited and punished by domestic law.[267] Furthermore, illegal traffic leads to the obligation of the responsible Party to ensure the re-importation of the hazardous wastes by the exporter or, if necessary, by the State Party itself.[268]

During the years following the entry into force of the Basel Convention, the Conferences of the Parties of the Basel Convention (COPs) agreed on further enhancements to strengthen the importance of the Convention. Particularly, in September 1995, the Third Conference of the Parties (COP3) decided to add a new Article 4a comprising a prohibition of any export of hazardous wastes from developed countries[269] to developing countries.[270] This so-called "Basel Ban Amendment" or simply the "Basel Ban" applies to hazardous wastes that are destined both for final disposal and for recycling and recovery operations. It has not yet entered into force.[271] The reluctance of States Parties to ratify the Basel Ban Amendment may be explained by the fact that this issue represents one of the most controversial disputes, if not the crucial question surrounding the Basel Convention. The ban-issue centres upon the very fundamental decision as between the two opposing approaches of either pursuing a complete ban or following a prior informed consent principle (PIC). In other words, it concerns the question of

---

[266] *Ibid.*, Article 9(1).

[267] *Ibid.*, Article 4(3) and 9(5). This rule is implemented, for example, in Germany by Section 326 of the German Criminal Code.

[268] *Ibid.*, Article 9(2) and (3). See also the following section.

[269] The new Article 4a of the Basel Convention distinguishes between Parties listed in a new Annex VII and those not listed there. The new Annex VII of the Basel Convention reads as follows: "Parties and other States which are members of OECD, EC, Liechtenstein".

[270] Already in 1994 Decision II/12, (Doc. UNEP/CHW.2/30) was adopted by the Conference of the Parties (COP2), according to which the export from OECD to non-OECD States of hazardous wastes destined for disposal was immediately banned. Furthermore, the States agreed to ban the export from OECD to non-OECD States of hazardous wastes intended for recovery and recycling as of 31 December 1997. Since, however, this Decision was not incorporated into the text of the Convention, there was a dispute about its legal force. As a result, the States adopted Decision III/1 (Doc. UNEP/CHW.3/35) at the Conference of the Parties in 1995 (COP3), according to which this ban had to be formally incorporated by means of an amendment in the Convention. See also Kummer, *The Basel Convention* (1995), at 64.

[271] A dispute existed about the number of ratifications required for the entry into force of the Basel Ban Amendment pursuant to Article 17(5) of the Basel Convention. The ambiguity is whether the required three-fourths of the "Parties who accepted them" applies to those Parties present and voting at the time of deposit of the ratification or to those Parties present and voting at the time the amendment was adopted. By Decision X/3 (Doc. UNEP/CHW.10/28) adopted by COP10 in 2011, the Parties to the Basel Convention agreed to interpret Article 17(5) to mean "that the acceptance of three-fourths of those Parties that were Parties at the time of the adoption of the amendment is required for the entry into force of such amendment". Today, 78 Parties have ratified the Ban Amendment, only a part of which were Parties at the time of the adoption of the Amendment.

whether a complete prohibition or, alternatively, an admission of a limited trade in hazardous wastes within a controlled and regulated procedure is more suitable to achieve the Convention's principal aim of reducing undesirable hazardous waste movements into economic inferior States.[272] In the end, however, this controversy may be of less practical significance owing to the subsequent practice of States after the adoption of the Basel Convention. In consequence of the absence of a total ban of hazardous waste exports from developed to developing countries in the Basel Convention the number of regional conventions providing for a total ban of hazardous waste transports or imports, instead of pursuing the PIC-approach, increased.[273] In this respect the already mentioned 1991 Bamako Convention, the 1995 Waigani Convention and the 1996 Izmir Protocol to the Barcelona Convention can be mentioned as examples. It is furthermore to be mentioned that the Basel Ban has *de facto* already been implemented into the EU legislation on the exportation of hazardous wastes.[274]

A further defect of the Basel Convention was seen in the absence of a distinct regime of liability and compensation for damages occurring in the course of permitted or illegal hazardous waste transports. During the negotiation process of the Convention, this issue of liability and compensation turned out to be highly contested among the different delegations. Therefore, it was decided to put this "hot potato" aside for a later consideration, and to concentrate on achieving an agreement upon the procedural and other substantive rules and aspects of hazardous waste transportation.[275] To this end, Article 12 was introduced into the Convention and calls for the States Parties to co-operate in the elaboration of a Protocol to the Convention that sets out appropriate rules and procedures in the field of liability and compensation. In the course of those negotiations, finally at the Fifth Conference of the Parties (COP5) in December 1999 the Parties reached

---

[272] This dispute cannot be illustrated at this point. For a further discussion, see e.g.: Clapp, 'Africa, NGOs, and the International Toxic Waste Trade', 3 *J. Env. & Dev.* (1994), at 17 *et seq.*; Clapp, 'The Toxic Waste Trade', 15 *TWQ* (1994), at 512–516; Friedrich-Ebert-Stiftung (ed.), *Zehn Jahre Basler Übereinkommen* (1999), at 28–33; Kitt, 'Waste Exports to the Developing World', 7 *Geo. Int'l Envtl. L. Rev.* (1994/1995), at 504–512; Krueger, 'Prior Informed Consent and the Basel Convention', 7 *J. Env. & Dev.* (1998), at 115 *et seq.*; Kummer, 'The Basel Convention: Ten Years On', 7 *RECIEL* (1998), at 229–230; O'Neill, 'Out of the Backyard', 7 *J. Env. & Dev.* (1998), at 138 *et seq.*; Poulakidas, 'Waste Trade and Disposal in the Americas', 21 *Vt. L. Rev.* (1996/1997), at 894–901; Schneider, 'The Basel Convention Ban on Hazardous Waste Exports', 20 *Suffolk Transnat'l L. Rev.* (1996/1997), at 247 *et seq.*; Widawsky, 'In My Backyard', 38 *Envtl. L.* (2008), at 577 *et seq.* and particularly at 610–615; Wirth, 'Trade Implications of the Basel Ban', 7 *RECIEL* (1998), at 237 *et seq.*

[273] Anand, *International Environmental Justice* (2004), at 82–84; Birnie/Boyle/Redgwell, *International Law and the Environment* (3rd ed., 2009), at 473–475.

[274] As to the EU Waste Shipment Regulation see *infra*, Sect. "European Union Legislation".

[275] See Birnie/Boyle/Redgwell, *International Law and the Environment* (3rd ed., 2009), at 482–483; 232; Gwam, 'Travaux Preparatoires of the Basel Convention', 18 *J. Nat. Resources & Envtl. L.* (2003/2004), at 66–67; Kummer, *The Basel Convention* (1995), at 72.

agreement upon a Protocol on Liability and Compensation. It is conceived not as an amendment, but as a supplement to the Basel Conventions and needs to be ratified separately.[276]

(2) Obligations Imposed on States

It has already been outlined that the Basel Convention imposes some general obligations on States concerning a reduction of the generation and transportation of hazardous wastes as well as specific obligations imposing procedural requirements with regard to the conduct of transboundary hazardous waste movements, these being configures in a three-tiered framework of requirements. These obligations are examined in this section in more detail with a view to determining their precise nature and content.

(i) General Obligation: Minimisation of Generation and Transportation of Hazardous Wastes

The first general obligation imposed by the Basel Convention on its States Parties is the obligation to "take the appropriate measures to [... e]nsure that the generation of hazardous wastes and other wastes within it is reduced to a minimum".[277] The scope of this obligation is limited in two respects. First, States are only obliged to "take appropriate measures" and, second, the obligation is ambiguous in respect of the concrete result to be achieved. The generation of hazardous wastes is to be "reduced to a minimum", which by itself is an unequivocal objective. However, according to the further wording, "social, technological and economic aspects" need to be taken into account. Thus, the concrete "minimum" of hazardous waste generation to be achieved is relaxed in its definition and will differ from one case to another depending on the respective circumstances. These particulars have an influence on the character of that obligation. The wording "to take appropriate measures to ensure" is usually indicative of an obligation of result since it states nothing other than the very content of an obligation of result. In the present case, however, the obligation does not define a concrete outcome. Therefore, it must rather be concluded that it represents an obligation of conduct that requires the State to take actions towards the achievement of a certain legal aim. Such endeavours of the State are thus prevailing. Finally, it should be stressed that the general standard of due diligence applies to this obligation.

In order to establish a breach of this obligation, it is necessary to first define the particular "minimum" to which the hazardous waste generation must be reduced in the individual case. In a subsequent step, it must be ascertained whether the measures actually taken by the State are sufficient to achieve this particular "minimum". It consequently becomes apparent that, with regard to both issues, the content of this

---

[276] See in detail *infra*, Sect. "1999 Protocol to the Basel Convention".

[277] Basel Convention, Article 4(2)(a).

obligation is far from being definite. First, "social, technological and economic aspects" need to be taken into account when defining the particular "minimum" to be achieved. And, second, the State has the discretion to decide which particular measures it considers appropriate and implements to achieve this particular "minimum". From a practical perspective, therefore, it will hardly be possible to establish that the failure of a State to adopt a certain measure constitutes a breach of this obligation. A different conclusion may result only in very obvious constellations, especially where a State fails to take measures which are considered simply indispensable for achieving the minimisation of hazardous waste generation.

The obligation to reduce the generation of hazardous wastes is supplemented by further related rules. The States Parties are under an obligation to co-operate in "the development and implementation of new environmentally sound low-waste technologies and the improvement of existing technologies with a view to eliminating, as far as practicable, the generation of hazardous wastes".[278] Furthermore, States Parties "should" establish regional or sub-regional centres for training and technology transfer regarding, in particular, the minimisation of the generation of hazardous wastes.[279] And finally, the Preamble of the Convention establishes the reduction of the generation of hazardous wastes as a target serving to protect the human health and the environment.[280] However, these supplementary rules are of even less binding force. They represent programmatic declarations of political intent or the determination of legal aims rather than legally binding primary obligations. Concrete rules of conduct, therefore, do not arise from these provisions.[281]

In addition to the obligations regarding the minimisation of the generation of hazardous wastes, the Basel Convention also contains general obligations concerning the minimisation of their exportation and transportation. In accordance with the principles of proximity and self-sufficiency,[282] the States Parties "shall take appropriate measures to [... e]nsure the availability of adequate disposal facilities [...] that shall be located, to the extent possible, within it, whatever the place of their disposal".[283] Furthermore, States "shall take appropriate measures to [... e]nsure that the transboundary movement of hazardous wastes and other wastes is reduced to the minimum consistent with the environmentally sound and efficient management of such wastes".[284] Concerning the regulatory character of these obligations, the findings outlined in respect of the obligations to minimise the generation of hazardous wastes apply accordingly. Since the primary intention of these obligations,

---

[278] *Ibid.*, Article 10(2)(c).

[279] *Ibid.*, Article 14(1).

[280] *Ibid.*, Preamble, Paragraphs 3, 17.

[281] Apart from that, the preamble of an international convention does not form part of the legally binding corpus of a convention. See Dahm/Delbrück/Wolfrum, *Völkerrecht, vol. I/3* (2nd ed., 2002), at 516.

[282] See Louka, *International Environmental Law* (2006), at 429.

[283] Basel Convention, Article 4(2)(b).

[284] *Ibid*, Article 4(2)(d). See also *ibid.*, Preamble, Paragraph 8.

again, is to require the States to take actions towards the achievement of a certain legal aim, they must be regarded obligations of conduct. Due to the inclusion of the term "to the extent possible", the particular circumstances of the individual case, such as the relevant economic, social and technological aspects, need to be taken into account even though this is not expressly mentioned.

(ii) First Tier: Absolute Trade Restrictions

The first tier of the Basel Convention's three-tier concept of requirements for the permissibility of transboundary movements of hazardous wastes involves general export restrictions to certain States and areas. According to the Convention, a Party "shall not permit hazardous wastes or other wastes to be exported to a non-Party or to be imported from a non-Party".[285] The Parties are, furthermore, under the obligation "not to allow the export of hazardous wastes or other wastes for disposal within the area south of 60° South latitude".[286] And finally, the Parties "shall prohibit or shall not permit the export of hazardous wastes or other wastes to the Parties which have prohibited the import of such wastes".[287] These obligations establish particular trade bans for the transportation of hazardous wastes to or from certain States or areas. If one considers the precise wording, it becomes apparent that they are conceived as "relative" obligations[288] that are expressly addressed to the States Parties, whereas the intended legal result, i.e. the non-transportation to certain areas and States, relates to the conduct of the involved private parties. But one has to be precise: These obligations do not impose on States the obligation to ensure that the respective conduct of private parties, which is described by these obligations, is not taking place in the first instance. Such a wording would suggest an obligation of result. In this case, the obligations instead directly focus on the legal and administrative measures to be taken by the State and, therefore, emphasise the particular conduct owed by the State. In conclusion, they must be considered obligations of conduct, according to which the (only) obligation of the States Parties is to domestically prohibit such transports. Finally, it should be stated that the general standard of due diligence applies. A breach of these obligations, thus, is conditioned upon a failure in the State's conduct regarding the legislative and administrative implementation of these obligations.

Following from their character as relative obligations, the content of these obligations changes after they have been implemented into the various national legal regimes. The act of implementation forms a *caesura* by which the State's original obligation to implement evolves into the obligation to further regulate, control and ensure compliance and enforcement.[289]

---

[285] *Ibid.*, Article 4(5).
[286] *Ibid.*, Article 4(6).
[287] *Ibid.*, Article 4(1)(b).
[288] As to this term see *supra*, Sect. "Relevance of the Respective Primary Obligation".
[289] See *ibid.*

(iii) Second Tier: Environmentally Sound Management

The second tier of requirements for the permissibility of transboundary hazardous waste movements under the Basel Convention is concerned with the general requirement of an environmentally sound management of hazardous wastes. According to this, the States parties "shall take appropriate measures to [... n]ot allow the export", or, as the case may be, "[p]revent the import of hazardous wastes and other wastes if it has reason to believe that the wastes in question will not be managed in an environmentally sound manner".[290] In addition to this, the States "shall take the appropriate measures to ensure that the transboundary movement of hazardous wastes and other wastes only be allowed if: (a) The State of export does not have the technical capacity and the necessary facilities, capacity or suitable disposal sites in order to dispose of the wastes in question in an environmentally sound and efficient manner; or (b) The wastes in question are required as a raw material for recycling or recovery industries in the State of import".[291] From the wording and the regulatory content of these obligations it becomes apparent that they are conceived as obligations of conduct. Although the States are basically obliged to "take appropriate measures", which generally indicates an obligation of result, the very core action required by the State is the legal or administrative act of non-allowance or prevention of certain transports, which consists in a particular conduct by the State rather than in a result to be achieved. The phrase "take appropriate measures" in this context may be understood as referring to the applicable standard of conduct, which is one of due diligence.[292]

The precise content of these obligations remains indefinite in some respect. The central term "environmentally sound management" (ESM) is defined by the Convention as "taking all practicable steps to ensure that hazardous wastes or other wastes are managed in a manner which will protect human health and the environment against the adverse effects which may result from such wastes".[293] This definition actually is not of great help since it again relies on indefinite terms, as there are "practical steps", "adverse effects" and the "manner which will protect human health and the environment".[294] Consequently, in the following years the Secretariat of the Basel Convention elaborated a number of Technical Guidelines for the ESM regarding certain types of hazardous wastes, which provide manageable criteria for an environmentally sound treatment tailored to the particular types of wastes.[295] In the end, it can therefore be said that there is a set

---

[290] Basel Convention, Article 4(2)(e) and (g).

[291] *Ibid.*, Article 4(9).

[292] See also Kummer, *The Basel Convention* (1995), at 56.

[293] Basel Convention, Article 2(8).

[294] This issue was subject to broad criticism already at an early stage of the Convention, see e.g. Abrams, 'Regulating the International Hazardous Waste Trade', 28 *Colum. J. Transnat'l L.* (1990), at 828; Kummer, *The Basel Convention* (1995), at 57–60; Rublack, 'Fighting Transboundary Waste Streams', 22 *VRÜ* (1989), at 376.

[295] These Technical Guidelines can be found at www.basel.int.

of general criteria which has been developed to provide in more detail a definition of what must be considered an ESM of the particular hazardous wastes. Another question, with regard to the first obligation mentioned above, is under which conditions an exporting State has "reason to believe" that hazardous wastes are not disposed of in an environmentally sound manner in the importing State. A corresponding right of the exporting State to actively investigate and to collect data exists neither under the Basel Convention nor according to customary international law.[296] Therefore, in accord with the applicable standard of conduct, i.e. due diligence, the "reason to believe" can be determined only on the basis of the information received from the importing State. With regard to the second obligation mentioned above, concerns have been voiced about the inclusion of the criterion of "efficient" domestic disposal as a factor in the authorisation of transboundary hazardous waste movements. It was argued that by virtue of this term, hazardous waste exports could be justified solely by lower disposal costs abroad, provided that minimum environmental standards are observed.[297] This contention, however, is not especially convincing. The term "efficient" must be understood in a broader, global context relating to the social, economic and technological circumstances of a State as a whole. Considering this, it is appropriate to permit hazardous waste transports if it is not possible for the State of export to keep available sufficient disposal and treatment facilities for any particular kind of hazardous wastes.

A breach of the ESM provisions, thus, may be established if a State allows the export or, alternatively, the import of hazardous wastes although, based on the information in its possession, it ought to have reason to believe that the wastes in question will not be managed in an environmentally sound manner, or it ought to know that neither are sufficient capacities for an ESM lacked by the State of export nor are the wastes required for recycling or recovery operations by the State of import. However, due to the fact that these obligations involve several indefinite legal terms, the particular requirements for a breach of these obligations are pending further clarification on a case-by-case basis by international case law.

(iv) Third Tier: Prior Informed Consent Principle

The third and final tier of the Basel Convention's three-tier concept of requirements for the permissibility of hazardous waste transports concerns the procedural rules of notification and approval to be observed during the performance of each individual transport. According to this, the "State of export shall notify, or shall require the generator or exporter to notify, in writing [...] the competent authority of the States concerned of any proposed transboundary movement of hazardous

---

[296] Abrams, 'Regulating the International Hazardous Waste Trade', 28 *Colum. J. Transnat'l L.* (1990), at 828; Kummer, *The Basel Convention* (1995), at 57.

[297] Lang, *The International Waste Regime*, in: Lang/Neuhold/Zemanek (ed.) (1991), at 153; see also Kummer, *The Basel Convention* (1995), at 56.

wastes or other wastes".[298] In turn, the State of import "shall respond to the notifier in writing, consenting to the movement with or without conditions, denying permission for the movement, or requesting additional information."[299] The same duty is incumbent upon each State of transit.[300] Until the point at which the State of export has received written confirmation from the notifier that it has received the written consent of the State of import and each State of transit, the State "shall not allow the generator or exporter to commence the transboundary movement".[301] Finally, any transport of hazardous wastes "shall be covered" by insurance, bond or other financial guarantee.[302] As soon as the disposer receives the wastes, as well as upon completion of the disposal, the disposer is required to inform both the exporter and the exporting States. If no such information is received by the exporting State or the exporter, they "shall so notify" the State of import.[303] These provisions, as a whole referred to as the Prior Informed Consent (PIC) Principle, establish procedural rules that in detail prescribe the particular conduct required by States with regard to the conduct of private parties.[304] According to the wording that precisely describes the factual, administrative or legal acts required by States, these obligations represent obligations of conduct, to which the general standard of due diligence applies. This outcome also applies as far as these rules impose a duty to ensure that a certain action is performed by private parties. This results from the fact that within its scope of discretion, the State may decide whether the act of notification is to be performed by the State itself or by the generator or exporter under its jurisdiction.

A lively debate has emerged during the drafting process of the Basel Convention concerning the extent of the application of the PIC procedures to coastal States. This debate particularly includes the question whether a coastal State is to be considered a State of transit within the meaning of the Basel Convention if a

---

[298] Basel Convention, Article 6(1). This applies both to States of import and States of transit. According to Article 7, this provision, furthermore, applies *mutatis mutandis* to States of transit which are not Parties to the Basel Convention.

[299] *Ibid.*, Article 6(2).

[300] *Ibid.*, Article 6(4). In contrast to the duty to notify each transport to States of transit, a duty which, according to Article 7, applies *mutatis mutandis* to States of transit that are not States Parties of the Convention, the Convention does not contain a comparable rule concerning the consideration of any objections or non-approvals regarding such transports by States of transit that are not States Parties to the Convention. See Kummer, *The Basel Convention* (1995), at 68–69, who tries to resolve this conflict by referring to the *telos* of the Convention, according to which the position of non-Parties may not be lower than that of States Parties, which leads to the conclusion that the express consent also of non-Parties is necessary.

[301] Basel Convention, Article 6(3)(a) and (4).

[302] *Ibid.*, Article 6(11).

[303] *Ibid.*, Article 6(9).

[304] For a detailed analysis of these obligations see Kummer, *The Basel Convention* (1995), at 65–70.

vessel carrying hazardous wastes is only passing through its EEZ.³⁰⁵ According to Article 6(1) and (4) of the Basel Convention prior notification and consent is required with regard to "the States concerned", which also comprises any State of transit. The legal definition of the term "State of transit", however, is ambiguous as regards the coverage of maritime zones and has left room for different interpretations by the signatory States.³⁰⁶ This definition covers any State "through which" a hazardous waste movement takes place.³⁰⁷ By using this wording this definition avoids making a clear reference either to the territory of a coastal State, or to the "area under the national jurisdiction of a State" as defined in Article 2(9) of the Convention.³⁰⁸ However, due to the avoidance of an explicit reference to the EEZ or, at least, to the "area under the national jurisdiction" of the coastal State it must be concluded that an inclusion of the EEZ into the definition of a "State of transit" is not intended. What remains is that a movement must have taken place "through" a State, which can thus only be understood as referring to the territory of the State. Since, according to Article 2(1) of the UNCLOS, the sovereignty of a coastal State is extended beyond its land territory to its territorial sea, also the territorial sea of a coastal State must be considered as a part of its territory.

---

³⁰⁵ For a detailed discussion see Pineschi, *The Transit of Ships Carrying Hazardous Wastes*, in: Francioni/Scovazzi (ed.) (1991), at 299 *et seq.*; Rummel-Bulska, *The Basel Convention and the UNCLOS*, in: Ringbom (ed.) (1997), at 83 *et seq.*

³⁰⁶ See Kummer, *The Basel Convention* (1995), at 52–54; Pineschi, *The Transit of Ships Carrying Hazardous Wastes*, in: Francioni/Scovazzi (ed.) (1991), at 302–304, 310–315; Rummel-Bulska, *The Basel Convention and the UNCLOS*, in: Ringbom (ed.) (1997), at 90–92, 98–101.

No clarification of this issue could be achieved by means of the subsequent State practice. The Bamako and Waigani Conventions and the Izmir Protocol to the Barcelona Convention provide no clear rules in this regard. With regard to the Mediterranean Sea, Scovazzi, 'Transboundary Movement of Hazardous Waste in the Mediterranean', 19 *UCLA J. Envtl. L. & Pol'y* (2000–2002), at 241–243, argues for the "notification without authorization" scheme of the Izmir Protocol to the Barcelona Convention (Article 6(4)) that applies to the territorial seas of coastal States. According to this, prior notification of any movement must be given to the coastal State if a vessel carrying hazardous wastes is crossing the coastal State's territorial sea. The authorisation of this movement by the coastal is, however, not a requirement for the movement to commence. Although there are some good reasons speaking in favour of such solution, it is of no relevance for the Basel Convention, since a corresponding provision does not exist. Due to the fact that no EEZs are established in the Mediterranean Sea, there is no provision in the Izmir Protocol regarding movements through the EEZ of a coastal State.

³⁰⁷ Article 2(12) of the Basel Convention defines the State of transit to be "any State, other than the State of export or import, through which a movement of hazardous wastes or other wastes is planned or takes place". A similar wording is used by Article 7 of the Basel Convention, according to which Paragraph 1 of Article 6 applies *mutatis mutandis* to hazardous waste movements from a Party "through" a State or States which are not Parties.

³⁰⁸ This term is defined by Article 2(9) of the Basel Convention to be "any land, marine area or airspace within which a State exercises administrative and regulatory responsibility in accordance with international law in regard to the protection of human health or the environment". The question whether or not the EEZ is covered by this definition is discussed *infra*, Sect. "Transboundary Movement" in Chap. 5.

Consequently, the definition of the term "State of transit" must be understood to cover the territorial sea of a coastal State, but not the EEZ of that State.

In consequence of the different positions of States concerning the coverage of maritime zones a further "disclaimer clause" was included in the Convention text at the end of the negotiation process. Article 4(12) of the Basel Convention, however, has no substantive, regulatory content; it merely intends to clarify that the obligations contained in the Convention have no effect on the existing rights and obligations of States established in international law and, specifically, in the international law of the sea.[309] In this context, it needs to be highlighted that the interpretation of the term "coastal State" in a way that it includes the territorial sea and excludes the EEZ of a coastal State is consistent with the provisions of the law of the sea. The competence of the coastal State to establish the PIC requirements of the Basel Convention in its territorial sea follows from Article 21(1)(f) of the UNCLOS. The corresponding obligation of foreign ships to comply with those obligations is laid down in Article 21(4) of the UNCLOS. This outcome, however, is different with regard to the EZZ. Article 56(1)(b)(iii) of the UNCLOS establishes jurisdiction of the coastal State with regard to the protection and preservation of the marine environment only as far as this is provided for in the relevant provisions of the UNCLOS. Those relevant provisions could be seen in Article 211 (5) and in Article 220(3), (5) and (6) of the UNCLOS. However, these provisions refer to international rules and standards for the prevention, reduction and control of pollution particularly from vessels. It is doubtful that the establishment of a notification and prior consent requirement regarding the mere transit of vessels carrying hazardous wastes through the EEZ of a coastal State can be related to the prevention, reduction and control of pollution from vessels. This is at least true in cases where there is no indication whatsoever that there is a specific threat of damage to the marine environment emanating from the vessel carrying hazardous wastes. It must be concluded, therefore, that coastal States do not possess the necessary competence under the international law of the sea to require prior notification and consent in accordance with the rules of the Basel Convention as far as ships are concerned that are merely crossing their EEZ. The exclusion of the EEZ from the definition of the term "State of transit", thus, corresponds with the competences of coastal States under the international law of the sea.

In conclusion of this section, it should be stressed that the establishment of a breach of the PIC requirements is relatively simple. Since the particular act to be performed by the State is explicitly described by the Convention, the determination of whether or not these rules have been observed does not require an additional legal assessment or the consideration of any relevant social, economic or

---

[309] Basel Convention, Article 4(12), states that "[n]othing in this Convention shall affect in any way the sovereignty of States over their territorial sea established in accordance with international law, and the sovereign rights and the jurisdiction which States have in their exclusive economic zones and their continental shelves in accordance with international law, and the exercise by ships and aircraft of all States of navigational rights and freedoms as provided for in international law and as reflected in relevant international instruments".

technological circumstances. An internationally wrongful act thus exists, for example, if a State of export permits the commencement of a hazardous wastes shipment without having received the prior confirmation of the notifier that it has received the prior written consent of all States concerned. According to the view posited by this work, a prior consent of coastal States is not required as long as the shipment is merely conducted through the territorial sea or the EEZ of such States.

(v) Further Obligations Imposed by the Basel Convention

In addition to the above-mentioned general obligations and the core requirements of the Convention regarding the prior informed consent of the States involved when performing a transboundary hazardous waste movement, the Basel Convention also imposes additional or supplementary obligations on the States Parties.

This first concerns the obligation to re-import hazardous wastes in case a transboundary movement cannot be completed in accordance with the terms of the contract, as laid down by Article 8 of the Convention, as well as in case the transport has been carried out in contravention of the requirements imposed by the Convention, as stipulated by Article 9. It has been shown that these rules represent ordinary primary rules of international law and cannot be ascribed to the body of rules of State responsibility or State liability.[310] According to the common wording of Articles 8 and 9, States "shall ensure" that the wastes in question "are taken back". Furthermore, a secondary duty to perform the re-importation is imposed on the States themselves in case the private parties concerned are unavailable or incapable of doing so. From this it becomes apparent that these obligations do not confine themselves to requiring the performance of a certain conduct, but rather oblige the States to ensure the occurrence of a particular result. Thus, they represent obligations of result, to which the general standard of due diligence applies. A breach of these obligations is established solely on the condition that in consequence of an illegal or uncompleted shipment, the hazardous wastes in question are not shipped back even though so required by Articles 8 and 9 of the Convention.

Further obligations concern interstate co-operation and the exchange of information. States Parties are obliged (i) to co-operate in order to improve and achieve environmentally sound management of hazardous wastes[311] and, for this purpose, (ii) to make available information with a view to promoting the harmonisation of technical standards and practices and the development, implementation and transfer of new technologies and management systems.[312] In addition, States are under the obligation to monitor and report information on various issues, such as

---

[310] *Supra*, Sects. "Article 8" and Article 9.
[311] Basel Convention, Article 10(1) and 4(2)(h).
[312] *Ibid.*, Article 10(2).

on accidents occurring during the transboundary movement of hazardous wastes, on particulars concerning the domestic regulatory system and on the number and extent of transboundary hazardous waste transports in which they have been involved.[313] These supplementary obligations complete the framework of duties imposed on States Parties by the Basel Convention and serve the purpose of putting the States and the Basel Secretariat in position to fully comply with their main obligations and functions under the Convention. They are not targeted at regulating the permissibility or performance of hazardous waste transports as such, so that a breach of these obligations will be of no importance as concerns State responsibility for damages occurring during the transport of hazardous wastes.

(3) The Application of the Basel Convention to End-of-Life Ships

A further issue relevant to the Basel Convention concerns its application to end-of-life ships that are intended to sail to another country in order to be dismantled. Since end-of-life ships often contain hazardous components like oils, asbestos and PCBs, which may be released to the environment during the dismantling process and may cause serious damage to health of the workforce, a likely debate has emerged regarding the application of the PIC and ESM requirements of the Basel Convention to those ships. This issue is of considerable practical importance, since States in whose ports ships are anchoring prior to its last journey might be under the obligation to prevent the departure of the vessel unless the procedural requirements of the Basel Convention are fulfilled.

(i) The Dismantling of End-of-Life Ships

The procedure of ship dismantling is usually initiated by the decision of the shipowner to terminate the commercial operation of the ship and, instead, to realise the value of the raw materials (predominantly steel) that are contained in the ship. To this end, the shipowner usually sells the ship to a cash buyer, who renames the ship, registers it in another registry and resells it to a shipbreaking yard. Large dismantling capacities for sea-going ships can today be found mainly in India, Bangladesh, China and Pakistan. After the ship has been sold to a cash buyer the ship usually sails with a last cargo of material, such as scrap metals, to the shipbreaking yard and then is run aground at the beach at high watermark. This procedure is commonly referred to as "beaching". Subsequently, the ship is often dismantled manually and without sufficient protective gear for workers and without any measures to protect the coastal environment. While the steel and the other raw metals are sold for large profits on the commodities market, the remaining waste materials are often simply burned or deposited without further

---

[313] *Ibid.*, Article 13. See also Shibata, *Ensuring Compliance with the Basel Convention*, in: Beyerlin/Stoll/Wolfrum (ed.) (2006), at 70–74.

treatment at dumping grounds. In this way, it is possible to generate proceeds from the dismantling of the ship of up to 20 million Dollars for large vessels.[314]

(ii) The Definition of "Waste"

In a situation where the shipowner intends to sell his ship for the purpose of dismantling in a third country and where this ship contains hazardous components it is argued by interest groups that such a ship generally falls within the scope of the Basel Convention, with the consequence that the port State would be obliged to prevent the departure of the ship unless the relevant ESM and PIC requirement have been complied with.[315] A less strict position is held by the Parties to the Basel Convention, who assume that the definition of "waste" under the Basel Convention is basically to be understood in a broad meaning. The Parties to the Convention, consequently, adopted at COP6, in 2002, the Technical Guidelines for the Environmentally Sound Management of the Full and Partial Dismantling of Ships.[316] Moreover, at COP7, in 2004, the Parties formally recognised that "a ship *may* become waste as defined in article 2 of the Basel Convention and that at the same time it may be defined as a ship under international rules".[317] The precise meaning of this statement, however, remains ambiguous. This applies all the more since the Basel Convention in its convention text explicitly provides that nothing in the Convention affects any rights, freedoms and obligations of States as provided for in international law, particularly in the international law of the sea.[318]

What is for sure is that the question of whether or not an end-of-life ship comes under the scope of the Basel Convention cannot be answered in general terms, but rather depends on the individual case. The focus of this consideration clearly lays

---

[314] As to the *modus operandi* of shipbreaking see EC Doc. COM (2007) 269 final, at 5–8; Lagoni/Albers, 'Schiffe als Abfall?', 30 *NuR* (2008), at 221–222; Matz-Lück, 'Safe and Sound Scrapping of "Rusty Buckets"?', 19 *RECIEL* (2010), at 95–97. The impacts of the global shipping crises on the ship dismantling industry since 2009 are outlined by Wägener, 'Dank der Schifffahrtskrise floriert die Abwrackindustrie', 149 *HANSA* (2012), at 74–78, who stresses the fact that as a result of the shipping crises the average age of dismantled ships has fallen, whereas the revenues from the sale of ships to "cash buyers" have increased up to 500 USD per light displacement ton (ldt).

[315] See e.g. Basel Action Network (BAN) and Greenpeace, press release of 10 January 2002, 'Shipbreaking and the Legal Obligations under the Basel Convention'. See also BAN and Greenpeace, 'Environmentally Sound Management of Ship Dismantling', OEWG Doc. UNEP/CHW/OEWG/4/INF/20, at 3.

[316] COP6 Decision VI/24 (Doc. UNEP/CHW.6/40).

[317] COP7 Decision VII/26 (Doc. UNEP/CHW.7/33).

[318] Basel Convention, Article 4(12) reads: "Nothing in this Convention shall affect in any way the sovereignty of States over their territorial sea established in accordance with international law, and the sovereign rights and the jurisdiction which States have in their exclusive economic zone and their continental shelves in accordance with international law, and the exercise by ships and aircraft of all States of navigational rights and freedoms as provided for in international law and as reflected in relevant international instruments."

on the interpretation and construction of the terms "wastes" and "hazardous wastes" as defined under the Basel Convention. Only if the ship as such is to be considered "waste" under the Convention, it is possible to subsequently determine whether this particular waste also comes under the definition of "hazardous wastes".

An end-of-life ship comes under the definition of "waste" of the Basel Convention if it is disposed of or is intended to be disposed of or is required to be disposed of by the provisions of national law.[319] Based on this definition, a ship becomes waste, at the latest, when it *is disposed of*. This is the case when the ship is actually being dismantled at the shipbreaking yard.[320] But also the previous period beginning in that moment when the ship is "beached" at the shore and is waiting to be scrapped represents a disposal operation,[321] so that the ship is to be considered waste as of this moment.[322]

A highly contentious issue involves the question whether a ship may become waste already before the moment when it is "beached". This would be the case if the ship is *intended to be disposed of* already at an earlier stage. In order to give an answer, one first need to determine who the person is that decides about the intended purpose of the ship. Since this issue is not addressed by the Basel Convention, it is necessary in this respect to resort to the relevant national laws. Most national laws consider the actual possessor of the wastes to be the person that decides about the intended purpose.[323] At this point it becomes apparent that the provisions of the Basel Convention are not properly aligned to the peculiarities of sea-going vessels and thus simply do not fit when it comes to cases of ship dismantling. It must be kept in mind that unlike usual wastes, sea-going ships are registered in a public registry allowing for an easy identification of the owner. The typical situation underlying regular wastes, according to which the actual owner of the wastes can hardly be identified, does not apply to ships. With regard to ships, it is, therefore, not necessary to focus on the actual possessor of the wastes in order to ensure effective risk prevention. Consequently, there can be no justification to

---

[319] Basel Convention, Article 2(1), defines "wastes" as "substances or objects which are disposed of or are intended to be disposed of or are required to be disposed of by the provisions of national law". The term "disposal" is defined by Article 2(4) of the Basel Convention, referring to any operation specified in Annex IV to the Convention.

[320] In Annex IV, Section B of the Basel Convention recycling and resource recovering operations are listed. Those include *inter alia* the recycling/reclamation of metals and metal compounds (R4) and the recycling/reclamation of other inorganic materials (R5).

[321] Basel Convention, Annex IV, Section B , R13 classes among the recycling operations also the accumulation of materials intended for any operation in Section B.

[322] See also Lagoni/Albers, 'Schiffe als Abfall?', 30 *NuR* (2008), at 223.

[323] For example, the EU law focuses in it definition of "waste" on the "holder", see Article 3 No. 1 of the Directive 2008/98/EC of the European Parliament and the Council of 19 November 2008 on Waste. The "waste holder" is defined as meaning "the waste producer or the natural or legal person who is in possession of the waste", see Article 3 No. 6 of this Directive.

restrict the shipowner's fundamental right to property by relying on the intention of the actual possessor of the ship instead of on the intention of the shipowner.[324] It must be concluded that only the shipowner is in a position to determine the purpose of the ship. Since, however, the shipowner is in general not the possessor of the ship, which is usually (bareboat) chartered to an operator, it is obvious that the provisions of the Basel Convention do not come to sufficient results regarding ships.

The most ambiguous issue, however, is to define the precise moment as of which the ship is intended to be disposed of by the shipowner. It is obvious that purely objective criteria like the age or a poor condition of the ship cannot be decisive, considering that a ship can easily be overhauled or rebuilt in order to serve another purpose, such as a museum or hotel ship. It is, therefore, necessary to focus on the subjective intention of the shipowner. Since such intention is often not explicitly expressed, it is argued that it is possible to extrapolate the intention to dispose of a ship from external indicators.[325] In particular, it is assumed that as of the moment the shipowner enters into a sales contract with a cash buyer or shipbreaker the ship is intended to be disposed of and, thus, represents waste.[326] Such approach, however, must fail from the outset, since it is nothing more than the fiction of intent to the detriment of the shipowner and the replacement of the subjective criterion of "intention" by objective circumstances. In cases where a sales contract is concluded or where other external indicators are present it is actually far from certain that the shipowner intends to dispose of the ship.[327] Such external indications can also be explained by other reason, such as to be seen during the recent shipping crises when hundreds of ships were laid up for an indefinite period. Those ships are surely not intended to be disposed of during this period.

Consequently, the subjective intention of the shipowner to dispose of a ship needs to be determined in each individual case. But does already the sale of a ship to a cash-buyer or a shipbreaker manifest this intention? When entering into a sales

---

[324] See on this issue Lagoni/Albers, 'Schiffe als Abfall?', 30 *NuR* (2008), at 224.

[325] Such external indicators are said to include e.g. the deletion of the ship from the national ship registry, the non-renewal of relevant certificates or classifications, the cancellation or modification of insurance, preparatory steps towards the later dismantling of the ship, or simply the fact that the ship is taken out of traffic on a permanent basis or is not maintained anymore for transport of cargo or passengers. See Claußen, *Die Abwrackung von Seeschiffen in Nicht-OECD-Staaten* (2009), at 44–45; Ulfstein, 'Legal Aspects of Scrapping of Vessels', report annexed to LWG Doc. UNEP/CHW/LWG/4/4, at 8; see also Basel Action Network (BAN) and Greenpeace, press release of 10 January 2002, 'Shipbreaking and the Legal Obligations under the Basel Convention', at 2.

[326] Claußen, *Die Abwrackung von Seeschiffen in Nicht-OECD-Staaten* (2009), at 42–44; Ulfstein, 'Legal Aspects of Scrapping of Vessels', report annexed to LWG Doc. UNEP/CHW/LWG/4/4, at 8.

[327] Lagoni/Albers, 'Schiffe als Abfall?', 30 *NuR* (2008), at 224; see also Engels, *European Ship Recycling Regulation* (2013), at 129–130, however, erroneously citing Lagoni/Albers for the opposite view at 129.

contract regarding the ship the shipowner's predominant aim is to realise the value of the raw materials contained in the ship. It is true that such aim to generate a (last) economic benefit from a disused item does, in general, not preclude that this item is intended to be disposed of.[328] However, as regards the sale of end-of-life ships the shipowner's intention is not to generate an economic benefit *from the disposal* of the ship, his intention amounts to nothing more than to the previous sale of the ship. In other words, the intention to sale an end-of-life ship cannot be equated with the intention to dispose of the ship in accordance with any of the disposal operations listed in Annex IV of the Basel Convention.[329]

It may now be argued that the intention of the shipowner is not restricted to the sale of the ship, but also comprises the later disposal of the ship.[330] At this point, again, it becomes apparent that the definition of "waste" under the Basel Convention does not take account of the factual circumstances relevant to end-of-life ships that are sold for the purpose of dismantling. Ships, even when they are sailing to their last destination at a shipbreaking yard, are used as a means of transport. Even if the ship is supposed to be dismantled after its arrival, it is still intended to serve as a vehicle enabling transportation. If such ship would have to be considered "waste" already at the moment the shipowner takes the decision to dismantle the ship at a later time, this ship would, as of this moment, be subject to the export restrictions of the Basel Convention and, thus, could effectively not be used as a means of transport anymore. It must be concluded, therefore, that only the current purpose given to the ship by the shipowner can be decisive with regard to the definition of the term "waste"; any later (or subsequent) purpose must remain irrelevant in this context. The intention of the shipowner to use the ship as a means of transport and his intention to dispose of the ship are thus mutually exclusive. Only in that moment when the ship is "beached" at the shipbreaking yard, the intention of the shipowner to use the ship as a means of transport ends and, in lieu thereof, the ship is intended to be disposed of.[331]

Finally, a ship also becomes waste if it is *required to be disposed of* by the provisions of national law. It is therefore possible that a State creates national law, according to which an end-of-life ship under certain objective conditions is required to be disposed of by the owner and, thus, would be considered "waste" under the Basel Convention. However, such national law can only be lawful and valid if it is not in contravention with the provisions of the UNCLOS or the corresponding rules of customary international law. Under the UNCLOS and the customary international law ships that hold the required IMO certificates enjoy the freedom of navigation and may only be detained in those cases provided for in the

---

[328] Annex IV, Section B of the Basel Convention enumerates disposal operations which may lead to resource recovery and recycling reclamation. The Convention, thus, also applies to wastes destined for recovery operations.

[329] See also Lagoni/Albers, 'Schiffe als Abfall?', 30 *NuR* (2008), at 224.

[330] Ulfstein, 'Legal Aspects of Scrapping of Vessels', report annexed to LWG Doc. UNEP/CHW/LWG/4/4, at 8.

[331] Lagoni/Albers, 'Schiffe als Abfall?', 30 *NuR* (2008), at 225.

UNCLOS.[332] The UNCLOS, however, does not allow the detention of a ship based on the ground that the owner did not comply with the procedural requirements of the Basel Convention.[333] As long as the ship is seaworthy and does not threaten damage to the marine environment it may not be restricted in the exercise of its navigational freedom.[334] Consequently, it is not possible to restrict the freedom of navigation by means of an obligation under national law to dispose of a sea-going vessel under certain conditions.

As a result, it can be concluded that an end-of-life ship can be considered "waste" under the Basel Convention only at that moment when it is "beached" at the place of dismantling. An earlier intention of the shipowner to dispose of the ship does not exist, since prior to the moment when it is "beached", it is still intended to be used as a means of transport. The intention to use a ship as a means of transport and the intention to dispose of the ship are mutually exclusive. In addition, any provision of national law, according to which a ship is required to be disposed of prior to this moment would conflict with the navigational rights and freedoms as guaranteed by international law and, thus, would be legally void.

(iii) Simultaneity of Ship and Waste?

The outcome of the previous section, according to which the intention of the shipowner to use the ship as a means of transport and his intention to dispose of the ship are mutually exclusive, seems to contradict the decision of the Parties to the Basel Convention, according to which "a ship *may* become waste as defined in article 2 of the Basel Convention and that at the same time it may be defined as a ship under international rules".[335] In order to assess whether a conflict actually exists, again, one needs to be precise.

At first, this statement needs to be seen in the light of the other provisions of the Basel Convention, including particularly Article 4(12). According to this, it is established that no provision of this Convention affects the exercise of navigational rights and freedoms by ships of all States as provided for in international law and as reflected in relevant international instruments. This, however, would be the case if the definition of the term "waste" is interpreted in a way that a ship, which still serves the purpose as a means of transport, is at the same time considered to be "waste" so that the export restrictions of the Basel Convention would apply. It has

---

[332] See UNCLOS, Articles 87, 90.

[333] Lagoni/Albers, 'Schiffe als Abfall?', 30 *NuR* (2008), at 225–226; Wolfrum, in: Wolfrum/Matz, *Conflicts in International Environmental Law* (2003), at 115.

[334] See UNCLOS, Article 219.

[335] COP7 Decision VII/26 (Doc. UNEP/CHW.7/33). A similar declaration has been included in Paragraph 35 of the Preamble to the Regulation (EC) No 1013/2006 of the European Parliament and of the Council of 14 June 2006 on Shipments of Waste, which reads: "Furthermore, it should be noted that a ship may become waste as defined in Article 2 of the Basel Convention and at the same time it may be defined as a ship under other international rules". See also Ulfstein, 'Legal Aspects of Scrapping of Vessels', report annexed to LWG Doc. UNEP/CHW/LWG/4/4, at 7.

been outlined in the previous section that as long as a ship holds the required IMO certificates, is seaworthy and does not threaten damage to the marine environment, it enjoys the freedom of navigation as laid down in the UNCLOS and as accepted by customary international law. Therefore, it may not be detained by coastal States.[336] The declaration of the Parties to the Basel Convention, according to which a ship may at the same time be considered waste, would amount to a restriction of the navigational rights and freedoms of ships as established by international law and, thus, also contradicts Article 4(12) of the Basel Convention.

In addition, it should be noted that a decision made by the Conference of the Parties to the Basel Convention is not legally binding as such. It may be considered as a political intention and as a recommendation for the interpretation of legal terms.[337] But even in this regard this decision is hardly useful. It does not define under which conditions a ship may at the same time be considered waste, but confines itself to describing the intended outcome of a legal interpretation.[338]

It can be concluded, therefore, that the decision of the Parties to the Basel Convention, according to which a ship may at the same time be considered waste is not in conformity with the international law establishing and guaranteeing navigational rights and freedoms of ships of all States. Consequently, this decision should not be taken into account when determining whether or not end-of-life ships come under the definition of "wastes" of the Basel Convention. It thus remains true that the intention of the shipowner to use a ship as a means of transport and the intention to dispose of this ship are mutually exclusive.

(4) Summary

In summary, it can be said that the Basel Convention imposes a wide range of particular obligations on the States Parties, most of which represent obligations of conduct. To these belong the general obligations to reduce to a minimum the generation and transportation of hazardous wastes. In order to achieve this goal the States are obliged to take "appropriate measures" while taking into account the respective social, technological and economic circumstances of the States concerned. Since, therefore, the particular measures to be taken have to be determined on a case-by-case basis, a breach of these obligations will be hard to establish in practice. Furthermore, obligations of conduct are to be found in the context of the Convention's three-tiered concept of requirements for the permissibility of transboundary movements of hazardous wastes. First, a State Party is responsible for an infringement if it fails to prohibit either by law or by administrative act the transport of hazardous waste to non-Parties or to the Antarctic. Secondly, a State may be responsible for the non-compliance with the requirements of the environmentally

---

[336] See *supra*, Sect. "The Definition of "Waste"".

[337] See also Ulfstein, 'Legal Aspects of Scrapping of Vessels', report annexed to LWG Doc. UNEP/CHW/LWG/4/4, at 6.

[338] Lagoni/Albers, 'Schiffe als Abfall?', 30 *NuR* (2008), at 226.

sound management principle. In this respect, again, it is necessary to first determine what is the precise content and meaning of the "environmentally soundness" of waste treatment in the particular case and which measures therefore are deemed to be "appropriate". In addition, within the context of the applicable due diligence standard, it needs to be taken into account what the State could have reasonably assumed as to the manner in which the wastes will be treated in the State of import. Therefore, in practice it will hardly be possible to establish that a State has breached the obligations deriving from the environmentally sound management principle. Finally, a State may be internationally responsible for non-compliance with the requirements of the prior informed consent principle. A breach of these obligations is established in case the State either itself acts in contravention of these rules or fails to legally or administratively transpose these rules into domestic law applicable to the private parties concerned. Apart from the general obligations and the core obligations of the Basel Convention concerning the permissibility of transboundary movements of hazardous wastes, which represent obligations of conduct, the Basel Convention also lays down obligations of result. These involve the obligations to re-import the hazardous wastes in case their transport is deemed illegal or cannot be completed according to the terms of the contract. A breach of these obligations is established solely on the condition that the hazardous wastes in question are not shipped back despite such action being required by the obligation applicable. Finally, it should be stressed that as to all of these obligations the general standard of due diligence applies, meaning that in each particular case the concrete actions available to the State need to be taken into account in order to assess a potential breach of these obligations.

(bb) Other Conventions and Regulations Relevant to the Trade in and Transport of Hazardous Wastes

In addition to the Basel Convention, further international conventions provide for substantive rules relevant to the trade in and transport of hazardous wastes.

(1) The Bamako Convention

(i) Background of the Convention

In response to the waste dumping incidents which occurred in Africa during the 1980s, the former Organisation of African Unity (OAU)[339] in 1988 adopted Resolution 1153, by which it declared that "the dumping of nuclear and industrial

---

[339] The OAU was disbanded in 2002 and replaced by the African Union (AU), which today consists of 54 States, i.e. all of Africa's countries except for Morocco.

wastes in Africa is a crime against Africa and the African people", and by which it emphasised its determination to pursue the approach of a total ban of hazardous waste imports from non-African countries to Africa.[340] Consequently, the OAU did not perceive its position as being sufficiently taken into account in the course of the subsequent negotiations of the Basel Convention, during which it transpired that the African countries were not able to push through their position concerning a total ban of hazardous waste exports to developing countries.[341] The Basel Convention, instead, adopted a general principle of prior informed consent whose global application makes no distinction between developed and developing countries. The OAU-countries, therefore, initially refrained from signing the Basel Convention and instead advocated the elaboration of an African regional convention.[342] Finally, in 1991, the Bamako Convention on the Ban of the Import into Africa and the Control of Transboundary Movement and Management of Hazardous Wastes within Africa was set up under the auspices of the OAU. It entered into force on 22 April 1998 and, as of today, has been ratified by 25 States.[343]

The Bamako Convention is designed as a regional convention which is open for accession only by African countries.[344] Despite its regional character, the Bamako Convention nevertheless has a significant global impact. This is due to its interactions with the Basel Convention, as a result of which the trade restrictions of the Bamako Convention indirectly apply also to the Parties of the Basel Convention.[345] Since States may become Parties to both Conventions,[346] the import ban

---

[340] OAU Resolution CM/Res/1153 of 23 May 1988 on Dumping of Nuclear and Industrial Waste in Africa.

[341] Gudofsky, 'Transboundary Shipments of Hazardous Waste', 34 *Stan. J. Int'l L.* (1998), at 246; Kummer, *The Basel Convention* (1995), at 44; Marbury, 'Hazardous Waste Exportation', 28 *Vand. J. Transnat'l L.* (1995), at 269; Ovink, 'Transboundary Shipments of Toxic Waste', 13 *Dick. J. Int'l L.* (1994/1995), at 287–288.

[342] Eze, 'The Bamako Convention', 15 *Afr. J. Int'l & Comp. L.* (2007), at 214; Gudofsky, 'Transboundary Shipments of Hazardous Waste', 34 *Stan. J. Int'l L.* (1998), at 246; Jones, 'The Evolution of the Bamako Convention', 4 *Colo. J. Int'l Envtl. L. & Pol'y* (1993), at 324 *et seq.*; Kaminsky, 'Assessment of the Bamako Convention', 5 *Geo. Int'l Envtl. L. Rev.* (1992/1993), at 88.

[343] The Parties to the Convention today are: Benin, Burkina Faso, Burundi, Cameroon, Chad, Côte d'Ivoire, Comoros, Republic of the Congo, Democratic Republic of the Congo, Egypt, Ethiopia, Gabon, Gambia, Libya, Mali, Mozambique, Mauritius, Niger, Senegal, Sudan, Tanzania, Togo, Tunisia, Uganda and Zimbabwe.

[344] According to its Articles 21–23, only Member States of the AU, which is the successor to the OAU, may accede to the Convention.

[345] See Donald, 'The Bamako Convention', 17 *Colum. J. Envtl. L.* (1992), at 434.

[346] Today, all Signatory States of the Bamako Convention are as well Parties to the Basel Convention. This dual membership is permitted by Articles 11 of both the Bamako and the Basel Convention and will lead to a precedence of the Bamako Convention. See to this issue Kummer, *The Basel Convention* (1995), at 104–107; Shearer, 'Comparative Analysis of the Basel and Bamako Conventions', 23 *Envtl. L.* (1993), at 173–174.

established by the Bamako Convention is recognised by the Basel Convention, thus imposing upon its Parties the corresponding obligation to prohibit any export of hazardous wastes to the Contracting States of the Bamako Convention.[347]

(ii) The Obligations Imposed by the Convention

Regarding its scope of application the Bamako Convention provides for broader provisions than the Basel Convention. This applies especially to the catalogue of covered wastes[348] as well as to the activities[349] and the means of transport[350] covered by the Convention.

Apart from that the Bamako Convention basically follows the model of the Basel Convention and adopts its basic structure and regulatory approach. But also in respect of its regulatory content, the Bamako Convention contains in some respects stricter provisions which are designed to remedy the perceived shortcomings of the Basel Convention.[351] It basically distinguishes between hazardous wastes generated outside of Africa and those generated within Africa.[352] In accordance with this distinction the Bamako Convention pursues two different goals: First, the complete prohibition of the importation of hazardous wastes generated outside of Africa to African countries; and, second, the establishment of a regime of prior informed consent and environmentally sound management in respect of the transboundary movement and treatment of hazardous wastes generated within Africa.[353]

With regard to the first goal, the Bamako Convention imposes on its Member States the obligation to "take appropriate legal, administrative and other measures

---

[347] According to Article 4(1)(a) of the Basel Convention a Party may notify the other Parties of their decision to prohibit the import of hazardous wastes, which by virtue of Article 4(1)(b), (2)(e) entails the corresponding obligation of the other Parties to prohibit any export to that State. This would in principle also apply to all shipments that take place among Parties to the Bamako Convention. However, since the rules of the Bamako Convention prevail in respect of shipments among its Member States according to Article 11 of the Basel Convention, intra-African shipments would be permissible, whereas exports from non-members to members of the Bamako Convention would be prohibited also under the Basel Convention.

[348] The Bamako Convention covers a more comprehensive catalogue of wastes, even including radioactive wastes, see Eze, 'The Bamako Convention', 15 *Afr. J. Int'l & Comp. L.* (2007), at 217; Gudofsky, 'Transboundary Shipments of Hazardous Waste', 34 *Stan. J. Int'l L.* (1998), at 247–251; Kummer, *The Basel Convention* (1995), at 101.

[349] The Convention contains provisions banning the dumping of hazardous wastes at sea and in internal waters. See Bamako Convention, Article 4(2).

[350] The Convention applies to vessels flying the flag of a Contracting Party and aircrafts registered in the territory of a Contracting Party, see Article 11(3) of the Convention.

[351] Kummer, *The Basel Convention* (1995), at 100.

[352] Gudofsky, 'Transboundary Shipments of Hazardous Waste', 34 *Stan. J. Int'l L.* (1998), at 246; Kaminsky, 'Assessment of the Bamako Convention', 5 *Geo. Int'l Envtl. L. Rev.* (1992/1993), at 78.

[353] OAU Res. CM/Res/1225(L) of 22 July 1989 on Control of Transboundary Movements of Hazardous Wastes and their Disposal in Africa; see also Kummer, *The Basel Convention* (1995), at 100.

[...] to prohibit [...] for any reason" the transportation of hazardous wastes from non-Parties into Africa.[354] The Parties shall furthermore "[c]o-operate to ensure that no imports of hazardous wastes from a non-Party enter a Party to this Convention."[355] The application of these obligations is comprehensive and does not allow for any exemptions or less stringent rules in respect of wastes destined for recovery or recycling operations.[356] As a result, these rules amount to a complete import ban of all hazardous wastes shipped from a non-Party, irrespective of whether this is an African or non-African State, into the national territory of a State Party to this Convention. In contrast, the Convention does not prohibit the export of hazardous wastes from Africa to non-States Parties.[357] The transport of hazardous wastes from a Contracting State out of the Convention area is only limited insofar as the Parties "agree not to allow the export of hazardous wastes for disposal within the area South of 60° South Latitude"[358] The Convention finally provides that the Parties "shall, in the exercise of their jurisdiction [...] prohibit the dumping at sea of hazardous wastes, including their disposal in the seabed and sub-seabed".[359] The regulatory character of these obligations generally corresponds to the regulatory character of the absolute trade restrictions established under the Basel Convention. From the precise wording of these obligations, which require that the State "shall take appropriate measures", "shall co-operate" with a view to a certain result, and "shall not allow" or "shall prohibit" certain activities, it becomes obvious that the purpose of these obligations is not to prescribe the occurrence of a certain factual result that has to be achieved by the State by whatever means necessary. These obligations rather emphasise the legal or administrative conduct of the State with regard to the activities of private parties and, therefore, must be considered obligations of conduct, to which the general standard of due diligence applies. A breach of these obligations is constituted by any failure of the State to domestically prohibit such transports or to sufficiently control, monitor and enforce this prohibition.

In respect of its second goal, the Bamako Convention establishes a prior informed consent principle as regards transboundary movements of hazardous wastes among States Parties, reproducing in almost identical terms the relevant

---

[354] Bamako Convention, Article 4(1).
"Africa" is not defined as a legal term within the Bamako Convention and, therefore, this terms denotes the continent in its geographical expansion, irrespective of whether the particular State concerned is a Contracting Party to the Convention or not.

[355] Bamako Convention, Article (4)(1)(b).

[356] Kummer, *The Basel Convention* (1995), at 101; Marbury, 'Hazardous Waste Exportation', 28 *Vand. J. Transnat'l L.* (1995), at 270.

[357] See Kummer, *The Basel Convention* (1995), at 102; Louka, *International Environmental Law* (2006), at 431; van der Mensbrugghe, *Commentary*, in: Lang/Neuhold/Zemanek (ed.) (1991), at 162.

[358] Meaning Antarctica, Bamako Convention, Article 4(3)(l).

[359] *Ibid.*, Article 4(2).

provisions of the Basel Convention.[360] However, one difference is that the Bamako Convention requires that the written confirmation of the State of import is received directly by the State of export, instead of merely requiring that the notifier confirms with the State of export that it has received the written confirmation from the State of import.[361] Nevertheless, the same statements and findings as made in respect of the PIC-procedure of the Basel Convention apply to the PIC-requirements of the Bamako Convention. These obligations thus represent ones of conduct. Since the particular conduct required from the respective State is precisely described by these provisions, the determination of a breach of these provisions should not involve major difficulties.

Finally, the Bamako Convention prescribes that if a transport of hazardous wastes cannot be completed in accordance with the terms of the contract the exporting State "shall ensure that the wastes in question are taken back into the State of export, by the exporter". The same applies to cases of illegal traffic, where the State of export "shall ensure" that the wastes are taken back "by the exporter or generator of if necessary by itself" or, alternatively, that the State of import "shall ensure" that those wastes are returned to the exporter by the importer.[362] These provisions, again, are carried over almost verbatim from the relevant provisions of the Basel Convention and thus have to be considered obligations of result since the States have to ensure the occurrence of a particular result by whatever means necessary. The applicable standard of behaviour is the general standard of due diligence. A breach of these obligations of re-importation is thus established solely on the condition that hazardous wastes which have been shipped illegally or incompletely are not taken back notwithstanding the fact that the State knew or ought to have known about the circumstances rendering the transport illegal or incomplete.

(iii) Summary

It can be concluded that the obligations imposed by the Bamako Convention resemble to a large extent those of the Basel Convention. This basically also applies to the establishment of ban on the import of hazardous wastes into the Convention area. In fact, both Conventions establish a ban of hazardous wastes shipments from non-Contracting Parties to Contracting Parties. In fact, the Basel Convention goes even further and additionally bans any hazardous waste movements from Contracting Parties to non-Contracting Parties. The unique impact of the import ban established by the Bamako Convention only results from its locally restricted Convention area. Regarding the establishment of an import ban, the

---

[360] *Ibid.*, Articles 6, 7.

[361] See Articles 6(3) of both the Bamako and Basel Convention. The latter procedure is criticised as facilitating misdeclarations by the notifier, see Eze, 'The Bamako Convention', 15 *Afr. J. Int'l & Comp. L.* (2007), at 218.

[362] Bamako Convention, Articles 8, 9.

Bamako Convention thus does not provide stricter rules than the Basel Convention. In other respects, however, the Bamako Convention indeed provides for stricter rules and thus may in theory be considered as a suitable and effective supplement to the Basel Convention. From a practical perspective, however, the significance of the Bamako Convention is quite limited. This is to be explained by the inability of its Contracting States to either effectively implement the provisions or ensure compliance by local and national authorities.[363]

Regarding the legal character of the obligations, the findings made in respect of the Basel Convention apply accordingly. The obligations are ones of conduct to which the general standard of due diligence applies. Establishing a breach of these obligations is not likely to involve major difficulties as far as concerns a failure of the State to implement the procedural rules into the national law. The same basically applies to any failure of the State to comply with the sole obligation of result, namely to ensure that illegally or incomplete shipped hazardous wastes are re-imported by the State of export.

(2) The Waigani Convention

Acknowledging the need to prohibit the importation of hazardous and nuclear wastes into the South Pacific Region and the need to create a complementary regime to the Basel Convention that takes into consideration particular regional concerns, in September 1995 the Member States of the Pacific Island Forum (PIF) adopted the Waigani Convention to Ban the Importation into Forum Island Countries of Hazardous and Radioactive Wastes and to Control the Transboundary Movement and Management of Hazardous Wastes within the South Pacific Region. The Waigani Convention entered into force in 2001 and today has 13 Member States.[364]

The regulatory content of the Waigani Convention to a great degree corresponds to the obligations imposed by the Bamako Convention.[365] The Convention establishes a ban on the importation of hazardous wastes from outside the Convention Area into any Pacific Island Developing Party.[366] In contrast, shipments of hazardous wastes among Pacific Island Developing Parties as well as exports out of the Convention Area are not prohibited. In respect of those shipments the Convention provides for a distinct regime of PIC-requirements that are taken almost verbatim from the Basel and Bamako Conventions.[367] The Convention, finally, also stipulates an obligation of re-importation in cases of illegal traffic or where a transport cannot be completed in accordance with the terms of the

---

[363] Pratt, 'Decreasing Dirty Dumping?', 35 *Wm. & Mary Envtl. L.& Pol'y Rev.* (2011), at 603.

[364] Australia, Cook Islands, Federated States of Micronesia, Fiji, Kiribati, New Zealand, Niue, Papa New Guinea, Samoa, Solomon Islands, Tonga, Tuvalu and Vanuatu.

[365] See to the Waigani Convention in general: van Hoogstraten/Lawrence, 'The Waigani Convention', 7 *RECIEL* (1998), at 268 *et seq.*

[366] Waigani Convention, Article 4(1)(a). The "Convention Area" and the term "Pacific Island Developing Party" are defined in Article 1.

[367] *Ibid.*, Article 6.

contract.[368] In accordance with the corresponding findings in respect of the Basel and Bamako Convention, these represent obligations of conduct, except for the obligation of re-importation, which has to be considered an obligation of result. The elements of an internationally wrongful act under the Waigani Convention are thus largely the same as under the Bamako Convention.

(3) The Cotonou Agreement

In June 2000, the then 15 Member States of the European Union and 78 States of the African, Caribbean and Pacific Group of States (ACP Group) signed in Cotonou, Benin a partnership agreement between the European Union and the ACP Group intending to promote and expedite the political, economic, cultural and social development of the ACP Group.[369] The Agreement entered into force on 1 April 2003 and succeeded the previous Lomé IV Convention of 1989[370] that had expired in February 2000. Whereas the Lomé IV Convention contained a provision according to which the "Community shall prohibit all direct or indirect export" of hazardous wastes and "the ACP States shall prohibit the direct or indirect import into their territory",[371] the Cotonou Agreement abandons the approach of a trade ban and does not provide for any binding rules on the transboundary shipment of hazardous wastes. It rather focuses on the "[c]ooperation on environmental protection and suitable utilisation and management of natural resources" while "[t]aking into account issues relating to the transport and disposal of hazardous waste."[372] The Agreement furthermore encourages its Parties to ratify as quickly as possible the Basel Convention.[373] From this wording it can be inferred that these obligations are of mere programmatic character reflecting a political intent rather than a rule directly binding on the Contracting Parties.

(4) OECD Council Decisions

The Organisation for Economic Co-operation and Development (OECD) consists today of 34 highly developed countries and aims at the promotion of economic progress and world trade. The Environment Directorate of the OECD Secretariat has been concerned with the issue of transboundary movements of hazardous wastes since the 1970s, which led to a number of Decisions and mixed Decision-Recommendations of the Council in this field. A Decision of the OECD Council is

---

[368] *Ibid.*, Articles 8, 9.

[369] Partnership Agreement between the Members of the African, Caribbean and Pacific Group of States, of the One Part, and the European Community and its Member States, of the Other Part (Cotonou Agreement).

[370] The Fourth African, Caribbean and Pacific States—European Economic Community Convention of Lomé (Lomé IV Convention). It entered into force in 1990.

[371] Lomé IV Convention, Article 39.

[372] Cotonou Agreement, Article 32.

[373] *Ibid.*, Declaration IX para. 4, with a view to Article 49, para. 2.

binding on the OECD Member States; the same applies to a Decision-Recommendation as far as its mandatory part is concerned.[374]

The OECD first focused on hazardous wastes in general and imposed on its Member States the mere general obligation to "control the transfrontier movements of hazardous waste".[375] Particularly with a view to exports of hazardous wastes out of the OECD area for final disposal, the Council, in 1986, adopted a comprehensive set of requirements including obligations (i) to apply no less stringent controls than on movements involving only Member States, (ii) to prohibit movements absent the consent of the importing country and the prior notification of any transit country, and (iii) to prohibit movements unless the wastes are directed to an adequate disposal facility in the importing country.[376] In addition, a draft of an international convention among OECD countries on this issue was elaborated in 1988.[377] Since, however, in 1989 the Basel Convention was adopted and already contained all of the relevant provisions and principles envisaged by the draft convention, the OECD Council ultimately suspended its work on this convention.[378] The obligations imposed by the Council decisions of 1984 and 1986 remain binding for the OECD members; they do not, however, reach the distinct requirements of the Basel Convention nor its level of protection and, thus, do not come under Article 11 of the Basel Convention. These rules are consequently of no practical relevance.[379]

After the adoption of the Basel Convention the OECD focused on the issue of transfrontier movements of wastes for recovery operations. The Council, in 1992, adopted a Decision which applied to all wastes shipped between OECD countries and designed for recovery operations a set of different procedural requirements that varied depending on the respective category of wastes.[380] This set of requirements was revised in 2001, when in conformance with the respective lists of the Basel Convention the former classification comprising three categories was reduced to a system of two categories.[381] No specific trade restrictions apply to

---

[374] Convention on the Organisation for Economic Co-operation and Development, Articles 5, 6.

[375] Decision-Recommendation of the Council on Transfrontier Movements of Hazardous Waste of 1 February 1984, C(83)180/FINAL.

[376] Decision-Recommendation of the Council on Exports of Hazardous Wastes from the OECD area of 5 June 1986, C(86)64/FINAL.

[377] OECD Draft International Agreement on Control of Transfrontier Movements of Hazardous Wastes, 27 October 1988, (ENV(87)9(5th Revision)).

[378] OECD Council Resolution on the Control of Transfrontier Movements of Hazardous Wastes, 30 January 1989 (C(89)1/FINAL); OECD Council Resolution on the Control of Transfrontier Movements of Hazardous Wastes, 18–20 July 1989, (C(89)112/FINAL). See also Kummer, *The Basel Convention* (1995), at 159–161.

[379] Kummer, *The Basel Convention* (1995), at 165.

[380] OECD Council Decision Concerning the Control of Transfrontier Movements of Wastes Destined for Recovery Operations, 30 March 1992 (C(92)39/FINAL).

[381] Decision of the Council Concerning the Control of Transboundary Movements of Wastes Destined for Recovery Operations, of 14 June 2001 (C(2001)107/FINAL), as amended.

wastes on the green list,[382] whereas wastes on the amber list[383] are subject to a control system including, *inter alia*, the requirements of a written contract and prior notification and at least the tacit consent of the importing country. Regarding its practical significance, it should first be stressed that this OECD Decision applies only to waste shipments among OECD countries and is, therefore, of no relevance for the environmentally dangerous exports of wastes to developing counties. This is of particular importance considering that States which are not members to the OECD cannot invoke the breach of an individual or "subjective" right vis-à-vis an OECD Member State that is contravening the internal OECD rules. In addition, it may well be argued that the OECD rules do not meet the procedural requirements of the Basel Convention since they allow for a tacit consent of the State of import. The State of import may furthermore decide in general not to raise objections concerning movements to specific recovery facilities.[384] Thus, the OECD Decision of 2001 cannot be considered an agreement within the meaning of Article 11 of the Basel Convention.[385] In conclusion, it must be said that the Basel Convention prevails within its ambit over the OECD Decision. This applies to shipments of hazardous wastes covered by the amber list contained in the OECD Decision.

(5) European Union Legislation

The European Union (EU) legislation on the management of wastes is both very comprehensive and very distinct. The rules on the transboundary movement of hazardous wastes represent only one component of this overall framework.

The general basis for the European Union legislation on wastes is provided by Directive 2008/98/EC on Waste.[386] This Directive replaced former Directive 2006/12/EC[387] which had been newly created only 2 years earlier. The new Directive also includes general rules on hazardous wastes and on waste oils that previously had been subject to separate directives.[388] As regards its content the Directive on Waste stipulates basic legal definitions and lays down a set of general requirements to be observed by the domestic legislation of the EU Member States,

---

[382] Wastes on the green list are those contained in Annex IX of the Basel Convention plus additional wastes the OECD members agreed upon.

[383] Wastes on the amber list are those contained in Annexes II and VIII of the Basel Convention plus additional wastes the OECD members agreed upon.

[384] See in respect of the amber list already Kummer, *The Basel Convention* (1995), at 167.

[385] This would apply although the Basel Secretariat has been notified about the OECD Council Decision C(2001)107/FINAL in accordance with Article 11(2) of the Basel Convention.

[386] Directive 2008/98/EC of the European Parliament and the Council of 19 November 2008 on Waste.

[387] Directive 2006/12/EC of the European Parliament and of the Council of 5 April 2006 on Waste.

[388] Council Directive 91/689/EEC of 12 December 1991 on Hazardous Wastes; Council Directive 75/439/EEC of 16 June 1975 on the Disposal of Waste Oils.

including, for example, rules on the prevention, recycling and disposal of wastes.[389] Particularly with regard to hazardous wastes, it furthermore imposes obligations to control, not to mix and to properly label hazardous wastes.[390] This general Directive on Waste is supplemented by further specific EU regulations concerning particular hazardous wastes and particular types of waste. This involves, in respect of particular hazardous wastes, the EU directives on titanium dioxide,[391] on PCBs and PCTs,[392] and on POPs.[393] In respect of particular types of waste, specific regulations are provided by the EU directives on packaging and packaging waste,[394] on end-of-life vehicles,[395] on waste electrical and electronic equipment[396] and on spent batteries and accumulators.[397]

The EU regulations on the transboundary movement of wastes form part of the body of rules concerning particular types of waste treatment.[398] The central instrument is Waste Shipment Regulation (EC) No 1013/2006,[399] by which the requirements of the Basel Convention and Decision C(2001)107/FINAL of the OECD Council are implemented in the EU area.[400] This Regulation also takes into account the principles of origin and proximity as prescribed by general EU

---

[389] For a detailed discussion see Meßerschmidt, *Europäisches Umweltrecht* (2011), at 845 *et seq.*

[390] Directive 2008/98/EC, Articles 17–19.

[391] Council Directive 78/176/EEC of 20 February 1978 on Waste from the Titanium Dioxide Industry. See also Council Directive 82/883/EEC of 3 December 1982 on Procedures for the Surveillance and Monitoring of Environments Concerned by Waste from the Titanium Dioxide Industry and Council Directive 92/112/EEC of 15 December 1992 on Procedures for Harmonizing the Programmes for the Reduction and Eventual Elimination of Pollution Caused by Waste from the Titanium Dioxide Industry.

[392] Council Directive 96/59/EC of 16 September 1996 on the Disposal of Polychlorinated Biphenyls and Polychlorinated Terphenyls.

[393] Regulation (EC) No 850/2004 of the European Parliament and of the Council of 29 April 2004 on Persistent Organic Pollutants and Amending Directive 79/11/EEC.

[394] European Parliament and Council Directive 94/62/EC of 20 December 1994 on Packaging and Packaging Waste.

[395] Directive 2000/53/EC of the European Parliament and of the Council of 18 September 2000 on End-of-life Vehicles.

[396] Directive 2002/96/EC of the European Parliament and of the Council of 27 January 2003 on Waste Electrical and Electronic Equipment.

[397] Directive 2006/66/EC of the European Parliament and of the Council of 6 September 2006 on Batteries and Accumulators and Waste Batteries and Accumulators and Repealing Directive 91/157/EEC.

[398] Also belonging to this body of rules are Directive 2000/76/EC of the European Parliament and of the Council of 4 December 2000 on the Incineration of Waste and Council Directive 1999/31/EC of 26 April 1999 on the Landfill of Waste.

[399] Regulation (EC) No 1013/2006 of the European Parliament and of the Council of 14 June 2006 on Shipments of Waste.

[400] Levis, 'The EC's Internal Regime on Trade in Hazardous Wastes', 7 *RECIEL* (1998), at 286; Meßerschmidt, *Europäisches Umweltrecht* (2011), at 894.

environmental law.[401] It applies to hazardous and non-hazardous wastes as well and covers any involvement of EU Member States in transboundary waste shipments, to the extent there are exports, imports or transits of wastes, as well as transports solely among Member States. The specific obligations imposed by the Waste Shipment Regulation depend on several factors. First, a distinction is made as to whether the shipment is among Member States or concerns an export, import, or transit of wastes. Here again, within the respective set of rules, different obligations apply to wastes destined for disposal as opposed to wastes destined for recovery operations. With a view to waste exports to non-EU members (Articles 34 et seq.), a further distinction is made according to the respective State of destination and its accession to relevant international conventions. Finally, different rules apply to shipments of hazardous and non-hazardous wastes in accordance with the classification of wastes into either a green list (Annex III) or an amber list (Annex IV).

The particular content of these obligations cannot be examined in detail at this point.[402] However, as to the question of whether in the context of transboundary movements of hazardous wastes claims for damage compensation can be based on the customary principle of State responsibility—invoking a breach of the relevant EU legislation—the particular content of these obligations is of minor importance. This is due to the fact that non-compliance with existing EU legislation will not give rise to claims under international law. As far as the Waste Shipment Regulation contains rules relevant to the export of hazardous wastes from EU members to third States, the internal EU Directive does not constitute an international commitment of EU States vis-à-vis third States, and, accordingly, third States cannot invoke the infringement of a corresponding individual or "subjective" right. On the other hand, as far as concerns hazardous waste shipments exclusively among EU members, the EU Waste Shipment Regulation may basically be considered an arrangement within the ambit of Article 11 of the Basel Convention.[403] Since the Regulation also meets the regulatory standards required by Article 11, it takes, to this extent, precedence over the provisions of the Basel Convention.[404] Transboundary shipments of hazardous wastes within the EU area are thus exclusively governed by and judged against the EU Waste Shipment Regulation.

---

[401] EU Treaty, Article 191(2). These principles are laid down in Article 11(1)(a) and (g)(i) of Regulation (EC) No 1013/2006.

[402] See for a detailed discussion: Dieckmann, 'Die neue EG-Abfallverbringungsverordnung', 17 ZUR (2006), at 561 et seq.; Meßerschmidt, Europäisches Umweltrecht (2011), at 891 et seq.; Oexle, 'Rechtsfragen des neuen Verbringungsrechts', 18 ZUR (2007), at 460 et seq.

[403] See Kummer, The Basel Convention (1995), at 149–151; Rublack, 'Fighting Transboundary Waste Streams', 22 VRÜ (1989), at 372.

[404] Unlike its predecessors, the 2006 EU Waste Shipment Regulation no longer provides for less stringent rules on the shipment of hazardous wastes for recovery operations. See Dieckmann, 'Die neue EG-Abfallverbringungsverordnung', 17 ZUR (2006), at 564; Meßerschmidt, Europäisches Umweltrecht (2011), at 896–897, and to the previous Regulation, which was in contravention of the standards of the Basel Convention, Kummer, The Basel Convention (1995), at 155–157.

Its breach, however, does not entail the application of the principle of State responsibility. The application of this general rule of customary international law is overridden by the internal EU non-compliance rules that represent *leges speciales*.[405]

In summary, it can be concluded that a breach of the relevant EU legislation on the transboundary shipment of hazardous wastes under no circumstances constitutes an internationally wrongful act within the meaning of the principle of State responsibility.

(cc) The Law of the Sea Convention

The most important global legal instrument on the law of the sea is the 1994 United Nations Convention for the Law of the Sea (UNCLOS), which provides a comprehensive coverage of maritime issues and consolidates the main general principles and obligations in this legal field. It is, therefore, often described as "the general part" as regards the legal framework of the law of the sea.[406]

(1) Background and Basic Legal Features of the Convention

The UNCLOS entered into force on 16 November 1994, 12 years after its adoption in Montego Bay in Jamaica,[407] this following 9 years of negotiations undertaken during the Third UN Conference on the Law of the Seas between 1973 and 1982. For its States Parties this Convention replaced the four Geneva Treaties on the Law of the Sea[408] which had been signed in 1958.[409] As of today, 165 States and the European Union have become Contracting Parties to the UNCLOS.[410] Because of both its widespread acceptance as well as its universal coverage of maritime legal issues, the Convention has since its creation been commonly referred to as the "constitution of the oceans".[411]

---

[405] Relevant rules are laid down in Articles 259 *et seq.* of the EU Treaty.

[406] Lagoni, 'Die Abwehr von Gefahren für die marine Umwelt', 32 *BDGVR* (1992), at 94.

[407] An incentive for its adoption was finally given after major parts of Part XI of the Convention had been modified by the 1994 Agreement Relating to the Implementation of Part XI of the UNCLOS. See Anderson, 'Implementation of Part XI of the UNCLOS', 55 *ZaöRV* (1995), at 275 *et seq.*; Rattray, 'Implementation of Part XI of the UNCLOS', 55 *ZaöRV* (1995), at 298 *et seq.*

[408] 1958 Convention on the Territorial Sea and the Contiguous Zone 1958 Convention on the Continental Shelf, 1958 Convention on the High Seas and 1958 Convention on Fishing and Conversation of Living Resources of the High Seas.

[409] The development of the UNLCOS regime is described by Churchill/Lowe, *The Law of the Sea* (3rd ed., 1999), at 13–22.

[410] The only major shipping nation not Member to the UNCLOS is the USA.

[411] This description was used already by the President of the Third UN Conference on the Law of the Sea, *Tommy Koh*, in his remarks under the title 'A Constitution for the Oceans'. See also Beyerlin/Marauhn, *International Environmental Law* (2011), at 120.

Unlike the earlier Geneva Treaties on the Law of the Sea, the UNCLOS pursues a comprehensive and universal approach by restating the relevant rules as to almost all aspects of the law of the sea. To a large extent, it codifies pre-existing principles of customary international law and also incorporates the essential rules of relevant international treaty law. The UNCLOS, furthermore, reflects a change of sentiment with regard to the oceans representing a global resource that is in need of protection. Pollution is no longer seen as an implicit right covered by the freedom of the seas. The Convention rather acknowledges a fundamental obligation to control and prevent those impacts on the marine environment caused by human activities.[412] It moreover introduces a balance of powers system, juxtaposing the interests of exploitation and preservation.[413] To this end, it monopolises the economic utilisation of marine resources[414] by allocating this right to the respective coastal States.[415] Conversely, it guarantees to the benefit of flag States and the community as a whole the freedoms of navigation and of conducting activities not related to the exploitation of natural recourses. The UNCLOS thus implements both the *mare clausum* and the *mare liberum* concepts.[416] The realisation of the Convention's attempt to strike a reasonable balance between rational exploitation and conservation of the sea's living and non-living recourses was made possible only by means of a comprehensive package deal.[417] The Convention is on the one hand comprehensive in scope and protected against deviating party agreements; on the other it is limited to general obligations and to those mandates given to competent institutions to elaborate specific rules. Thus, the UNCLOS itself largely abstains from defining substantive rights and obligations, instead establishing a general regulatory framework that requires a subsequent specification of these general rules by substantive provisions elaborated by global or regional legal instruments in the respective fields.[418]

The UNCLOS consists of 17 chapters (Parts). Parts II to VII deal with rights and obligations of States in respect of certain marine areas, such as in the territorial sea, in the exclusive economic zone (EEZ), or on the high seas. Parts VIII to X are

---

[412] Birnie/Boyle/Redgwell, *International Law and the Environment* (3rd ed., 2009), at 383.

[413] Beyerlin/Marauhn, *International Environmental Law* (2011), at 118; Birnie/Boyle/Redgwell, *International Law and the Environment* (3rd ed., 2009), at 383; Graf Vitzthum, in: Graf Vitzthum (ed.), *Handbuch des Seerechts* (2006), at 41.

[414] Almost 90 % of the global fish stocks can be found in the 200 nautical miles zone of the EEZ; the same is true for non-living resources that are found on the continental shelf, see Graf Vitzthum, in: Graf Vitzthum (ed.), *Handbuch des Seerechts* (2006), at 41.

[415] The relevant rules on the EEZ can be found in Part V of the UNCLOS.

[416] Grotius, *The Free Sea* (transl. ed., 2004). Selden, *Of the Dominion, or, Ownership of the Sea* (transl. ed., 1972). As to the "war of books" see Graf Vitzthum, in: Graf Vitzthum (ed.), *Handbuch des Seerechts* (2006), at 32–35.

[417] Birnie/Boyle/Redgwell, *International Law and the Environment* (3rd ed., 2009), at 382; Schneider, 'Environmental Aspects of the UNCLOS', 20 *Colum. J. Transnat'l L.* (1981), at 244.

[418] Beyerlin/Marauhn, *International Environmental Law* (2011), at 119; Lagoni, 'Die Abwehr von Gefahren für die marine Umwelt', 32 *BDGVR* (1992), at 94–95.

concerned with particular geographic circumstances. The regime of the deep seabed, the "Area", is contained in Part XI. The seabed is declared to be part of the common heritage of mankind,[419] and its exploitation is made subject to the supervision of the newly created International Seabed Authority (ISA).[420] The protection and preservation of the marine environment is dealt with in Part XII of the UNCLOS. Part XII starts with some general obligations of States to protect and preserve the marine environment and this applicable to any kind of marine-related activities (Articles 192–206). These general obligations are followed by more specialised rules dealing with particular sources of pollution (Articles 207–212). These rules largely do not contain detailed substantive obligations, but refer to existing or future legal instruments that are supplemented with expanded rules of enforcement (Articles 213–233).

(2) Obligations Imposed on States

The following sections shall outline the particular obligations of States under the UNLCOS which are of relevance in the context of transboundary movements of hazardous wastes.

(i) The Obligation to Protect and Preserve the Marine Environment, Articles 192, 194(1)

Article 192 of the UNCLOS states that "[s]tates have the obligation to protect and preserve the marine environment". This provision embodies the core principle of the protection of the marine environment and to this extent restates customary international law.[421] However, the content of this principle as such is not sufficiently prescribed in order to be applied directly. It rather announces the programmatic goal of the entire set of rules and therefore has to be defined in more detail by further provisions.[422]

The most important concretisation of this general principle is made by Article 194(1) of the UNCLOS. It obliges States to "take [...] all measures [...] that are necessary to prevent, reduce and control pollution of the maritime environment from any source". It becomes apparent that this obligation protects the international community as a whole and, therefore, is to be considered an obligation *erga*

---

[419] UNCLOS, Article 136 reads: "The Area and its recourses are the common heritage of mankind". The "Area" is defined in Article 1 as "the seabed and ocean floor and subsoil thereof, beyond the limits of national jurisdiction".

[420] The ISA is established by Article 156 of the UNCLOS, see on its history Wood, 'International Seabed Authority', 3 *Max Planck YBUNL* (1999), at 175–179.

[421] UN Doc. A/44/461, at para. 27; Birnie/Boyle/Redgwell, *International Law and the Environment* (3rd ed., 2009), at 387; As regards the customary principle see *infra*, Sect. "The Obligation to Protect and Preserve the Marine Environment".

[422] Proelß, *Meeresschutz im Völker- und Europarecht* (2004), at 77; Hafner, in: Graf Vitzthum (ed.), *Handbuch des Seerechts* (2006), at 366–367.

omnes.[423] It covers the entire maritime space including the high seas and does not depend on the violation of the national territory or any other individual or "subjective" right of another State.[424] On the other hand, this obligation is restricted in respect of the applicable standard of conduct. It requires the States to use "the best practicable means at their disposal and in accordance with their capabilities". The applicable standard of conduct, therefore, is not an absolute one, but a standard of due diligence, which is further modified by the said restriction, according to which the specific circumstances of the case and the particularities of the State concerned are to be taken into consideration. With regard to the legal character of this obligation, the wording "to take all necessary measures to" is, as an initial matter, indicative of an obligation of result because this wording only reproduces the very content of an obligation of result. However, in this case one has to take into account that this obligation does not prescribe a particular final result to be achieved either. It rather stipulates certain criteria of conduct consisting in the prevention, reduction and control of pollution. The central purpose of this obligation, thus, is to require States to take measures towards the achievement of these criteria of conduct. It therefore represents an obligation of conduct.

A breach of this obligation, however, will be difficult to establish. Although Articles 192, 194(1) stipulate a cohesive obligation that is directly applicable to States, it nevertheless remains a mere general obligation without any precise contours.

It is subject to the discretion of States which measures they consider necessary and appropriate to prevent, reduce and control pollution. Since, furthermore, the circumstances of the case need to be taken into account, there are two conditions that, in the end, must be fulfilled in order establish a breach of this obligation: First, it must be established that—in the particular context at hand—the State is required to undertake any measures in the first instance. Only in a second step can it be established whether the particular measures taken by the State are, from an *ex ante* point of view and under consideration of the given circumstances, sufficient to meet the substantive requirements of the obligation. With regard to the transboundary movement of hazardous wastes by sea, a breach of this obligation is conceivable only where the coastal State allows for the discharge of imported hazardous wastes into the marine and coastal areas, knowing that this will cause substantial damage to the marine or coastal area and being aware that this activity does not represent a use of the best practicable means at its disposal and is not in accordance with its capabilities. By contrast, any unintended discharge of hazardous wastes into the sea will not constitute an internationally wrongful act of the coastal or flag State. Finally, it must be considered that in respect of various

---

[423] Hafner, in: Graf Vitzthum (ed.), *Handbuch des Seerechts* (2006), at 367.

[424] Birnie/Boyle/Redgwell, *International Law and the Environment* (3rd ed., 2009), at 387; Lagoni, 'Die Abwehr von Gefahren für die marine Umwelt', 32 *BDGVR* (1992), at 121–122; Nordquist/Rosenne/Yankov/Grandy (ed.), *UNCLOS—Commentary Vol. IV* (1991), at 40.

marine activities, specific international agreements or conventions apply and take precedence over the general obligation of Articles 192, 194(1) of the UNCLOS.

In respect of the enforcement of this obligation, it should be noted that this obligation represents an obligation *erga omnes*, the performance of which is owed to the international community as a whole. A breach of this obligation, thus, may be invoked by any State, irrespective of whether it has actually been injured or even affected by the polluting activity.[425] Since this obligation also represents customary international law, those States entitled to invoke a breach of this obligation are not limited to the States Parties of the UNCLOS.[426] By invoking a breach of an obligation *erga omnes* the State is not acting in its individual capacity as an injured State, but acts on behalf of the international community as a whole.[427] In this capacity it may demand the cessation of this act and a corresponding guarantee of non-repetition.[428] Moreover, it is entitled to claim the reparation and restitution to the benefit of the international community as a whole.[429]

Besides the obligation to pursue necessary measures according to Article 194(1), there are also further concretisations of the general principle to protect and preserve the marine environment contained in Article 92 of the UNCLOS.[430] Thus, according to Article 197, States "shall cooperate [...] in formulating and elaborating international rules, standards and recommended practices and procedures [...] for the protection and preservation of the marine environment". Such envisaged agreements and conventions may provide for more stringent rules imposed on States.[431] By means of this provision the character of the UNCLOS as a framework convention becomes obvious. Part XII of the UNCLOS is designed to be complemented by further global and regional conventions and agreements that are concerned with particular sources of pollution or that apply to particular regions taking account of the respective regional peculiarities.[432] As far as those

---

[425] See ILC Draft Articles on State Responsibility, Article 48. See also *Barcelona TractionCase*, ICJ Reports 1970, at 32, para. 33.

[426] Lagoni, 'Die Abwehr von Gefahren für die marine Umwelt', 32 *BDGVR* (1992), at 123.

[427] Commentaries to the ILC Draft Articles on State Responsibility, Article 48, para. 1; Beyerlin/Marauhn, *International Environmental Law* (2011), at 363; Graf Vitzthum, in: Graf Vitzthum (ed.), *Handbuch des Seerechts* (2006), at 144.

[428] ILC Draft Articles on State Responsibility, Article 48(2)(a), see also *The S.S. Wimbledon Case*, PCIJ Series A No. 1 (1923), at 30; *South West Africa Cases, Preliminary Objections*, ICJ Reports 1962, at 322 *et seq.*

[429] ILC Draft Articles on State Responsibility, Article 48(2)(b). This latter obligation, however, has yet not been applied by international courts and thus cannot be considered part of customary international law. Commentary to the ILC Draft Articles on State Responsibility, Article 48, para. 9.

[430] The relation between Article 192 and the related specific obligations is described by Lagoni, 'Die Abwehr von Gefahren für die marine Umwelt', 32 *BDGVR* (1992), at 121–135.

[431] UNCLOS, Article 237.

[432] Such as the MARPOL 73/78 Convention, the London Dumping Convention or the conventions elaborated within the context of the UNEP Regional Seas Programme.

specific conventions provide for more distinct rules in the respective regulatory field, they will override the application of the general obligation to protect and preserve the marine environment as imposed by Article 192 of the UNCLOS. The UNCLOS, however, remains applicable to the extent that it sets the basic standard and the minimum degree of environmental protection and, in addition, provides for a general framework of procedural and enforcement rules.

The general principle in Article 192 of the UNCLOS, finally, is further defined by the requirement to co-operate in various regards.[433] The States are under the obligation to elaborate international rules and adopt national laws and regulations to prevent, reduce and control pollution of the marine environment in respect of certain sources of pollution,[434] as well as to enforce these laws and regulations.[435] However, these rules are of minor direct significance for the establishment of an internationally wrongful act in the context of transboundary movements of hazardous wastes by sea.

(ii) The Obligation to Prevent Transboundary Pollution to the Marine Environment, Article 194(2)

According to Article 194(2) of the UNCLOS, the States "shall take all measures necessary to ensure that activities under their jurisdiction or control are so conducted as not to cause damage by pollution to other States and their environment, and that pollution [...] does not spread beyond the areas where they exercise sovereign rights". This obligation, at first sight, strongly resembles the obligation to protect and preserve the marine environment as outlined in the previous section. However, whereas Articles 192, 194(1) of the UNLCOS concern the protection and preservation of the marine environment as such, the central issue of this obligation is the transboundary effect of pollution and thus the territorial impairment of the affected State or the international community as a whole. In order to examine the precise content of this obligation a distinction should first be made between the initial and latter part of this obligation.

As far as its first part is concerned, this obligation originates from the customary principle of territorial integrity that protects a State from being exposed to interferences perpetrated by other States.[436] The content of this obligation is much more specific and sophisticated than the content of the obligation to protect and preserve the marine environment as outlined above. In geographical terms it requires a transboundary infringement. This means that, first, there must be an activity that is conducted under the jurisdiction or control of a State. This includes any activities in the territorial sea of a State, but also in the EEZ or on the

---

[433] See e.g. UNCLOS, Articles 198, 199, 200, 201, 202, 226, and 235.
[434] *Ibid.*, Articles 207–212.
[435] *Ibid.*, Articles 213–222.
[436] *Infra*, Sect. "The Origin of this Obligation".

continental shelf.[437] Because of the inclusion of the term "control", this also comprises activities of private parties having a foreign nationality that are conducted in an area where the coastal State is entitled to exercise certain control rights (e.g. vessels flying the flag of a foreign country that are exercising either its right of innocent passage in the territorial sea or its freedom of navigation in the coastal State's EEZ).[438] Furthermore, this also covers activities on the high seas or in the EEZ of another State that are conducted under the control of the flag State.[439] The second requirement is that this activity must have caused damage to another State and its environment. The interests of the affected State protected by this obligation are comprehensive and correspond to the rights of the coastal State in the respective marine areas. This includes, in the territorial sea, all marine-related activities and general interests in the coastal environment, e.g. tourism, public health or a sustainable conservation of marine resources. In the EEZ and on the continental shelf this includes all sovereign rights that are exercised by the coastal State in the respective area.[440] Finally, this also involves individual or "subjective" rights and legally protected interests of any States in common areas beyond any national jurisdiction, such as on the high seas. This obligation is also in another regard much stricter than the obligation to protect and preserve the marine environment as laid down in Articles 192, 194(1) of the UNCLOS, namely the applicable standard of conduct is ot restricted such that the States have only to use the best practical means at their disposal and in accordance with their capabilities. Thus, the standard of due diligence to be performed by the State with regard to this obligation is interpreted more stringently than in respect of the obligation to protect and preserve the marine environment and also needs to be determined consistently on the international level. As regards the character of this obligation, the wording "shall take all measures necessary to ensure" indicates an obligation of result since this wording restates the very content of an obligation of result. The particular result that needs to be achieved by the States is that the activities of private parties under their jurisdiction are conducted in a way that ultimately this will not result in an infringement of the territorial rights of another State. The way in which this result is achieved by the States is, conversely, left to the discretion of the States.

With regard to the transboundary movement of hazardous wastes by sea, this obligation may be of particular significance. Shipping must be considered an "activity" within the meaning of Article 194(2) of the UNCLOS. Therefore, this obligation will basically be applicable to the State of export if hazardous wastes are shipped by a State-owned company or another company under the jurisdiction of this State. It will also apply to the flag State of a vessel carrying hazardous wastes or to the coastal State if hazardous wastes are shipped through or from its

---

[437] As to the latter, see UNLOS, Article 76.
[438] Lagoni, 'Die Abwehr von Gefahren für die marine Umwelt', 32 *BDGVR* (1992), at 137.
[439] This, however, applies only as far as the "control" of the flag State reaches.
[440] Lagoni, 'Die Abwehr von Gefahren für die marine Umwelt', 32 *BDGVR* (1992), at 137–138.

territorial sea or its EEZ.[441] This obligation applies if in such constellations the hazardous wastes are later discharged into the coastal area of another State with the consequence of adverse effects on the coastal environment of that State. It also applies if such wastes are discharged or dumped within the EEZ of another State, adversely affecting, for example, the economic interest of that State in exploiting fish stocks. From all this it follows that this obligation basically covers a broad scope of activities related to the transboundary movement of hazardous wastes by sea. However, in all these constellations it needs to be considered that the State is responsible only within the limits of due diligence, taking account of what activity the State could have reasonably been aware of as well as its ability to control and prevent the activity conducted by private persons. The State is basically obligated to take the necessary measures to ensure that such infringement is prevented, which means that the State is required to enact and implement sufficient legal instruments and enforcement measures[442] as well as undertake an effective regime of administrative supervision so as to ensure compliance. If there is no obvious indication for the source State that a particular activity conducted by private parties will produce an actual threat of damage by pollution to another State, the source State does not have any legal power to restrict, in particular, the freedom of navigation of flag States by taking sovereign measures. Thus, as long as there is no obvious violation of laws or no actual threat of damage to another State's territory, the coastal State is not allowed to restrict the shipment of hazardous wastes on the basis of Article 194(2) of the UNCLOS.

The second part of this obligation is concerned with pollution of the high seas caused by activities conducted in the EEZ or on the continental shelf of a coastal State. It is doubtful whether this obligation to prevent pollution of the high seas is already recognised as customary international law.[443] It is more likely that today this obligation must be considered an extension of the customary law that is established within the legal framework of the UNCLOS.[444] As regards it character, this obligation is an obligation *erga omnes* that is owed to the international community as a whole; but at the same time, since it is not acknowledged as customary international law, it is valid and applicable only in relation among the Parties of the UNCLOS. In consequence, any State that is Party to the UNCLOS is entitled to invoke a breach of this obligation by another Party to the Convention, irrespective of whether it has actually suffered a violation of its subjective rights.[445] As far as transboundary movements of hazardous wastes by sea are concerned, the same constellations as outlined with regard to the first part of the obligation apply with regard to shipments

---

[441] See Article 56(1)(b)(iii) of the UNCLOS.

[442] The regulatory power to implement such legal instruments arises from Articles 207–222 of the UNCLOS.

[443] To this issue see also *infra*, Sect. "Geographical Scope".

[444] Lagoni, 'Die Abwehr von Gefahren für die marine Umwelt', 32 *BDGVR* (1992), at 138; see also von Gadow-Stephani, *Der Zugang zu Nothäfen* (2006), at 75.

[445] ILC Draft Articles on State Responsibility, Article 48(1)(a).

of hazardous wastes causing pollution to the marine environment of the high seas. The particular significance of this second part of Article 194(2) lies in the fact that any pollution of the high seas is covered, irrespective of whether individual or "subjective" rights or interests of individual States are affected.

In summary, it can be concluded that both parts of this obligation form a cohesive whole that basically covers any transportation of hazardous wastes both across the high seas and through the territorial sea and the EEZ of another State. A breach of this obligation may be invoked by any State; albeit in respect of shipments crossing the high seas, only by Parties to the UNCLOS. If a State knew or ought to have known about a shipment of hazardous wastes taking place within its jurisdiction or under its control and if the State also knew or ought to have known that this would likely cause an imminent threat of damage by pollution to another State or the high seas, a claim may be lodged against the responsible State for the cessation of the wrongful act as well as for reparation and compensation. It becomes apparent that the main difficulty in establishing the international responsibility of States due to a breach of Article 194(2) of the UNCLOS is the requirement of establishing that the State failed to act with the required degree of diligence. It should furthermore be noted that Article 194(2) of the UNCLOS will be of direct relevance for the establishment of an internationally wrongful act only where the particular source of pollution is not covered by a more specific regulation.

(iii) Further Obligations

The UNCLOS contains further provisions apart from Part XII that may become relevant in the context of transboundary movements of hazardous wastes by sea. However, these provisions do not impose obligations on States that are relevant for the establishment of an internationally wrongful act of the State with regard to damage caused by wastes shipments.

According to Article 19(2)(g) and (h) of the UNCLOS, a passage of a foreign ship through the territorial sea of a coastal State is considered to be non-innocent if the ship unloads any commodity contrary to the laws and regulations of the coastal State, or if on the occasion of the passage a serious pollution is caused. In such cases the coastal State is, according to Article 25 of the UNCLOS, entitled to take the necessary steps to prevent a non-innocent passage. It becomes clear that these provisions establish a defensive right of the coastal State rather than an obligation of the flag State or coastal State to prevent such non-innocent passage.[446]

According to Article 139(1) of the UNCLOS, States are under the responsibility to ensure that activities conducted in the Area, irrespective of whether they are carried out by States or by private parties, comply with the requirements set out in Part XI of the Convention. "Activities in the Area" are defined in Article 1(1)(3) as "all activities or exploration for, and exploitation of, the resources of the Area". Resources are defined by Article 133(a) as "all solid, liquid or gaseous mineral

---

[446] See Graf Vitzthum, in: Graf Vitzthum (ed.), *Handbuch des Seerechts* (2006), at 124–125.

resources *in situ*". Based on this definition of deep-sea resources, it is hardly possible to consider the capability of depositing hazardous wastes on the deep-sea floor to be "resources of the Area" within the meaning of Part XI of the UNCLOS.

(3) Summary

The UNCLOS represents a framework convention that provides for a comprehensive set of rules to protect and preserve the marine environment. However, these rules, most of which are contained in Part XII, are of general character and can be applied directly to a particular transport of hazardous wastes only on the condition that no other convention provides for more specialised and distinct rules. In case the provisions of the UNCLOS apply, a breach of the obligation to take the necessary measures to protect and preserve the marine environment according to Articles 192, 194(1) is, from a practical perspective, unlikely to be established. This is mainly due to the standard of due diligence applicable to the States. When determining the particular measures that are required to be undertaken in a certain situation, the capabilities of the respective State and the discretion left to the State to determine the best practical means need to be taken into account. By contrast, a breach of the obligation to prevent transboundary pollution to the marine environment of another State's territory or on the high seas as laid down in Article 194(2) is, as a practical matter, more readily established. In this case the claimant State must prove, first, that an incident or an activity carried out in an area under the control of the respondent State has caused the transboundary pollution of either the territory of the claimant State or of the high seas. And, second, it needs to bring evidence that the respondent State in fact knew of the circumstances of the case as well as the threat of causing pollution with the result that it did not satisfy the due diligence standard.

(dd) Further Conventions Relevant to the Protection of the Marine Environment

(1) MARPOL 73/78 Convention

The 1973/1978 MARPOL Convention for the Prevention of Pollution from Ships (MARPOL 73/78)[447] provides for a set of distinct rules regulating on a global level the pollution of the marine environment caused by the operation of vessels and, therefore, lies within the ambit of Articles 197, 211(1) and 237(1) of the UNCLOS. MARPOL 73/78 succeeds the 1954 OILPOL Convention[448] and extends the protection of the marine environment to various sources of pollution in the context

---

[447] MARPOL Convention for the Prevention of Pollution from Ships of 2 November 1973 as amended by the 1978 Protocol of 17 February 1978.

[448] 1954 London International Convention for the Prevention of Pollution of the Sea by Oil (OILPOL).

of the operation of a vessel. The Convention was adopted in 1973 and, after an amendment of the initial text by the 1978 Protocol, it entered into force on 2 October 1983. Today, the Convention has 152 Contracting States.[449] It applies to vessels flying the flag of a Contracting State or operating under the authority of such State.[450] It is conceived as a framework convention that itself only provides a set of general provisions regulating the issues of control and enforcement of the substantive rules contained in the Annexes to the Convention. Annexes I and II are binding on all Contracting States; the further Annexes III to V are optional and binding only after a separate act of ratification. Today, all Annexes, including an additional Annex VI, which was adopted in 1997,[451] are in force and have found widespread acceptance.[452] For the transboundary shipment of hazardous wastes, Annexes II and III need to be considered.

Annex II of MARPOL 73/78 contains regulations for the control of pollution caused by noxious liquid substances (NLS) carried in bulk. It stipulates requirements for the construction and equipment of chemical and NLS tankers and lays out a set of strict and detailed conditions which prescribe when the discharge into the sea of cargo residues, ballast water, tank washings and other mixtures containing noxious liquid substances is permitted.[453] This Annex, furthermore, requires a specific prewashing procedure of tanks from which noxious substances have been unloaded, before the ship leaves the port of unloading in order to conduct a tank washing on the sea.[454] In addition to the general obligation to transpose these rules into national law, flag States and port States are under the obligation (i) to ensure that their ports and terminals are equipped with adequate reception facilities,[455] (ii) to conduct surveys of vessels carrying noxious liquid substances in bulk on a regular basis and to issue and maintain an "international oil pollution prevention certificate",[456] and (iii) to execute control over discharge procedures.[457] As long as a vessel does not comply with the standards set out in this annex and represents a threat of harm to the marine environment, or if there

---

[449] As of 28 February 2014 they represent 99.20 % of the gross tonnage of the world's merchant fleet. A list of the Contracting States can be retrieved from www.imo.org.
[450] MARPOL 73/78, Article 3.
[451] Protocol of 1997 to amend the 73/78 MARPOL Convention.
[452] Annex III today has 138 Parties representing 97.59 % of the gross tonnage of the world's merchant fleet; Annex IV has 131 Parties, representing 89.65 %; Annex V has 144 Parties representing 98.47 %; and Annex VI has 75 Parties representing 94.77 %.
[453] According to MARPOL 73/78, Annex II, Regulation 13, when discharging such substances into the sea the vessel must proceed at a speed of at least 7 knots, the water depth must be at least 25 m and the distance to the nearest land must be not less than 25 nautical miles.
[454] *Ibid.*, a tank from which noxious substances have been unloaded must be prewashed in the port of unloading until a certain concentration has been achieved and the residues must be discharged into a reception facility.
[455] *Ibid.*, Annex II, Regulation 18.
[456] *Ibid.*, Annex II, Regulations 7–10.
[457] *Ibid.*, Annex II, Regulations 13–16, in connection with Article 5.

are clear grounds for believing that the master or crew is not familiar with the essential shipboard procedures, the State has to ensure that the vessel does not continue her voyage.[458]

Annex III of MARPOL 73/78 is concerned with the prevention of pollution by harmful substances carried by sea in packaged form.[459] It basically prohibits the carriage of harmful substances, except in accordance with the provisions set out in this Annex.[460] Such provisions concern particular standards to be complied with regarding the packing, marking, labelling, documentation and stowage of harmful substances as well as certain quality restrictions. A further obligation of the Contracting States in addition to the transposition of these rules into national law is to conduct surveys and controls and ensure that the vessel does not sail if there are clear grounds for believing that the master or crew is not familiar with the essential shipboard procedures.[461]

Annexes II and III of the MARPOL 73/78 Convention provide for regulations that impact upon certain aspects of transboundary hazardous waste movements. Hazardous wastes are shipped both in liquid and solid form, but only seldom in gaseous form. Where hazardous wastes are collected prior to their transport to be shipped by sea they are usually filled in IBCs or barrels as far as liquid, fluid, pourable or granulated materials are concerned. Gaseous substances are mostly filled in MEGCs. Solid hazardous wastes are mostly stuffed in big bags, crates, flat racks and all forms of containers, but they may also be shipped in bulk depending on the amount and the hazard level of the wastes. Annex III of MARPOL 73/78 applies in respect of all hazardous wastes that are shipped in packaged form, obligating the port State to conduct controls and surveys. If the port State comes to know of any concrete reasons for believing that the master or crew is not familiar with the particular requirements but nonetheless allows the vessel to leave the port, this must be deemed a breach of the obligations arising under Annex III of MARPOL 73/78 and thus an internationally wrongful act. However, hazardous wastes also are generated during the sea transport or in connection with the transportation. This comprises in particular residues of liquid bulk cargo collected in the vessel's slop tanks. Even though the liquid cargo as such may not come under Annex II of MARPOL 73/78, the mixed residues of various shipments collected in the slop tanks predominantly must be considered noxious liquid substances in bulk within the meaning of Annex II.[462] The port States thus are under an obligation to take delivery of such substances in appropriate discharge facilities and to not allow the vessel to leave the port if the vessel has not

---

[458] *Ibid.*, Annex II, Regulations 8, 16.

[459] "Packaged form" means any form of containment specified for harmful substances in the IMDG Code, such as IBCs, MEGCs, closed bulk containers or large packaging.

[460] MARPOL 73/78, Annex III, Regulation 1.

[461] *Ibid.*, Annex III, Regulation 8.

[462] In respect of such hazardous wastes which derive from the normal operations of a ship, the Basel Convention is not applicable on account of such wastes being excluded from its scope of application; see Basel Convention, Article 1(4).

conducted a required prewashing procedure or if there are any clear grounds for believing that the master or crew is not familiar with the respective shipboard procedures. Any failure to perform these obligations must be considered an internationally wrongful act. The same basically applies if the flag State fails to conduct inspections of the vessel on a regular basis and issue a corresponding certificate. It is only in obvious constellation, however, that establishing the breach of any of such obligations by the State might prove possible.

Against this background, it becomes apparent that the practical relevance of the MARPOL 73/78 for establishing the responsibility of States is limited.[463] The obligations established by MARPOL 73/78 cannot be considered obligations *erga omnes*. Therefore, it is not the case that any State may invoke the wrongful act of another State; rather, only those States may claim for compensation that have actually incurred damage caused by the wrongful act. Since, however, in most cases it is the coastal or port State itself that—on account of its own (in)actions—suffers damage from an improper discharge of cargo residues or an insufficient prewashing, compensation from another State may not be claimed under the principle of State responsibility. The same applies to the improper discharge of cargo residues or ballast water or to the wrongful conduct of tank washings on the high seas, where there is no State that is affected in respect of its individual or "subjective" rights. The only constellation conceivable in which a State may be entitled to claim compensation from another State under the principle of State responsibility is that such activities are conducted in the EEZ of another State and adversely affect that coastal State's economic interests in the form of actual damage.

(2) OPRC Convention

An additional international instrument for the prevention and combating of pollution of the marine environment is the 1990 OPRC Convention,[464] requiring its contracting States to take effective measures to prepare for and to respond to oil pollution incidents. On the part of the carriers such measures include obligations to carry a shipboard oil pollution emergency plan and to report any incident to the authorities of the coastal States; the latter are obliged to keep oil spill combating equipment available and to elaborate emergency plans for combating oil spills.[465] Moreover, in 2000, a HNS Protocol to the OPRC Convention was adopted

---

[463] For a general criticism of MARPOL 73/78 due to its lack of including non-compliance procedures as well as the reluctance of Parties to either fulfil their reporting obligations or establish sufficient port discharge facilities see Birnie/Boyle/Redgwell, *International Law and the Environment* (3rd ed., 2009), at 409–413.

[464] 1990 International Convention on Oil Pollution Preparedness, Response and Cooperation (OPRC Convention). The Convention entered into force on 13 May 1995.

[465] See for a detailed discussion: Geisler, *Das IPRC Übereinkommen* (2004).

extending the preparedness and response regime also to HNS incidents.[466] Since neither instrument excludes wastes from its scope, they basically also apply to hazardous wastes shipped by sea. The nature of the obligations established by the OPRC Convention closely resembles the nature of the obligations imposed by the MARPOL 73/78 Convention. Therefore, it is essentially the case that a failure to comply with these obligations could lead to the international responsibility of that State. This may, for example, apply where damage to another State's territory has been exacerbated due to an obvious failure to keep sufficient pollution combating equipment at hand.

(3) London Dumping Convention

The deliberate disposal into the sea of wastes and other matter as well as the incineration of wastes and other matter at sea is at a global level regulated by the 1972 London Convention[467] and the 1996 Protocol to the London Convention.[468] The London Convention and its Protocol modify the general provisions of the UNCLOS and, thus, come under the ambit of Articles 197, 210, 237(1) of the UNCLOS.[469] According to the Protocol, Contracting States are required to prohibit the dumping of any wastes or other matter, except for those wastes listed in Annex I that may only be dumped subject to the conditions laid down in Annex II.[470] Under these rules, wastes considered hazardous are in most cases not eligible for being dumped. Furthermore, States are under the obligation to prohibit any incineration of wastes or other matter at sea and also to not allow the export of such matter to other countries for dumping or incineration at sea, regardless of whether they are Parties to the Convention or not.[471]

Since a deliberate action is required, the London Dumping Regime will not be applicable in cases where the intent is to ship hazardous wastes by sea to another country for onshore disposal. In such cases, depending on the subjective intention of the cargo owner, the London Dumping Convention and the Basel Convention are mutually exclusive. To this extent the provisions of the London Dumping

---

[466] 2000 Protocol on Preparedness, Response and Co-operation to Pollution Incidents by Hazardous and Noxious Substances (OPRC-HNS Protocol). The Protocol entered into force on 14 June 2007.

[467] 1972 London International Convention on the Prevention of Marine Pollution by Dumping of Wastes and Other Matter (London Dumping Convention). The Convention entered into force on 30 August 1975 and today has 87 States Parties.

[468] 1996 Protocol to the London Dumping Convention. It entered into force on 24 March 2006 and today has 44 States Parties.

[469] As to the London Convention regime in general see Beyerlin/Marauhn, *International Environmental Law* (2011), at 128–129; Churchill/Lowe, *The Law of the Sea* (3rd ed., 1999), at 363–370; Coenen, 'The 1996 Protocol to the London Convention', 6 *RECIEL* (1997), at 54 *et seq.*; Rothwell/Stephens, *The International Law of the Sea* (2010), at 373–378; Hafner, in: Graf Vitzthum (ed.), *Handbuch des Seerechts* (2006), at 389–391.

[470] Protocol to the London Convention, Article 4(1).

[471] *Ibid.*, Articles 5 and 6.

Convention and its Protocol do not come under the scope of the present work and, therefore, will not be examined in more detail. The London Convention regime, however, remains relevant for the present consideration as far as it prohibits the export of hazardous wastes to other countries with the intention to subsequently dump or incinerate these wastes at sea. It is in such a constellation that the legal regimes of both the London Convention and the Basel Convention apply concurrently and provide for applicable rules to be observed by the States.[472] In the context of transboundary shipments of hazardous wastes, the Protocol to the London Dumping Convention thus can be relevant for the establishment of an internationally wrongful act entailing the responsibility of that State if the State of loading, in contravention with the Protocol, permits the shipment of hazardous wastes, although it knows or ought to know that the intention of the shipper of the wastes is to dump or incinerate them at sea after the shipment. In such constellation, a breach of the relevant PIC and ESM requirements of the Basel Convention is also likely to be established.

(4) The UNEP Regional Seas Programme

In 1974, in the aftermath of the United Nations Conference on the Human Environment held in 1972 in Stockholm, UNEP launched the Regional Seas Programme in order to promote the creation of regional action plans and the elaboration of conventions for the protection of marine and coastal environments, taking account of the particularities and special requirements of the respective regional seas.[473]

(i) Regional Seas Conventions

Within the UNEP Regional Seas Programme, 13 individual Programmes have been established under the auspices of UNEP. Six of these are also administered by UNEP, whereas the other seven are under the administration of other organisations. Five further Programmes have been set up independently and not under the aegis of UNEP. 14 of the 18 total Programmes have adopted conventions aiming at the protection of the marine environment at a regional level.[474]

These Regional Seas Conventions pursue a comprehensive regulatory approach addressing all sources of pollution; at the same time, they are limited in their geographical application to specific regions. Since they provide for regulations

---

[472] This, of course, only applies in case the respective State is a Party to both Conventions. For a very detailed analysis of the areas of conflict and overlap between the Basel Convention and the 1972 London Dumping Convention see Kummer, *The Basel Convention* (1995), at 184–189.

[473] See on the Regional Seas Programme in general: Akiwumi/Melvasalo, 'UNEP's Regional Seas Programme', 22 *Mar. Pol'y* (1998), at 229 *et seq.*; Rothwell/Stephens, *The International Law of the Sea* (2010), at 344–347; Hafner, in: Graf Vitzthum (ed.), *Handbuch des Seerechts* (2006), at 417–421.

[474] A list of the 14 Regional Seas Conventions can be found *supra*, section "Other Conventions Relevant to the Protection of the Marine Environment".

adapted for the particular circumstances and needs of specific regional seas, they represent, to this extent, *lex specialis* in relation to the UNCLOS. This is acknowledged and even provided for by the UNCLOS itself. The Regional Seas Conventions come under the ambit of Articles 197, 237 and 207 to 212 of the UNCLOS and, therefore, take precedence over the provisions of the UNCLOS so long as they meet the regulatory standards established by the UNCLOS.

As regards their structure and regulatory content, the Regional Seas Conventions resemble the general structure of Part XII of the UNCLOS. They pursue the "framework convention" approach, which means that a general umbrella convention lays down a set of basic rules, including the general obligation of States to protect and preserve the marine environment, and, a requirement to implement the respective regional Action Plans; the umbrella convention additionally provides a set of specific rules addressing particular sources of marine pollution. Due to the "framework convention" approach, these conventions largely contain programmatic commitments rather than directly applicable rules, and they are designed to be complemented by separate Protocols or Annexes addressing in detail the specific sources of pollution.[475]

In each case using more or less the same words, the Regional Seas Conventions first restate the general obligation of States to take all appropriate measures to prevent, abate, combat and control pollution of the respective regional sea and to ensure the sound environmental management of natural resources.[476] This obligation corresponds to the general obligation to protect and preserve the marine environment as laid down in Articles 192, 194(1) of the UNCLOS. Hence, the same conclusions that were drawn in respect of those provisions apply here.[477] Since this general obligation is specified further by the subsequent provisions regarding the particular sources of pollution, including the respective Protocols and Annexes to the Conventions, the application of this general obligation will in most cases be rejected in favour of the more specific provisions. What is remarkable is that only in few of the Regional Seas Convention there is an equivalent to the obligation to prevent transboundary marine pollution as laid down in Article 194(2) of the UNCLOS.[478] In most cases, the general obligation is

---

[475] A deviating structure is provided only by the 1991 Madrid Protocol to the Antarctic Treaty, which owes to the specific need for protection of the Antarctic area.

[476] Barcelona Convention, Article 4(1); Kuwait Convention, Article III(a); Abidjan Convention, Article 4(1); Lima Convention, Article 3(1); Jeddah Convention, Article III(1); Cartagena Convention, Article 4(1); Nairobi Convention, Article 4(1); Noumea Convention, Article 5(1); Bucharest Convention, Article V(2); Helsinki Convention, Article 3(1); OSPAR Convention, Article 2(1); Antigua Convention, Article 5(1); Tehran Convention, Article 4. A different wording is provided only by the Madrid Protocol, Article 3.

[477] See *supra*, section "The Obligation to Protect and Preserve the Marine Environment, Articles 192, 194(1)".

[478] Moreover, these obligations are often of rather general nature. See: Lima Convention, Article 3(5); Cartagena Convention, Article 4(2); Nairobi Convention, Article 4(5); Noumea Convention, Article 5(2); Helsinki Convention, Article 3(6); OSPAR Convention, Article 2(4); Antigua Convention, Article 5(5).

followed by a more or less comprehensive set of provisions addressing the particular sources of pollution. Such provisions contain the obligation of States to take appropriate measures to prevent pollution caused by land-based sources,[479] ships,[480] dumping and incineration,[481] transportation of hazardous substances,[482] exploitation of the deep seabed[483] and other activities.[484] These obligations, however, do not contain any substantive standards or procedural rules which must be complied with; and only in some cases do they stipulate general principles that need to be taken into account in the elaboration of associated Protocols and Annexes.[485] Protocols relevant to the transboundary movement of hazardous wastes have been elaborated in the context of the Barcelona Convention and the Kuwait Convention.

(ii) The 1996 Izmir Protocol to the Barcelona Convention

With regard to the Mediterranean region, in 1996 the Parties to the Barcelona Convention adopted the Izmir Protocol on the transboundary movement of hazardous wastes.[486] As far as its content is concerned, it resembles to a large extent the general structure and provisions of the Basel and Bamako Conventions, but in some respect contains stricter rules.[487] The Protocol has been elaborated under the

---

[479] Barcelona: Art. 8; Kuwait: Art. VI; Abidjan: Art. 7; Lima: Art. 4; Jeddah: Art. VI; Cartagena: Art. 7; Nairobi: Art. 7; Noumea: Art. 7; Bucharest: Art. VII; HELCOM: Art. 6; OSPAR: Art. 3; Antigua: Art. 6; Tehran: Art. 7.

[480] Barcelona: Art. 6; Kuwait: Art. IV; Abidjan: Art. 5; Lima: Art. 4; Jeddah: Art. IV; Cartagena: Art. 5; Nairobi: Art. 5; Noumea: Art. 6; Bucharest: Art. VIII; HELCOM: Art. 8, 9; Antigua: Art. 6; Tehran: Art. 9.

[481] Barcelona: Art 5; Kuwait: Art. V; Abidjan: Art. 6; Lima: Art. 4; Jeddah: Art. V; Cartagena: Art. 6; Nairobi: Art. 6; Noumea: Art. 10, 11; Bucharest: Art. VI, X; HELCOM: Art. 10, 11; OSPAR: Art. 4; Antigua: Art. 6; Tehran: Art. 10.

[482] Barcelona: Art. 11; Nairobi: Art. 9; Bucharest: Art. VI, XIV; HELCOM: Art. 5.

[483] Barcelona: Art. 7; Kuwait: Art. VII; Abidjan: Art. 8; Jeddah: Art. VII; Cartagena: Art. 8; Nairobi: Art. 8; Noumea: Art. 10; Bucharest: Art. XI; HELCOM: Art. 12; Tehran: Art. 8.

[484] Kuwait: Art. VIII; Abidjan: Art. 9; Lima: Art. 4; Jeddah: Art. VIII; Cartagena: Art. 9; Noumea: Art. 9, 12, 13; Bucharest: Art. XII; OSPAR: Art. 5, 7; Antigua: Art. 6, 7; Tehran: Art. 11.

[485] See also Lefeber, *The Liability Provisions of Regional Sea Conventions*, in: Vidas/Østreng (ed.) (1999), at 510–513.

[486] Izmir Protocol on the Prevention of Pollution of the Mediterranean Sea by Transboundary Movements of Hazardous Wastes and their Disposal, of 1 October 1996. It entered into force on 19 January 2008.

[487] For a detailed discussion see Cubel, 'Transboundary Movements of Hazardous Wastes in the Mediterranean Area', 12 *IJMCL* (1997), at 460 *et seq.*; Scovazzi, 'The 1996 Mediterranean Protocol', 7 *RECIEL* (1998), at 264 *et seq.*; Scovazzi, 'Transboundary Movement of Hazardous Waste in the Mediterranean', 19 *UCLA J. Envtl. L. & Pol'y* (2000–2002), at 231 *et seq.*; Scovazzi, 'The Developments within the "Barcelona System"', 26 *Ann. Dr. Mar. Océanique* (2008), at 212–213.

framework of the Barcelona Convention and thus lies within the ambit both of Articles 4(5), 11 of the Barcelona Convention as well as Article 11 of the Basel Convention.[488]

It first provides for general obligations, including obligations to take all appropriate measures to prevent pollution by transboundary movements of hazardous wastes, to reduce the generation and transportation of such wastes to a minimum and, in particular, to prohibit not only the export of such wastes to developing countries but also the import or transit by States which are not members of the EU.[489] It becomes apparent that especially with regard to the imposition of absolute bans on hazardous waste shipments as occurring between developed and developing countries, as well as those among non-EU countries, the Izmir Protocol represents a stricter regulation than the Basel Convention.[490] This, however, does not mean that the Protocol imposes different or stricter obligations. Rather, it extends the application of the absolute trade ban to a larger number of constellations of waste shipments among States. Since States are obligated to "take appropriate measures" to prohibit such shipments, this is to be considered an obligation of conduct. The States are thus required to prohibit such shipments as well as to establish and maintain an administrative system of control and enforcement that is generally suitable and effective. This means that if a State fails to either enact such a prohibition or establish and maintain the required system of control and enforcement, this may constitute an internationally wrongful act of the State. Consequently, a State that is in a position to establish that it suffered damage due to a breach of this obligation has a cognisable claim on the merits for compensation under the principle of State responsibility.

Where a transboundary shipment of hazardous wastes is not generally prohibited according to an absolute trade ban of this nature, it remains permissible only in exceptional cases. This is the case if (1) disposal is not possible in the country of origin, (2) the requirements of a PIC procedure involving both the State of import and all States of transit[491] has been fulfilled, and (3) an environmentally sound management of these wastes in the State of destination is ensured.[492] An internationally wrongful act of the State of origin comes into consideration particularly if the State permits a transboundary shipment although it knows or ought to know that these particular requirements are not fulfilled in the given case.

According to the Izmir Protocol the State of export is, furthermore, under an obligation to re-import hazardous wastes either in case of illegal shipments or where a shipment of hazardous wastes could not be completed according to the

---

[488] Correspondingly, the Secretariat to the Basel Convention has been notified of this Arrangement according to Article 11(2) of the Basel Convention.

[489] Izmir Protocol, Article 5(1).

[490] See also Cubel, 'Transboundary Movements of Hazardous Wastes in the Mediterranean Area', 12 *IJMCL* (1997), at 466–467.

[491] As to this issue see Scovazzi, 'Transboundary Movement of Hazardous Waste in the Mediterranean', 19 *UCLA J. Envtl. L. & Pol'y* (2000–2002), at 239–244.

[492] Izmir Protocol, Article 6.

terms of the contract.[493] In this respect the conclusions reached with regard to the corresponding obligation under the Basel Convention apply.[494] Finally, also the Izmir Protocol contains a *"pactum de negotiando"* clause, according to which the Parties shall elaborate and adopt appropriate guidelines for the evaluation of damage as well as rules and procedures in the field of liability and compensation for damage resulting from the transboundary shipment of hazardous wastes.[495]

In conclusion, it can therefore be said that the 1996 Izmir Protocol, also taking account of the relevant umbrella norms of the Barcelona Convention, represents a legal instrument that acts as an intermediary between global hazardous waste legislation (Basel Convention) and the law of the sea (UNCLOS and Barcelona Convention). By means of this Protocol the basic norms and procedures established by the Basel Convention are applied to marine matters and are adapted to the specific regional conditions of the Mediterranean region. The Protocol does not constitute new obligations, but extends the existing obligations to certain constellations of States involved in transboundary shipments. Thus, for the present examination of the different obligations applicable in the context of transboundary hazardous waste movements, this Protocol is of no prominent significance.

(iii) The 1998 Tehran Protocol to the Kuwait Convention

With regard to the marine area covering the Persian Gulf and the Gulf of Oman, in 1998 the Parties to the Kuwait Convention adopted the Tehran Protocol on the Marine Transboundary Movements and Disposal of Hazardous Wastes.[496]

The structure of this Protocol differs from the Basel Convention pattern, and its regulatory content deviates from that of the Izmir Protocol to the Barcelona Convention. Not only does it address the issue of transboundary movements of hazardous wastes, it also includes rules on dumping into the sea. Regarding the general obligations imposed by the Tehran Protocol, it is remarkable that it states an obligation of minimisation that applies to the generation but not to the transportation of hazardous wastes.[497] The Protocol goes on to establish a ban on the import of hazardous wastes into the protocol area for the purpose of final disposal by any Contracting State; it allows, conversely importation for the purpose of recovery and recycling operations, provided that the State of import has the facilities and technical capacities to manage the hazardous wastes in an environmentally sound

---

[493] *Ibid.*, Articles 9, 7.
[494] *Supra*, at Sect. "Further Obligations Imposed by the Basel Convention".
[495] Izmir Protocol, Article 14.
[496] Tehran Protocol to the Kuwait Convention on the Control of Marine Transboundary Movements and Disposal of Hazardous Wastes, of 17 March 1998. It entered into force on 6 September 2005.
[497] Tehran Protocol, Article 4(1).

manner and the State of export lacks such facilities and capacities.[498] An exportation of hazardous wastes to a non-Contracting State is permitted under the Tehran Protocol only on the condition that the applicable requirements of the Basel Convention are complied with.[499] Movements of hazardous wastes between Contracting States which are destined for operations listed in Annex IV section B are generally permitted.[500] Movements of hazardous wastes destined for operations not listed in Annex IV section B are permitted only on the condition that the State of import possesses the facilities and technical capacities to manage these wastes in an environmentally sound manner and, in addition, that the State of export lacks such facilities and capacities.[501] For shipments among Contracting States the Protocol, furthermore, requires a PIC procedure as well as coverage by insurance, bond or other guarantee.[502] Finally, the Protocol sets out provisions on the illegal trafficking and re-importation of hazardous wastes.[503]

These above described obligations of the Tehran Protocol shall not be examined in more detail at this point. Even though this regime provides a differentiated arrangement concerning the authorisation and prohibition of imports into, exports from and transport among Contracting States, the basic obligations of States nevertheless resemble those of the Basel Convention. Therefore, the conclusions reached with regard to the obligations under the Basel Convention apply here as well.

(ee) Conventions Relevant to the Trade and Transport of Hazardous Substances

In the context of transboundary shipments of hazardous wastes by sea, relevant obligations of States may also arise from international conventions regulating the trade and transport of hazardous substances.

(1) Rotterdam PIC-Convention

The 1998 Rotterdam PIC Convention[504] addresses unwanted importation of hazardous chemical substances, and to this end it establishes a PIC procedure applicable to the trade in certain hazardous chemicals and pesticides.[505] However, since

---

[498] *Ibid.*, Article 5(1) and (2).
[499] *Ibid.*, Article 7. The further requirements mentioned in Article 7 merge in the procedural rules of the Basel Convention.
[500] *Ibid.*, Article 8(2).
[501] *Ibid.*, Article 8(1).
[502] *Ibid.*, Article 8(3) to (7).
[503] *Ibid.*, Articles 10 and 11.
[504] 1998 Rotterdam PIC Convention on the Prior Informed Consent Procedure for Certain Hazardous Chemicals and Pesticides in International Trade. It entered into force on 24 February 2004.
[505] For a detailed discussion see e.g. Barrios, 'The Rotterdam Convention', 16 *Geo. Int'l Envtl. L. Rev.* (2004), at 679 *et seq.*; Kummer, 'The Rotterdam Convention', 8 *RECIEL* (1999), at 323 *et seq.*; McDorman, 'The Rotterdam Convention', 13 *RECIEL* (2004), at 187 *et seq.*

wastes are explicitly excluded from the scope of the Convention,[506] it does not apply to the trade in or transport of hazardous wastes even where they consist of chemicals or pesticides otherwise covered by the Rotterdam Convention.

(2) Stockholm POPs Convention

Depending on the chemical composition of the hazardous wastes in question, the 2001 Stockholm POPs Convention[507] may provide relevant rules as to the responsibility of States in the context of transboundary hazardous waste shipments by sea.[508] The Convention applies to an enumerated list of 21 persistent organic pollutants (POPs) without providing an exclusion of wastes. It aims at the elimination or reduction of the production, use and transportation of the covered POPs. Accordingly, the first general obligation imposed on the Contracting States is "to prohibit" and to "take the legal and administrative measures necessary to eliminate" the production and use as well as the import and export of POPs.[509] With regard to the importation of POPs, the States "shall take measures to ensure" that it is only performed for specific types of uses or for the purpose of an environmentally sound final disposal that is in accordance especially with the provisions of the Basel Convention.[510] Furthermore, the Convention provides that States shall "take measures to ensure" that an exportation of POPs only takes place for the purpose of environmentally sound final disposal and shall "tak[e] into account any relevant provisions in existing international prior informed consent instruments",[511] to which particularly the Basel and the Rotterdam Conventions belong. Further obligations imposed by the Convention relevant to the use and handling of POPs are of minor relevance in the context of transboundary movements of such chemicals by sea.

The Stockholm Convention is the third pillar in the UNEP legal framework governing the transboundary movement of hazardous substances. It complements the Basel Convention and the Rotterdam Convention and provides for specific rules regarding the generation, use and transportation of substances that are considered to be of particular harmfulness to the environment and human health. The Stockholm Convention does not conflict with the rules of the Basel Convention, but rather relates to these rules with regard to the requirements for the exportation of POPs.

---

[506] Rotterdam Convention, Article 3(2)(c).

[507] 2001 Stockholm Convention on Persistent Organic Pollutants (POPs). It entered into force on 17 May 2004.

[508] For further details see Lallas, 'The Stockholm Convention', 95 *AJIL* (2001), at 692 *et seq.*; Vanden Bilcke, 'The Stockholm Convention', 11 *RECIEL* (2002), at 328 *et seq.*

[509] Stockholm Convention, Article 3(1)(a). The production and use of chemicals listed in Annex B, however, is subject to the provisions of that Annex.

[510] *Ibid.*, Article 3(2)(a). The standard of "environmentally sound disposal" is specified in Article 6(1)(d), which refers to the relevant international rules, standards and guidelines (iv).

[511] *Ibid.*, Article 3(2)(b).

Relevant obligations of States, therefore, may arise from both conventions.[512] As regards the particular content of the obligations established by the Stockholm Convention, it should be noted that they are all conceived as obligations of conduct. The State is in each case obliged to take a particular action which consists in either a legislative or administrative act to implement the requirements set out in detail by the Stockholm Convention. However, these obligations are not restricted to a mere obligation to implement. The State also has to conform its legislative and administrative acts to new circumstances and has to maintain a consistent system of control and enforcement of these rules. A Contracting State may invoke an internationally wrongful act only in those instances where it has suffered damage and is able to establish that this damage was caused by a failure of another Contracting State to comply with its obligations under the Stockholm Convention. Particularly the establishment of this causal relationship faces difficulties in practice.

(3) Hong Kong Convention

In 2009, at a Diplomatic Conference held in Hong Kong, the International Convention for the Safe and Environmentally Sound Recycling of Ships was adopted. The Convention has been elaborated under the auspices of the International Maritime Organization (IMO) in co-operation with the International Labour Organization (ILO), the Parties to the Basel Convention and several NGOs. It is not in force yet and it remains to be seen if and when this Convention will enter into force.[513]

The Hong Kong Convention acknowledges the risk that emanates from the dismantling and recycling of end-of-life ships, which are in most cases contaminated with hazardous materials like asbestos, heavy metals, hydrocarbons and oil residues.[514] It aims at combating the primary risks associated with the low-cost dismantling of ships, namely adverse effects on human health and unmitigated pollution and environmental degradation. To this end, the Convention imposes requirements on the construction, operation and preparation of ships for recycling operations, as well as on the operation and equipment of ship recycling facilities.[515]

Since the Hong Kong Convention covers the entire life span of a ship, it is not primarily concerned with the transboundary movement of hazardous wastes. It neither presupposes the character of the ship as being hazardous waste nor

---

[512] See also Vanden Bilcke, 'The Stockholm Convention', 11 *RECIEL* (2002), at 336.

[513] The Convention enters into force 24 months after ratification by 15 States, provided these States represent at least 40 % of the world merchant fleet by gross tonnage, and provided the combined maximum annual ship recycling volume of these States constitutes at least 3 % of their combined gross tonnage. See Hong Kong Convention, Article 17. To date, it has been ratified only by Norway.

[514] The usual *modus operandi* of dismantling end-of-life ships is outlined *supra*, Sect. "The Dismantling of End-of-Life Ships".

[515] Hong Kong Convention, Article 4, in connection with the Annex to the Convention.

constitutes this character by means of its provisions. Since the Basel Convention does not cover cases in which an end-of-life ship is sent to a foreign country for dismantling,[516] the Hong Kong Convention also fills a gap in the scope of application of the Basel Convention. It does not prohibit the export of end-of-life ships to foreign countries for dismantling, but rather imposes certain requirements with a view to a safe and environmentally sound dismantling of ships. By this means, it takes account of the existing realities of the global ship dismantling industry. Since sufficient capacities to dismantle large ships only exist in certain areas of the world, it is necessary to allow a global trade in end-of-life ships and, at the same time, to ensure sufficient standards for the protection of human health and the environment. The Hong Kong Convention consequently can be considered as a supplement to the Basel Convention regarding the dismantling of end-of-life ships.

A particular significance with regard to the transboundary movement of hazardous wastes by sea, however, is not attached to the Hong Kong Convention.

(4) Agreements and Conventions Concerning the Safe Transport and Handling of Dangerous Goods

There are several agreements and conventions at the international and regional level establishing rules on the safe transport and handling of harmful and dangerous goods. As far as the transportation of dangerous goods by sea is concerned, relevant rules can be found in the International Maritime Dangerous Goods Code (IMDG Code) that was elaborated under the auspices of the International Maritime Organization (IMO). These rules are made mandatory by way of reference in Chapter VII of the International Convention for the Safety of Life at Sea (SOLAS).[517] Further international rules on the safe transport and handling of dangerous goods have been elaborated with regard to air transportation.[518] In respect of other modes of transport, relevant rules have been adopted at a regional level, particularly under the auspices of the United Nations Economic Commission for Europe (UN-ECE). Those rules establish safety requirements for the carriage of

---

[516] See *supra*, Sect. "The Application of the Basel Convention to End-of-Life Ships". For a detailed analysis of the Convention see Engels, *European Ship Recycling Regulation* (2013), at 34–42; see also Matz-Lück, 'Safe and Sound Scrapping of "Rusty Buckets"?', 19 *RECIEL* (2010), at 95 *et seq.*

[517] The 1974 International Convention for the Safety of Life at Sea entered into force on 25 May 1980. See also IMDG Code, Chapter 1.1.1.5.

[518] The International Civil Aviation Organization (ICAO) published Technical Instructions for the Safe Transport of Dangerous Goods that were adopted by the International Air Transport Association (IATA) into the Dangerous Good Regulations (DGR). These rules, however, represent non-binding general business terms.

dangerous goods by road,[519] rail[520] and inland waterways[521] and apply to intra-European transports as well as to transports between EU members and third countries.[522] The procedural rules established by these agreements and conventions set out in detail the particular requirements to be complied with by the carrier and other persons involved in a shipment of dangerous goods. They do not impose direct obligations on States, which are thus only under the obligation to implement these rules into their domestic legislation. Consequently, these rules are only of minor relevance for the present consideration as to whether an internationally wrongful act of a State may exist.

(c) Obligations Arising from General International Law

Obligations of States relevant in the context of transboundary movements of hazardous wastes by sea may also arise from general international law. In this respect the relevant question is not only whether or not such customary obligations may apply to the cases under consideration, but also whether an independent significance can be attached to them. This is due to the fact that obligations of States established by international conventions represent *leges speciales* in relation to obligations arising from the customary international law, provided the explicit rules can be understood as a concrete specification and adaption of the general principles to the particular circumstances.

(aa) The Obligation not to Cause Significant Harm to the Environment of Another State's Territory

The first substantive obligation that comes into consideration is the obligation not to cause significant harm to the environment of another State's territory. The same obligation is sometimes also referred to as the obligation to prevent, reduce and control environmental harm.

(1) Recognition in International Law

The origins of this obligation go back to cases of river water diversion and air pollution by industrial plants which caused damage to the territory of adjacent States. These cases have ushered in a series of court decisions and arbitral awards

---

[519] European Agreement Concerning the International Carriage of Dangerous Goods by Road (ADR), of 30 September 1957.

[520] Regulations Concerning the International Carriage of Dangerous Goods by Rail (RID), embodied as Appendix C to the 1980 Convention Concerning International Carriage by Rail (COTIF), as amended on 3 June 1999.

[521] European Agreement concerning the International Carriage of Dangerous Goods by Inland Waterways (ADN), of 26 May 2000.

[522] Directive 2008/68/EC, Article 4.

developing and acknowledging a general obligation of States to prevent transboundary environmental harm.[523] This obligation, moreover, found its way into the international practice of States[524] and is largely supported by legal literature.[525] It was incorporated in Principle 2 of the 1992 Rio Declaration[526] and underlies the work of the ILC which resulted in the 2001 ILC Draft Articles on the Prevention of Transboundary Harm from Hazardous Activities.[527] Hence, there is today no serious doubt about the customary character of this obligation.

(i) The Origin of this Obligation

The doctrinal origin of this obligation, however, is still disputed. Some views, which refer to the principle of good neighbourhood or the principle of using your own property so as not to injure another's (*sic utere tuo ut alienum non laedas*),[528] are not especially convincing.[529] A more persuasive approach emanates from the principle of the sovereign equality of all States. This principle consists of two elements, the principle of territorial sovereignty, which entitles the State to use its territory *ad libitum*, and the principle of territorial integrity, which protects a State from being exposed to any interference perpetrated by other States. According to

---

[523] Some of the most prominent cases are: *Case Relating to the Territorial Jurisdiction of the International Commission of the River Oder*, PCIJ Series A No. 23 (1929), at 4 *et seq.*; *The Diversion of Water from the Meuse*, PCIJ Series A/B No. 70, at 26; *Trail Smelter Arbitration Award*, 3 RIAA (1949), at 1963–1969 *et seq.*; *The Corfu Channel Case*, ICJ Reports 1949, at 22; *Legality of the Threat or Use of Nuclear Weapons*, ICJ Reports 1996, at 241–242; *The Gabčíkovo-Nagymaros Project Case*, ICJ Reports 1997, at 41.

[524] Detailed references can be found, for example, in Lefeber, *Transboundary Environmental Interference* (1996), at 21–22.

[525] References are made by Epiney, 'Das Verbot erheblicher grenzüberschreitender Umweltbeeinträchtigungen', 33 *ArchVR* (1995), at 318–319, and Smith, *State Responsibility and the Marine Environment* (1988), at 76.

[526] It had already been included in Principle 21 of the 1972 Stockholm Declaration.

[527] See Commentaries to the ILC Draft Articles on Prevention of Transboundary Harm from Hazardous Activities, general commentary, para. 1–5. On the work of the ILC see *supra*, Sect. "State Liability".

[528] See e.g. Commentaries to the 2001 Draft Articles on Prevention of Transboundary Harm from Hazardous Activities, Article 3, para. 1; Handl, 'Territorial Sovereignty and the Problem of Transnational Pollution', 69 *AJIL* (1975), at 55.

[529] See Beyerlin, *Grenzüberschreitender Umweltschutz*, in: Hailbronner/Ress/Stein (ed.) (1989), at 54–59; Epiney, 'Das Verbot erheblicher grenzüberschreitender Umweltbeeinträchtigungen', 33 *ArchVR* (1995), at 324; Heintschel von Heinegg, in: Ipsen (ed.), *Völkerrecht* (5th ed., 2004), at 1055; Kasten, *Europarechtliche und völkerrechtliche Aspekte der grenzüberschreitenden Abfallverbringung* (1997), at 234–238. Particularly the latter principle presupposes a conflict of rights, which—as to be shown below—does not exist.

an early understanding, both elements overlap in case of transboundary interferences. Since they are also of equal importance, they need to be reconciled, which necessarily results in a mutual restriction of each element.[530] Running counter to this understanding, it was objected that the right of a State to use its territory can, *a priori*, only reach as far as the equivalent right of another State is not affected; by no means can the principle of territorial sovereignty constitute a right of a State to infringe the same right of another State.[531] Therefore, a later understanding argues for the two elements having a non-overlapping scope. Territorial sovereignty is rather limited to the State borders, thus leaving no room for a conflict between both elements. Consequently, a rule/exception relationship exists. Under this approach, as a general rule the territorial integrity of the impaired State basically overrides any interest of other States in conducting activities that result in impairments to this territory. Such harmful activities may be permitted only exceptionally, provided they are considered necessary for gaining the socio-economic benefits indispensable for the acting State.[532]

In the end, a significant (procedural) difference between both understandings exists only with regard to the onus of proof. Based on the general rule that the burden of proof lies on the State that raises a claim or invokes a defence or excuse, the earlier understanding means that the impaired State would have to show evidence that the actual impairment is beyond the threshold of what is considered to be a reasonable balance of interests. In contrast, assuming a rule/exception relationship with a basic prevalence of the territorial integrity, the burden to prove that the actual impairment is still covered by an accepted exception lies with the polluting State.

Since the latter understanding results in a fairer distribution of the onus of proof that corresponds to the actual risks involved, and since it also provides a more manageable criterion when referring to internationally acknowledged exceptions from a basic rule, the rule/exception approach will underlie the following examination.

---

[530] Dahm/Delbrück/Wolfrum, *Völkerrecht*, vol. *I/1* (2nd ed., 1989), at 445–446; Wolfrum, 'Purposes and Principles of International Environmental Law', 33 *GYIL* (1990), at 310–311.

[531] In the words of *Oppenheim*: "A state, in spite of its territorial supremacy, is not allowed to alter the natural conditions of its own territory to the disadvantage of the natural conditions of a neighbouring country". Oppenheim, *International Law* (2nd ed., 1912), at 182.

[532] Beyerlin, *Grenzüberschreitender Umweltschutz*, in: Hailbronner/Ress/Stein (ed.) (1989), at 40–42; Bryde, 'Umweltschutz durch allgemeines Völkerrecht?', 31 *ArchVR* (1993), at 4–5; Bryde, *Völker- und Europarecht als Alibi für Umweltschutzdefizite?*, in: Selmer/von Münch (ed.) (1987), at 781–784; Epiney, 'Das Verbot erheblicher grenzüberschreitender Umweltbeeinträchtigungen', 33 *ArchVR* (1995), at 320–324; Fröhler/Zehetner, *Rechtsschutzprobleme bei grenzüberschreitenden Umweltbeeinträchtigungen, Band I* (1979), at 74.

(ii) Exception from the Prevalence of Territorial Integrity in Case of Insignificant Harm

Contrary to a view that has been expressed occasionally,[533] it must today be considered a generally accepted rule that transboundary interferences which do not exceed the threshold of substantiality or seriousness constitute an exception from the basic prevalence of territorial integrity in interstate relations. Thus, the obligation not to cause environmental harm to areas beyond a State's own territory does not apply to insignificant harm.[534]

This immediately leads to the question of how to define the respective threshold of substantiality or seriousness. The international jurisprudence and practice of States is not clear on this issue.[535] Moreover, it might not be possible at all to define a universal threshold in general terms, such determination instead depending on a case-by-case consideration. A further specification of the respective thresholds to be applied is beyond the scope of the present work. It should be stressed, however, that in each case the same set of criteria need to be regarded and weighed according to the particular circumstances. The main criterion surely must be the interference level, which is composed of the quality and quantity of the impairment, a possible pre-existing impairment, or a particular sensitivity of the ecosystem. Furthermore, due regard needs to be paid to the degree of interest in conducting the particular activity.[536] The latter, however, may only serve as an additional criterion, but may not be attached an independent significance to the effect that it eventually allows for a balancing of interests. By allowing for a pure balancing of interests or a consideration of an equitable sharing of resources without referring to the objective requirement of substantiality, the exceptional nature of permitted impairments of territorial integrity would be eliminated and the obligation to prevent transboundary environmental damage would at the outset be limited by a criterion of socio-economic feasibility.[537]

---

[533] de la Fayette, 'The ILC and International Liability', 6 *RECIEL* (1997), at 324–326, referring to Principle 21 of the 1972 Stockholm Declaration and Principle 2 of the 1992 Rio Declaration. See also *Legality of the Threat or Use of Nuclear Weapons*, ICR Reports 1996, at 241–242; *The Gabčíkovo-Nagymaros Project Case*, ICJ Reports 1997, at 53.

[534] Beyerlin/Marauhn, *International Environmental Law* (2011), at 41; Epiney, 'Das Verbot erheblicher grenzüberschreitender Umweltbeeinträchtigungen', 33 *ArchVR* (1995), at 334–351; Lefeber, *Transboundary Environmental Interference* (1996), at 24–25; Wolfrum, 'Purposes and Principles of International Environmental Law', 33 *GYIL* (1990), at 311; and in particular see Article 3 of the 2001 ILC Draft Articles on Prevention of Transboundary Harm from Hazardous Activities.

[535] See Birnie/Boyle/Redgwell, *International Law and the Environment* (3rd ed., 2009), at 186–187.

[536] For a very detailed discussion see Epiney, 'Das Verbot erheblicher grenzüberschreitender Umweltbeeinträchtigungen', 33 *ArchVR* (1995), at 334–351.

[537] Beyerlin/Marauhn, *International Environmental Law* (2011), at 42; Birnie/Boyle/Redgwell, *International Law and the Environment* (3rd ed., 2009), at 188; Rao, ILC Doc. A/CN.4/510 (YBILC 2000 II/1), at 120.

(2) The Content of this Obligation

(i) Geographical Scope

Regarding its geographical scope, it is generally accepted that the obligation not to cause significant harm to the environment of another State's territory covers pollution to the territory of any State regardless of its geographical location in relation to the emitting State.[538]

There is also a group of authors arguing for an application of this obligation to areas beyond the limits of national jurisdiction, including Antarctica, the high seas, the deep seabed and outer space.[539] It is, however, doubtful whether this opinion today prevails in international legal literature and State practice.[540] It needs to be considered that such an extension to common areas can hardly be made consistent with the doctrinal origin of this customary obligation, which is deduced from the principle of territorial integrity. Since the principle of the territorial integrity does not apply to common areas like the high seas, an extension of this customary obligation to common areas would presuppose that such areas enjoy a level of legal protection that corresponds to that of the principle of territorial integrity of States.[541] This is recognised for the deep seabed and the moon, as well as for the Antarctic, and it is subject to the principle of common heritage of mankind.[542] One could conclude that these particular areas, hence, require a level of protection equivalent to the principle of territorial integrity, so that this could justify the application of the obligation not to cause significant environmental harm to these

---

[538] Lagoni, *Umweltvölkerrecht : Entwicklungen*, in: Thieme (ed.) (1988), at 246–247.

[539] Beyerlin/Marauhn, *International Environmental Law* (2011), at 39–40; Birnie/Boyle/Redgwell, *International Law and the Environment* (3rd ed., 2009), at 145; Gründling, 'Verantwortlichkeit der Staaten für grenzüberschreitende Umweltbeeinträchtigungen', 45 *ZaöRV* (1985), at 269; Smith, *State Responsibility and the Marine Environment* (1988), at 89–94. These authors mainly refer to the wording of Principle 21 of the Stockholm Declaration, Principle 2 of the Rio Declaration and also invoke Article 194(2) of the UNCLOS.

[540] The contrary opinion is held by, for instance, Epiney, 'Das Verbot erheblicher grenzüberschreitender Umweltbeeinträchtigungen', 33 *ArchVR* (1995), at 331–332; Lagoni, 'Die Abwehr von Gefahren für die marine Umwelt', 32 *BDGVR* (1992), at 136; Lagoni, *Umweltvölkerrecht : Entwicklungen*, in: Thieme (ed.) (1988), at 244. Also of major importance is that the 2001 ILC Draft Articles on Prevention of Transboundary Harm from Hazardous Activities apply only to harm caused to the territory or corresponding interest of another State (Commentaries to the ILC Draft Articles on Prevention of Transboundary Harm from Hazardous Activities, Article 2, para. 9).

[541] Such claim to protection of common areas may not arise from the interest in protecting and preserving the (marine) environment as such, but must have its origin in a right to avert any infringement of or adverse effects to that particular common area. In other words, it must be related to the protection of the territory rather than to the protection of the environment.

[542] See Article 136 of the UNCLOS, Article 11 of the Moon Treaty, and the Preamble of the Antarctic Treaty. See also Tuerk, *The Idea of the Common Heritage of Mankind*, in: Martínez Gutiérrez (ed.) (2010), at 156 *et seq.*; Wolfrum, 'The Principle of the Common Heritage of Mankind', 43 *ZaöRV* (1983), at 312 *et seq.*

particular common areas. However, since in particular the high seas do not yet belong to the common heritage of mankind,[543] and since further principles, like the precautionary principle or the polluter pays principle, do not provide an equivalent level of protection, it must be concluded that today this obligation does not apply to harm caused to the high seas. Accordingly, this obligation cannot be regarded as being owed to the international community as a whole, but is rather to be seen as an obligation *inter partes*.

(ii) Prohibitive and Preventive Elements

The obligation not to cause significant harm to the environment of another State's territory is understood as a compound obligation comprising prohibitive and preventive elements. It thus not only obligates States to refrain from actually causing significant transboundary harm, but also requires States "to take adequate measures to control and regulate in advance sources of potential significant transboundary harm".[544] Furthermore, this obligation implies related procedural duties, such as consultation, information exchange, notification of emergency situations, impact assessment and co-operation.[545]

As far as the preventive part of this compound obligation is concerned, one could argue that this represents a manifestation of the precautionary principle as laid down in Article 15 of the Rio Declaration. The question that arises in this context is under which conditions prior to the actual occurrence of harm the duty of diligent control and regulation arises. An answer cannot be given in general terms, requiring instead a case-by-case consideration that takes into account the respective degree of probability and the expected magnitude of harm.[546] This inquiry involves, in the first place, determining whether the State was or could reasonably have been aware of the activity conducted and the potential harmfulness of that activity.[547]

---

[543] Wolfrum, 'Purposes and Principles of International Environmental Law', 33 *GYIL* (1990), at 323.

[544] Beyerlin/Marauhn, *International Environmental Law* (2011), at 40–41; Lefeber, *Transboundary Environmental Interference* (1996), at 34; see also Commentaries to the 2001 ILC Draft Articles on Prevention of Transboundary Harm from Hazardous Activities, general commentary, para. 1–5.

[545] Beyerlin/Marauhn, *International Environmental Law* (2011), at 44–45; Birnie/Boyle/Redgwell, *International Law and the Environment* (3rd ed., 2009), at 137.

[546] Birnie/Boyle/Redgwell, *International Law and the Environment* (3rd ed., 2009), at 153; Lefeber, *Transboundary Environmental Interference* (1996), at 29–30; see also the definition in Article 2(a) of the 2001 ILC Draft Articles on the Prevention of Transboundary Harm from Hazardous Activities.

[547] Birnie/Boyle/Redgwell, *International Law and the Environment* (3rd ed., 2009), at 153.

(iii) Legal Character

As regards the determination of the legal character of this obligation, the particular difficulty lies in the fact that this customary obligation is conceived and denoted in various forms. However, considering that this obligation emanates from the principle of territorial integrity which protects any State territory from actual damage or infringements, preventing pollution represents the core content of this obligation. This is consistent also with the wording of the codification of this customary obligation in international treaty law.[548] Thus, this obligation is to be regarded as an obligation of result.[549] The standard of conduct required from the State is one of due diligence.[550] The particular degree of diligence owed by the State depends on the circumstances of the case, involving factors like the degree of risk of transboundary harm, the extent and duration of that activity as well as the scientific and technological capabilities. For example, activities considered to be ultra-hazardous require a standard of care that is much higher than with regard to activities entailing an average level of hazard.[551]

(3) The Burden of Proof

Regarding the procedural issue of who should bear the burden of proof, one first has to follow the general rule according to which the burden of proof lies with the State that raises a claim or invokes a defence or excuse.[552] This means that basically the claimant State has to bring evidence showing that damage occurred due to an activity that was conducted at least under the control of the respondent State. It furthermore has to establish that the polluting State failed to act with due diligence.[553] By contrast, if the polluting State invokes that—as an exception from the basic rule—the actually resulting pollution is not prohibited by this obligation

---

[548] See e.g. Article 3 of the 2001 ILC Draft Articles on the Prevention of Transboundary Harm from Hazardous Activities; Article 194(2) of the UNCLOS; Article 2(1) of the 1991 Espoo Convention on Transboundary Environmental Impact Assessment.

[549] Epiney, 'Das Verbot erheblicher grenzüberschreitender Umweltbeeinträchtigungen', 33 *ArchVR* (1995), at 355–356; de la Fayette, 'The ILC and International Liability', 6 *RECIEL* (1997), at 325–326; Scovazzi, 'State Responsibility for Environmental Harm', 12 *Yb. Int'l Env. L.* (2001), at 48–49.

[550] Commentaries to the ILC Draft Articles on Prevention of Transboundary Harm from Hazardous Activities, Article 3, para. 7–18; *The Corfu Channel Case*, ICJ Reports 1949, at 89; *Case Concerning United States Diplomatic and Consular Staff in Tehran*, ICJ Reports 1980, at 29–33; Beyerlin/Marauhn, *International Environmental Law* (2011), at 42–43; Birnie/Boyle/Redgwell, *International Law and the Environment* (3rd ed., 2009), at 147–148; see also Lefeber, *Transboundary Environmental Interference* (1996), at 62–69.

[551] Birnie/Boyle/Redgwell, *International Law and the Environment* (3rd ed., 2009), at 148–149.

[552] Commentaries to the ILC Draft Articles on State Responsibility, Article 19, para 8; Lefeber, *Transboundary Environmental Interference* (1996), at 107.

[553] See *Pulp Mills Case*, ICJ Reports 2010, at para. 97–109; *The MOX Plant Case*, Order No. 3 of 24 June 2003, at para. 53–55; Beyerlin/Marauhn, *International Environmental Law* (2011), at 43; Birnie/Boyle/Redgwell, *International Law and the Environment* (3rd ed., 2009), at 158.

since it remains below the threshold of what is considered significant harm, the onus of proving the circumstances relevant for this defence lies with the polluting State.

(4) Application to Transboundary Movements of Hazardous Wastes by Sea

The customary obligation not to cause significant harm to the environment of another State's territory has been put in explicit terms in Article 194(2) of the UNCLOS. From the ambit of the UNCLOS it follows that this provision only comprises pollution to the marine and coastal environment of the affected State. In this respect Article 194(2) of the UNCLOS represents *lex specialis* and supersedes the customary obligation. The latter, therefore, remains relevant only with regard to inland pollution. In this context, however, the question arises whether the provisions of the Basel Convention (or those of the respective regional conventions) as a whole may be considered *lex specialis* precluding the application of the customary obligation. Several aspects speak against this view: Even though the Basel Convention aims at the protection of the environment, its provisions are not explicitly concerned with the territorial aspect of preventing interference with the territory of other States. Thus, there is no direct counterpart to the customary obligation in the Basel Convention. Whereas the aim of the Basel Convention is to establish procedural rules including a PIC requirement, the customary obligation is primarily directed towards the defence of alleged interferences. Finally, the customary obligation requires the achievement of a certain result, whereas the provisions of the Basel Convention are conceived as obligations of conduct. As a result, it therefore must be concluded that the customary obligation not to cause significant harm to the environment of another State's territory is not fully absorbed by the provisions of the Basel Convention. Both regimes, rather, apply concurrently.

Notwithstanding this concurrent application of both regimes, the customary obligation is not of great practical relevance for transboundary movements of hazardous wastes by sea. This is due to the fact that the regimes influence each other, which leads to a certain synchronisation of both regimes. The Basel Convention specifies the legal requirements under which a transboundary movement is allowed to take place and, thus, determines what is deemed to be lawful at an international level. As a consequence, as long as a State complies with these procedural rules, it acts in conformance with what must be considered the relevant standard of diligence also with regard to the customary obligation to prevent transboundary harm. In practical terms this means that if an act is not in conformity with the customary obligation to prevent transboundary harm, the rules of the Basel Convention have simultaneously been breached. This applies with regard to the prohibitive function of the customary obligation since, if significant harm to the territory of another State has actually occurred, this must be considered a violation of the ESM requirements of the Basel Convention, rendering the transport illegal traffic and, accordingly, entailing the obligation of the State to ensure the re-importation of such wastes.[554] But this also applies with regard to the preventive function of the customary obligation. If the State knew or had reasonable grounds for knowing that a particular transport of hazardous wastes was likely to cause significant harm to the territory of another State, this must be considered a breach of the ESM requirements of the Basel Convention.[555]

As a result, it must be concluded, therefore, that although the customary obligation not to cause significant harm to the environment of another State's territory applies concurrently to the regime of the Basel Convention, its regulatory content does not go beyond the rules of the Basel Convention so that no independent relevance is attached to the customary obligation with regard to transboundary movements of hazardous wastes.

(bb) The Obligation to Protect and Preserve the Marine Environment

It is today acknowledged by State practice[556] and generally accepted by international literature[557] that the obligation to protect and preserve the marine environment forms part of customary international law. Unlike the customary obligation not to cause substantial harm to the environment of another State's territory, which arises from the principle of territorial integrity, this obligation is grounded on the conviction that the marine environment as such, including the natural ecosystem, must be safeguarded for present and future generations.[558] Consequently, this obligation applies to marine and coastal areas regardless of any legal classification, thus including the territorial seas, the EEZ and the high seas.[559]

However, this customary obligation is codified in full by Article 192 and further specified by Articles 194 *et seq.* of the UNCLOS. Thus, there remains no independent relevance of the customary obligation that goes beyond the regulatory content of Articles 192 *et seq.* of the UNCLOS. The customary obligation to protect and preserve the marine environment, therefore, does not need to be discussed further.

---

[554] Basel Convention, Article 9.

[555] *Ibid.*, Article 4(2)(e) and (g). The Basel Convention thus implements the precautionary approach, see Kummer, *The Basel Convention* (1995), at 283.

[556] This obligation is contained not only in Article 192, 194(1) of the UNCLOS but, amongst others, also in every Regional Seas Convention, see *supra*, Sect. "Regional Seas Conventions".

[557] Birnie/Boyle/Redgwell, *International Law and the Environment* (3rd ed., 2009), at 387; Dahm/Delbrück/Wolfrum, *Völkerrecht, vol. I/1* (2nd ed., 1989), at 451; Lagoni, 'Die Abwehr von Gefahren für die marine Umwelt', 32 *BDGVR* (1992), at 145–150.

[558] Dahm/Delbrück/Wolfrum, *Völkerrecht, vol. I/1* (2nd ed., 1989), at 451.

[559] Birnie/Boyle/Redgwell, *International Law and the Environment* (3rd ed., 2009), at 387; Dahm/Delbrück/Wolfrum, *Völkerrecht, vol. I/1* (2nd ed., 1989), at 451.

(d) Circumstances Precluding Wrongfulness

An act of a State that is in contravention with an international obligation is not considered internationally wrongful where the State can invoke circumstances that may justify or excuse the wrongfulness of that act.[560] Such circumstances precluding wrongfulness are laid down in Articles 20–25 of the ILC Draft Articles on State Responsibility; the presence of such circumstances must be shown and proven by the defendant State. A restriction of this principle applies to *ius cogens*, a breach of which cannot be justified by any consideration so that no circumstances precluding wrongfulness can be invoked if peremptory norms are breached.[561] As a result of such defences, the State's act cannot be considered internationally wrongful and does not entail the international responsibility of that State. It is, however, made clear by Article 27 of the ILC Draft Articles on State Responsibility that by virtue of such circumstances the acting State is relieved neither from the primary obligation that resumes as soon as the circumstances precluding wrongfulness cease, nor from an obligation to compensate the affected State that is subsequently negotiated and agreed upon by the States involved.[562]

In the context of transboundary movements of hazardous wastes by sea there are only few conceivable situations in which the State of export may invoke circumstances precluding wrongfulness. This includes, for example, the case where the master of a vessel decides due to a heavy storm or another peril of the sea to deviate from the designated route and to cross the territorial sea of a third State without the prior notification otherwise required by the provisions of the Basel Convention.[563] These situations further include cases in which the master of a vessel, in order to avert an imminent threat of a general average, decides to abandon or discharge the hazardous wastes into the sea causing damage to the marine or coastal environment of another State, or if in the same situation the vessel calls at a port of refuge to avoid being wrecked and discharges its hazardous cargo into the port State area.[564] Yet since the State of export is not responsible for acts of private parties, this applies only to vessels that are operated by State-owned companies. In this case it may be possible that a State of export which is otherwise in breach of its obligations arising from the Basel Convention, the UNCLOS or other relevant conventions is entitled to invoke the defences of force majeure or distress.[565] Finally, a possible circumstance precluding the wrongfulness of a

---

[560] For a detailed discussion see Dahm/Delbrück/Wolfrum, *Völkerrecht*, vol. I/3 (2nd ed., 2002), at 919–924; Lowe, 'Precluding Wrongfulness', 10 *EJIL* (1999), at 405 *et seq.*

[561] ILC Draft Articles on State Responsibility, Article 26.

[562] See Commentaries to the ILC Draft Articles on State Responsibility, Article 27, para. 1–6; Schweisfurth, *Völkerrecht* (2006), at 243.

[563] As to this obligation see *supra*, Sect. "Third Tier: Prior Informed Consent Principle".

[564] The regulatory framework regarding places of refuge for ships in distress are outlined by von Gadow-Stephani, *Der Zugang zu Nothäfen* (2006), at 213–269.

[565] ILC Draft Articles on State Responsibility, Articles 23 and 24.

State's act otherwise in contravention with an international obligation may be the consent of the affected State, e.g. where it consents to an interim storage of illegally exported hazardous wastes for a period greater than the 30 days specified in Article 9 of the Basel Convention.[566]

In summary, it must be concluded, however, that there are only very limited conceivable situations in which the act of a State in breach of an international obligation relevant to the exportation of hazardous wastes may be justified or excused by circumstances precluding the wrongfulness of that act.

(e) Summary

The second main element of an internationally wrongful act is the existence of a breach of an international obligation. Basically, each binding international obligation, regardless of its origin and character, is eligible to give rise to the international responsibility of States in case of non-conformity. Subjective elements are not in general required. International law rather follows an objective approach generally applying a standard of conduct to be performed by the acting person, which is one of due diligence. In individual cases this standard may be altered or an additional requirement of fault may be established by the respective substantive obligation.

In the context of transboundary movements of hazardous wastes, the international legislation on the management and transportation of hazardous wastes is of primary significance. The Basel Convention represents the major instrument at the international level. A breach of its general obligations to take appropriate measures towards the reduction of generation and exportation of hazardous wastes may be established only in case of obvious and flagrant infringements. This is due to the fact that these obligations merely require the States to take unspecified measures towards the achievement of a certain legal aim and because they contain a number of indefinite legal terms and other qualifications. The Basel Convention establishes absolute trade bans regarding hazardous waste transports to non-Parties, to the Antarctic and to States Parties that have prohibited its import. These trade bans are conceived as obligations of conduct that are breached when a State fails to implement or subsequently regulate, control or enforce these bans. The further rules on the ESM of hazardous wastes basically require the States to perform the legal or administrative act of preventing the commencement of transboundary shipments unless the particular ESM requirements are fulfilled. The establishment of a breach faces difficulties: First, the term ESM is still not sufficiently defined and, second, the capability of States to gather sufficient information is limited. Non-compliance with the standard of due diligence, thus, requires that the State allowed the commencement of a hazardous waste shipment although it knew or ought to have known of the lack of ESM capacities of the respective other States

---

[566] *Ibid.*, Article 20.

involved. The same basically applies to the PIC provisions of the Basel Convention that describe factual, administrative and legal measures to be taken by the State. These obligations are breached if the State of export permits the commencement of a hazardous waste shipment although it has not received the PIC of all States concerned. A PIC of coastal States, however, is not required where a vessel carrying hazardous wastes merely passes through the EEZ of that coastal State. Finally, the Basel Convention stipulates the obligations of result to ensure the re-importation of hazardous wastes that have been shipped either illegally or incompletely. A breach of these obligations exists solely on the condition that the wastes in question are not shipped back in accordance with the particular requirements laid down by Articles 8 and 9 of the Basel Convention. A similar set of obligations is provided by regional conventions like the Bamako and Waigani Convention as well as by the Izmir Protocol to the Barcelona Convention, and in parts also by the Tehran Protocol to the Kuwait Convention. These obligations mainly differ with regard to their application to the respective countries and areas; the general structure and the character of these obligations, however, is basically the same.

Since in the present work the shipment of hazardous wastes by sea is under particular consideration, also the international conventions on the law of the sea, and Part XII of the UNCLOS in particular, are of substantial importance. Article 192 first lays down the programmatic goal that States have to protect and preserve the marine environment. A binding obligation of States to take the respective necessary measures, however, is imposed only by Article 194(1). This obligation is conceived as an obligation *erga omnes*, so that any State is entitled to invoke a wrongful act and may claim for its cessation and for reparation and restitution to the benefit of the community as a whole. However, this obligation is a mere general obligation without specifying further the particular necessary measures. In addition, the degree of due diligence required by this obligation is further qualified by the restriction that the respective capabilities of the State and the best practical means at its disposal need to be taken into account. A breach of this obligation, therefore, may be established only in constellations in which the State has obviously and flagrantly exceeded the limit of its discretion. In the context of transboundary movements of hazardous wastes such situations are conceivable only in cases where a coastal State allows for the large-scale discharge of hazardous wastes into marine or coastal areas, knowing that this will cause substantial pollution and being aware that this activity is not covered by the use of the best practicable means at its disposal and in accordance with its capabilities. The same outcome basically applies to the general obligations established under the UNEP Regional Seas Conventions. A more sophisticated obligation is established by Article 194(2), which requires States to ensure that activities under their jurisdiction or control are conducted so as not to cause damage by pollution to other States and their environment as well as to the high seas. Since it is construed as an obligation of result, it applies only on the condition that damage to the environment of another State or the high seas has been caused. Due to this broad scope it basically covers the conduct of a coastal State with regard to any activities of private parties conducted in the territorial seas, in the EEZ or on the continental shelf, irrespective of their nationality. The responsibility of the coastal State for such activities is limited only by the applicable standard of due diligence. This means that a State bears responsibility where it knew about a particular activity of a private party and there was visible indication for the State that this activity would be in contravention with the applicable laws and would certainly cause damage to the environment of another State or the high seas.

In the context of transboundary hazardous waste movements by, sea a breach of the provisions of MARPOL 73/78 is unlikely to be established. This may be possible only if it can be established that the port State failed to take delivery of hazardous cargo residues or that it came to know of concrete grounds suggesting that the master or crew was not familiar with the respective rules but nonetheless allowed the vessel to leave the port. Partial significance is also attached to the provisions of the London Dumping Convention in those cases where a State permits the shipment of hazardous wastes although it knew or ought to have known that the intention of the shipowner was to dump or incinerate the wastes at sea after they were shipped to another State. A breach of the provisions of the Stockholm POPs Convention may be possible with regard to shipments of hazardous wastes that also come under this Convention. Relevant constellations in practice may be any kind of non-compliance with the established rules of procedure.

As a final remark to this complex of rules that may possibly be breached in the context of transboundary movements of hazardous wastes by sea, it should be mentioned that a breach of these obligation will not in any case entitle the injured State to claim compensation in full for any environmental damage. It is rather an additional issue, apart from the existence of an internationally wrongful act, whether or not there is a causal link between the particular wrongful act and the damage in question.[567]

3. Legal Consequences

According to Article 1 of the ILC Draft Articles on State Responsibility, which restates existing customary law, every internationally wrongful act of a State entails the international responsibility of that State.[568] The internationally wrongful act, thus, functions as a trigger to the emergence of a new legal relationship between the acting and the injured State and gives rise to the application of the secondary rules concerning State responsibility. Whereas the legal prerequisites, the "if", of State responsibility have been the subject of the previous section, in the present section the legal consequences, meaning the content or

---

[567] This issue will be addressed *infra*, Sect. "Causal Link".

[568] See *supra*, Sect. "The Principle of State Responsibility for Internationally Wrongful Acts".

substances, i.e. the "how", of State responsibility is outlined with regard to typical damage in the context of transboundary movements of hazardous wastes. The extent of the international responsibility of a State evolved in customary international law and has been developed further by the rules of the ILC Draft Articles on State Responsibility. These rules, however, do not preclude the possibility that express rules in international conventions may provide for different legal consequences of State responsibility.

(a) Continuation of the Primary Obligation

A first set of rules triggered by an internationally wrongful act is, in fact, not a consequence of the breach of that obligation, but rather arises from the respective primary obligation itself. Articles 29 and 30(a) of the ILC Draft Articles on State Responsibility, thus, have to be considered mere declaratory norms.[569] Article 29 restates the general principle that by means of an internationally wrongful act and due to the resulting application of the secondary rules of State responsibility the duty of the State to perform the primary obligation it has breached is not affected and still continues. Article 30(a) obliges the State to cease a wrongful conduct if that wrongful act or omission has a continuing character. Article 30(b), in addition, stipulates that an injured State may seek from the State in breach of an obligation that is gives appropriate assurances and guarantees of non-repetition. Such assurances and guarantees may be sought in case the injured State has reason to believe that a mere restoration of the pre-existing situation will not sufficiently prevent further infringements and that it is likely that the wrongfully acting State will repeat this violation. Particularly the right to claim the cessation of a continuing wrongful act and to claim assurances and guarantees of non-repetition is of special practical significance. The focus of international attention, besides the issue of reparation, often lies in the cessation of a wrongful act. Since, furthermore, many international obligations that might be breached in the context of hazardous waste shipments are conceived as obligations *erga omnes*, the right to claim cessation and related assurances and guarantees of non-repetition can be invoked not only by directly injured States, but by any State in accordance with Article 48. These obligations, thus, represent a "quite powerful instrument for the upkeep of the international legal order".[570]

---

[569] Ipsen, in: Ipsen (ed.), *Völkerrecht* (5th ed., 2004), at 655; Schweisfurth, *Völkerrecht* (2006), at 244.

[570] Wolfrum/Langenfeld/Minnerop, *Environmental Liability in International Law* (2005), at 477.

(b) Reparation

The central component of State responsibility is the obligation of the State that committed a wrongful act to make full reparation.[571] The term full reparation is in this context used in the most general sense, requiring, as far as possible, the eradication of all consequences of the wrongful act and re-establishment of the situation that would have existed if that wrongful act had not been committed.[572] Full reparation may encompass the forms of restitution, compensation and satisfaction, or a combination of these forms. These forms of reparation, however, do not have an equal status, but are ranked in order of priority. It is only where restitution in kind is inadequate that the responsible State is under the obligation to pay compensation or, in case of non-material injury, to pay satisfaction to the injured State.[573]

(aa) Causal Link

The breach of an international obligation may not, however, give rise to a duty of reparation for each and every consequence, but only for damage that was actually caused by the internationally wrongful act.[574] The establishment, in a first stage, of a causal link between the wrongful act and the injury requires that the wrongful act is to be considered *conditio sine qua non* for the actual injury. Since, however, such consideration based on a mere factual causation alone would lead to a boundless causal allocation, in a second stage a normative consideration of the causal link must take place and eliminate cases in which the injury is too remote or consequential. Normative criteria for the identification of extraordinary damage have been developed to a considerable extent. There is, however, no general rule as to how to apply these criteria to the different wrongful acts; rather, the different criteria must be resorted to with regard to the circumstances of the individual case and the respective obligation that has been breached.[575] In particular, this involves the criteria of directness as well as that of remoteness or proximity of damage, according to which a causal link is rejected when the damage is too indirect or remote. Further criteria are the foreseeability of damage and uncertainties in appraising the damage.[576] Particular importance is also attached to the question of

---

[571] ILC Draft Articles on State Responsibility, Article 31(1).

[572] Commentaries to the ILC Draft Articles on State Responsibility, Article 31, para. 1–3; *Factory at Chorzów Case*, PCIJ Series A No. 17 (1928), at 47.

[573] This follows from Articles 34 to 37 of the ILC Draft Articles on State Responsibility. See also *Factory at Chorzów Case*, PCIJ Series A No. 17 (1928), at 48.

[574] ILC Draft Articles on State Responsibility, Article 31(2).

[575] Commentaries to the ILC Draft Articles on State Responsibility, Article 31, para. 9–10.

[576] See *ibid.*, Article 31, para. 10, with further references to international case law.

whether or not the actual injury lies within the ambit of the protective purpose of the obligation that was breached.[577]

With regard to constellations in which either a combination of several conditions conjunctly or two or more conditions concurrently have caused the actual damage, international case law does not apply the principle of *causa proxima*,[578] instead applying a test according to which it is sufficient for establishing the causal link when at least one of the different conditions can be ascribed to the State and, when considering this particular wrongful act, the damage is not too remote or indirect and not beyond any experience and foreseeability.[579] An exception from this is only cases of contributory fault of the claimant State.[580]

The requirement of a causal link practically means that after the establishment of a factual causation in terms of a *conditio sine qua non*, a normative test must be applied. This involves, first, the assessment of whether the actual damage or injury is not too remote or indirect or could not have been envisioned regarding the typical course of events; and, second, a consideration of whether the damage or injury lies within the ambit of the protective purpose of the obligation that has been breached. For this normative consideration the respective primary obligation is of decisive importance. As a basic rule it can be said that the more general the primary rule is, the easier it is to establish a causal link to the wrongful act. This particularly applies to obligations of result which leave to the discretion of the State how the required result is to be achieved. If, for example, transboundary pollution to the marine environment has been caused by an activity under the jurisdiction or control of a State, the conduct or omission of that State is *a priori* causal in terms of a *conditio sine qua non* for the occurrence of pollution. But it is also hardly possible for this State to raise the defence that this damage was too remote, indirect, or could not have been foreseen, since the obligation is not linked to the activity conducted under the jurisdiction or control of that State, but rather solely on the actual result of pollution. This, at least, applies to the establishment of a causal link. In most conceivable scenarios which involve an entirely atypical course of events that could not reasonably have been predicted, the State's conduct will not be in breach of the required standard of due diligence so that there will be no internationally wrongful act in the first place. As a result, in respect of obligations of result the issue of causality remains relevant only if in consequence of a

---

[577] Commentaries to the ILC Draft Articles on State Responsibility, Article 31, para. 10, referring to *Cases No. A15(IV) and A24 (Iran v. United States)*, 11 WTAR (1999), at 45.

[578] Different however: Wolfrum/Langenfeld/Minnerop, *Environmental Liability in International Law* (2005), at 476.

[579] Commentaries to the ILC Draft Articles on State Responsibility, Article 31, para. 13; Dahm/Delbrück/Wolfrum, *Völkerrecht, vol. I/3* (2nd ed., 2002), at 950–951; see also *Zafiro Case*, 6 RIAA (1955), at 164–165; *Responsabilité de l'Allemagne (Portugal v. Germany)*, 2 RIAA (1949), at 1011 *et seq.*; *Newchwang Case*, 6 RIAA (1955), at 68.

[580] ILC Draft Articles on State Responsibility, Article 39; Moutier-Lopet, *Contribution to the Injury*, in: Crawford/Pellet/Olleson (ed.) (2010), at 639 *et seq.*

wrongful act of a State a large range of damage or injuries occurs, only part of which could not have been predicted.

In contrast, when a detailed and specified obligation of conduct has been breached, such as the PIC or ESM requirements of the Basel Convention that are directly addressed at the States, the normative assessment of the causal link plays a major role. Of particular importance is in this respect the ambit of the protective purpose of the primary obligation. If, for example, a State is in breach of the obligation to re-import hazardous wastes according to Articles 8 or 9 of the Basel Convention without having violated at the same time the PIC or ESM requirements, environmental damage is not eligible for unrestricted reparation, but only to the extent it can be attributed to the specific failure to perform the re-importation according to Article 8 or 9 of the Basel Convention. Another example involves the breach of the obligation to implement into national laws the rules of procedure which are contained in an international convention. If the State failed to implement such rules, pollution damage that occurred due to non-compliance with these rules may give rise to claims of reparation only if and to the extent that it can be established that the protective purpose of these procedural rules was to prevent this specific type of damage and that this particular damage would have been prevented if the State had actually complied with its duty of implementation and enforcement.

The issue of conjunctive or concurrent causes can also become relevant in the context of transboundary movements of hazardous wastes by sea. This may involve, for instance, cases in which a shipment of hazardous wastes takes place without having complied with the PIC or ESM requirements of the Basel Convention due to corruption and the acceptance of bribes of State officials involved. In such a case, a causal link between the wrongful act of a State involved and the consequent pollution damage can basically be established irrespective of any conjunctive or concurrent causes attributable to the wrongful act of another State. The claimant State, however, must deduct from its claim its own portion of contributory fault. In case an obligation *erga omnes* has been breached, any State is entitled to claim the cessation of that wrongful act as well as reparation to the benefit of the community as a whole in full amount from all States contributorily causing the damage as joint and several debtors.

Finally, the question arises whether a causal link can be established with regard to preventive measures. Preventive measures are taken after an internationally wrongful act has been conducted and prior to the actual occurrence of damage or injury. Required, however, is an actual and imminent threat of damage, the occurrence of which is considered to be only a matter of time. In this situation the threatened State is entitled to take appropriate measures in order to prevent the actual occurrence of damage by virtue of the principle of territorial integrity. The causal link between the act of the State that is in breach of an international obligation and the preventive measures is established by a hypothetical consideration of a causal link to the subsequent actual damage according to the general criteria outlined above. As long as such a causal link can be established, this applies *a maiore ad minus* to preventive measures taken in order to avoid the actual occurrence of damage. Conversely, prior to the conduct of the internationally wrongful act preventive measures can consist only of requiring appropriate assurances and guarantees of non-repetition in those instances where a wrongful act has already been conducted before.

(bb) Forms of Reparation

Full reparation covers the forms of restitution, compensation and satisfaction, or a combination of these forms.

Restitution (*restitutio in integrum*) is the first and preferred form of reparation. It aims at the re-establishment of the *status quo ante*, i.e. the situation that existed before the wrongful act was committed.[581] Restitution may encompass in principle any action that needs to be taken in order to undo damage of either a material or immaterial nature.[582] With regard to material damage this action may consist of the material restoration or return of territory, persons or property, such as in the repair of damaged objects, the restoration and decontamination of polluted areas and ecosystems, the release of an arrested vessel or the restitution of other seized property. With regard to immaterial damage, and particularly regarding "legal damage" ("*préjudice juridique*"), restitution may take the form of "juridical restitution". This consists in the modification of the internal laws of the responsible State or its legal relation to the injured State and may involve, for example, the revocation, annulment or amendment of a legislative act, or the rescinding or reconsideration of an unlawful administrative or judicial measure.[583] The obligation to make restitution, however, is excluded if restitution in kind is materially impossible or involves a burden fully disproportionate to the benefit.[584]

In those latter cases the responsible State is under the obligation to pay compensation for any financially assessable material damage including loss of profits (*lucrum cessans*)[585] and resulting interest.[586] The amount of compensation is basically calculated on the basis of the fair market value of the destroyed or lost goods.[587] Punitive damages may not be considered.[588] Particularly with regard to damage resulting in the context of transboundary movements of hazardous wastes by sea, compensation for material damage may include any kind of expenses

---

[581] ILC Draft Articles on State Responsibility, Article 35.

[582] *Ibid.*, Article 31(2).

[583] Commentaries to the ILC Draft Articles on State Responsibility, Article 35, para. 5.

[584] ILC Draft Articles on State Responsibility, Article 35(2).

[585] *Ibid.*, Article 36. See also Wolfrum/Langenfeld/Minnerop, *Environmental Liability in International Law* (2005), at 479.

[586] ILC Draft Articles on State Responsibility, Article 38.

[587] Commentary to the ILC Draft Articles on State Responsibility, Article 36, para. 22; further methods of evaluating the damage eligible for compensation are outlined *ibid.*, Article 36, para. 22–26; see also Wolfrum/Langenfeld/Minnerop, *Environmental Liability in International Law* (2005), at 479.

[588] Wolfrum/Langenfeld/Minnerop, *Environmental Liability in International Law* (2005), at 479.

incurred for the location, recovery and removal of hazardous wastes or its residues as well as costs for cleaning up affected areas.[589] A comprehensive list formulated by the United Nations Compensation Commission (UNCC) in 1991[590] may, furthermore, serve as a guideline as to what types of environmental damage are eligible for compensation. This list additionally mentions indirect damage and costs incurred for the abatement and prevention of environmental damage, the evaluation of harm, the restoration of the environment, medical measures to combat or prevent health risks as well as the depletion of or damage to natural resources.[591] All this shows that a large part of damages and costs incurred in the wake of transboundary hazardous waste shipments are eligible for compensation, including preventive measures and measures of re-instatement. However, environmental damage is often beyond what can be financially assessed as required by the concept of compensation. Particularly, damage to environmental values such as biodiversity and the composition of entire ecosystems cannot be compensated with money.

As far as restitution is not sufficient to remedy the consequences of a wrongful act and if these adverse consequences consist in moral or immaterial injury, the responsible State is under the obligation to give satisfaction.[592] Satisfaction, thus, is the appropriate remedy for those damages or injuries that cannot be financially assessed and mostly consist of an (political) affront of the State. Such injuries include, for example, insults to the symbols of the State, violations or attacks on diplomatic or consular staff or premises, or violations of sovereignty or territorial integrity.[593] The latter example is of significance particularly with regard to the transboundary pollution of another State's territory, including its marine and coastal environment, since in those cases restitution of the *status quo ante* is mostly impossible and compensation fails due to the impossibility of financially assessing the damage. Thus, satisfaction appears appropriate given the acknowledgment that this remedy can be sought also for purely ecologic damage that is not economically quantifiable but consists of damage to the environment as a value in itself.[594] Satisfaction can take several forms, such as a formal acknowledgement of the wrongful act or a corresponding declaration by a competent court or tribunal, an expression of regret or a formal apology, the award of symbolic damages, disciplinary or penal action against the acting individual, assurances or guarantees of non-repetition.[595]

---

[589] See, for instance, the claim for damage presented by Canada to the USSR for damage caused by the soviet Cosmos 954 satellite, 18 ILM (1979), at 899–930.

[590] See UNCC Doc. S/AC.26/1991/7/Rev.1, at para. 35.

[591] Wolfrum/Langenfeld/Minnerop, *Environmental Liability in International Law* (2005), at 481.

[592] ILC Draft Articles on State Responsibility, Article 37.

[593] Commentaries to the ILC Draft Articles on State Responsibility, Article 37, para. 4; *Rainbow Warrior Decision*, 20 RIAA (1990), at 272–273.

[594] Dahm/Delbrück/Wolfrum, *Völkerrecht*, vol. I/3 (2nd ed., 2002), at 977–978; Wolfrum/Langenfeld/Minnerop, *Environmental Liability in International Law* (2005), at 482.

[595] Commentaries to the ILC Draft Articles on State Responsibility, Article 37, para. 5; Wolfrum/Langenfeld/Minnerop, *Environmental Liability in International Law* (2005), at 482–483.

(c) Invocation by the Injured State

The extent of the right to invoke the international responsibility of a State depends on whether the injured State is directly or indirectly injured.

A State is "directly" injured if the obligation that has been breached is owed to this State individually, or if the State is part of a group of States to which the obligation is owed and the wrongful act specifically affects this State or its position.[596] From a practical perspective this primarily applies to importing States of hazardous wastes as well as to all coastal States involved, irrespective of whether they are on the exporting or importing side.[597] The directly injured State, or each directly injured State in case several States are directly injured, is entitled to invoke the responsibility of the State which has committed an internationally wrongful act.[598] If several States are responsible for the same internationally wrongful act, each State may be held responsible by the injured State or States.[599] The invocation of the responsibility of a State requires in all cases that a notice of claim be issued to the responsible State.[600] The directly injured State is furthermore entitled to take counter-measures (*reprisals* or *représailles*).[601]

A State is considered to be indirectly injured within the meaning of the concept of State responsibility if it is part of a group of States to which the obligation that has been breached is owed, or if this obligation is owed to the international community as a whole.[602] The invocation of international responsibility by an indirectly injured State, thus, requires the breach of an obligation *erga omnes partes* or *erga omnes*. In the context of hazardous waste shipments by sea this may involve a large number of States that have a collective interest in the affected area, or it may even involve all States as far as concerns obligations *erga omnes* aiming at the protection of the high seas.[603] The elements of State responsibility that may be invoked by an indirectly injured State are limited and include the cessation of the wrongful act, assurances and guarantees of non-repetition, and the performance of reparation in the interest of the injured State or to the benefit of the group of States or, alternatively, to the international community as a whole.[604] Unlike directly injured States, indirectly injured States are not entitled to take

---

[596] ILC Draft Articles on State Responsibility, Article 42.

[597] See for a more distinct discussion Rest, 'State Responsibility/Liability', 40 *Envtl. Pol'y & L.* (2010), at 299.

[598] ILC Draft Articles on State Responsibility, Articles 42 and 46.

[599] *Ibid.*, Article 47.

[600] *Ibid.*, Article 43.

[601] A detailed description of the legal prerequisites and contents of counter-measures is beyond the scope of the present consideration. See Articles 49–53 of the ILC Draft Articles on State Responsibility and the Commentary thereto. See also Dahm/Delbrück/Wolfrum, *Völkerrecht*, vol. I/3 (2nd ed., 2002), at 981–990.

[602] See ILC Draft Articles on State Responsibility, Article 48(1).

[603] Rest, 'State Responsibility/Liability', 40 *Envtl. Pol'y & L.* (2010), at 300.

[604] ILC Draft Articles on State Responsibility, Article 48(2).

counter-measures,[605] but only to take lawful measures against the responsible State (*rétorsions*) to ensure cessation and reparation in the interest of either the injured State or the beneficiaries of the obligation breached.[606]

Finally, it may also be possible that persons or entities other than States are entitled to invoke the responsibility of States. This particularly applies to cases in which the primary obligation that has been breached was established in order to protect the interests of individuals or non-State entities, like IGOs and NGOs.[607]

(d) Serious Breaches of Peremptory Obligations

Particular consequences arise from a serious breach of an obligation arising under a peremptory norm of general international law. This category of internationally wrongful acts as incorporated in Chapter III of the Final Draft of the ILC Articles on State Responsibility had previously been discussed under the topic of "international crimes".[608] This particular type of liability is, however, of minor significance for damage in the context of transboundary hazardous waste movements by sea. Two conditions must be fulfilled to result in these specific consequences: First, the obligation that has been breached must arise under a peremptory norm of general international law (*ius cogens*)[609] and, second, the breach of this obligation must be "serious", involving a gross or systematic failure by the responsible State.[610] In consequence of such a qualified internationally wrongful act, all States are under the obligation, first, to cooperate to bring to an end any serious breach, second, not to recognise as lawful the resulting situation and, third, not to render

---

[605] Commentary to the ILC Draft Articles on State Responsibility, Chapter II, para. 8; Wolfrum/Langenfeld/Minnerop, *Environmental Liability in International Law* (2005), at 486.

[606] ILC Draft Articles on State Responsibility, Article 54.

[607] See Commentaries to the ILC Draft Articles on State Responsibility, Article 33, para. 4; Brown Weiss, 'Invoking State Responsibility', 96 *AJIL* (2002), at 799; see also Cujo, *Invocation of Responsibility by International Organizations*, in: Crawford/Pellet/Olleson (ed.) (2010), at 969 *et seq.*; Tomuschat, *Individuals*, in: Crawford/Pellet/Olleson (ed.) (2010), at 985 *et seq.* As to the invocation of State responsibility by private entities in maritime matters, see Wendel, *State Responsibility for Interferences with the Freedom of Navigation* (2007), at 79–84.

[608] For a detailed discussion of Article 19 of the former 1996 ILC Draft Articles on State Responsibility see Graaff, *Staatenverantwortlichkeit* (2007).

[609] This condition refers to the definition of peremptory norms of general international law in Articles 53 and 64 of the 1969 Vienna Convention of the Law of Treaties. Applicable norms mainly include prohibitions of certain intolerable action against States and individuals, such as aggression, slavery, genocide, racial discrimination and torture, but may also include norms that are accepted and recognised as *ius cogens* by the international community of States. See Commentaries to the ILC Draft Articles on State Responsibility, Article 40, para. 1–9; Köck, *Staatenverantwortlichkeit und Staatenhaftung im Völkerrecht*, in: Köck/Lengauer/Ress (ed.) (2004), at 205–206.

[610] ILC Draft Articles on State Responsibility, Article 40(2).

aid or assistance in maintaining that situation.[611] In the context of transboundary hazardous waste movements the issue of serious breaches of peremptory norms may, however, be of rather limited practical relevance.[612]

4. Jurisdictional Issues

States are obliged by customary international law to settle disputes, including claims under the principle of State responsibility, peacefully and in accordance with Chapter VI of the UN-Charter. Possible means of peaceful dispute settlement are listed in Article 33 of the UN-Charter and involve, in particular, conciliation, arbitration and judicial settlement. Competent courts in the context of damage resulting from transboundary movements of hazardous wastes by sea may be the International Court of Justice (ICJ) and the International Tribunal for the Law of the Sea (ITLOS). Arbitral courts may be established *ad hoc* or on a contractual basis. A detailed description of the related procedural rules is beyond the scope of the present work, so that at this point reference shall be made only to the Statute of the ICJ and the Statute of the ITLOS, these legal instruments laying down the competence of these Courts and the related procedural rules.

III. Summary: State Responsibility

It has been outlined in the previous section that international law neither supports a customary principle of State liability for lawful but injurious activities, nor does it provide for explicit convention provisions on State liability relevant to the transboundary movement of hazardous wastes by sea. In case of damage resulting from those waste shipments, claims for compensation may, therefore, only be based on rules of State responsibility for internationally wrongful acts of the State in question. Relevant scenarios in the context of transboundary hazardous waste movements giving rise to the international responsibility of States are depicted in this section.

At the outset, this section has examined the fact that the relevant international conventions do not provide for explicit provisions regulating the international responsibility of States. Therefore, State responsibility can only be invoked under the customary principle of State responsibility. Two conditions must be fulfilled to this end. This involves, first, an act of the State, which is understood as an action or omission that is imputable to the State and, second, that this act of the State constitutes a breach of an international obligation of that State. Regarding the

---

[611] *Ibid.*, Article 41.

[612] See also Lammers, 'New Developments Concerning International Responsibility', 19 *Hague Y. B. Int'l L.* (2006), at 88.

imputation of an act to the State it can be summarised that, in general, the conduct of the official person in charge is attributable to the State, whereas with regard to the conduct of private persons the State can be responsible only in an indirect way, namely if at the same time the State is responsible for a failure to regulate or control such conduct or for a failure to ensure the non-occurrence of a particular prohibited result. The breach of an obligation may give rise to the international responsibility of the State regardless of the origin and character of this obligation. Subjective elements and physical damage are not generally required. Some conclusions as to when an obligation is deemed to be breached can be drawn from the character of the respective obligation as an obligation of conduct or as an obligation of result. In general, the State is obligated to conform to the standard of due diligence.

Subsequent to these general findings, this section examines which particular international obligations may be breached in typical scenarios involving the transboundary movement of hazardous wastes by sea. The relevant cases have already been summarised above,[613] so that at this point it will only be mentioned that the Basel Convention as well as the UNCLOS, including their respective regional counterparts, are of prime importance for the establishment of an internationally wrongful act. Furthermore, also the MARPOL 73/78 Convention, the London Dumping Convention and the Stockholm POPs Convention may become relevant in respect of certain constellations of hazardous waste movements by sea.

The state of being internationally responsible entails a set of legal consequences imposed on the responsible State. At first, it needs to be stressed that the respective primary obligation is not abolished in consequence of its breach, but remains in effect. The responsible State is obligated to cease the wrongful act and, where appropriate, to give adequate guarantees of non-repetition. A further legal consequence of State responsibility is the obligation to make full reparation. Reparation comprises the duty of restitution by taking any measures necessary to wipe out the factual consequences of the wrongful act and to restore the situation that existed before the wrongful act was committed. It is only in case restitution is substantively impossible or involves a burden fully disproportionate with the benefit that the responsible State either has to pay compensation as far as material damage is concerned, or that it has to give satisfaction with regard to immaterial damage. In the context of transboundary movements of hazardous wastes by sea, reparation may basically cover any direct and indirect damages and costs, including those costs necessary for locating, recovering and removing hazardous wastes and their residues, costs related to preventive measures and measures of reinstatement, costs for medical treatment of affected individuals, and costs estimating the amount of depletion of financially assessable natural resources. However, reparation can only be sought if in each particular case a causal link between the wrongful act and the actual damage has been established. This not only means that the damage may not be too remote and indirect, but also that the damage must

---

[613] *Supra*, Sect. "Summary" (p. 155).

come within the ambit of what particular kind of damage the respective obligation seeks to prevent.

In summary, it can be concluded, therefore, that the principle of State responsibility may basically provide for appropriate and effective solutions regarding the allocation of responsibilities and liabilities in international constellations involving different States and different territories and areas. The reason for this is that the principle does not contain a legal valuation of its own as regards the question under which conditions an internationally wrongful act exists, but rather resorts to the legal valuation of the respective primary obligations. Given the fact that especially sectoral and regional conventions provide for detailed and complex rules which are adjusted to the specific circumstances of the individual case, it becomes apparent that in general the principle of State responsibility is to be considered an appropriate reaction to any wrongful conduct of the State.[614]

Nevertheless, considerable difficulties arise if the principle of State responsibility is applied with regard to activities that are conducted by private parties, as is the case with regard to the transboundary movement of hazardous wastes. In such cases it is not the activity as such which is to be judged against the international law, but rather the conduct of the State with regard to this activity. Hence, the responsibility of the State and its obligation to make reparation cannot be judged simply by considering the lawfulness or wrongfulness of the activity conducted by private persons; rather, the inquiry must be undertaken only in an indirect way by judging the legislative or administrative act of the State with regard to this activity. This, however, entails considerable loopholes in respect of State responsibility. In practice, States lack reliable information about the circumstances under which hazardous waste transports are conducted. This particularly involves cases of unreported and illegal traffic, but it also concerns movements which are duly notified. Despite the fact that the State of export is responsible for the re-importation of hazardous wastes in case of illegal traffic according to Articles 8 and 9 of the Basel Convention, this does not mean that this State is financially responsible for the resulting damage according to the principle of State responsibility. Such consequence rather requires the additional finding that the State knew or ought to have known about the actual circumstances and failed to prevent the occurrence of damage despite having the opportunity to do so. Since, moreover, States are, for political reasons reluctant to bring interstate claims to international (arbitral) courts, State responsibility does not seem to be the appropriate instrument to respond to transnational damage caused by the activities of private parties.[615] The international practice of States thus tends towards the establishment of civil liability conventions,[616] which are outlined in the following section.

---

[614] See also Birnie/Boyle/Redgwell, *International Law and the Environment* (3rd ed., 2009), at 430–431; Smith, *State Responsibility and the Marine Environment* (1988), at 255.

[615] See Murphy, 'Prospective Liability Regimes', 88 *AJIL* (1994), at 38–45; Soares/Vargas, 'The Basel Liability Protocol', 12 *Yb. Int'l Env. L.* (2001), at 73.

[616] This has been established also by the ILC, *supra*, Sect. "State Liability".

D. Existing Civil Liability Conventions

Besides the principle of State responsibility, civil liability conventions play a major role for the establishment of liabilities for transnational damage caused by the conduct of private parties. Civil liability conventions establish a uniform legal regime that provides uniform standards and procedural rules regarding liability and compensation among the Contracting States of the respective convention. By means of civil liability conventions liability is directly attached not only to individuals and corporate entities, but also to States provided the State acts in the individual case as a private entity, meaning that it does not exercise public power while acting. This is the case, for example, if the State is deemed to be the civil operator of an activity that is governed by a specific civil liability convention.

In this section, the existing international civil liability conventions relevant to the transboundary movement of hazardous wastes by sea are outlined.

I. 1999 Protocol to the Basel Convention

Article 12 of the Basel Convention contains a *pactum de negotiando*, i.e. a mandate upon the Contracting States to elaborate and adopt a protocol on liability and compensation for damage resulting from the transboundary movement of hazardous wastes. This prospective protocol was not intended to be conceived as an amendment to the Convention, but rather to assume the form of an independent legal instrument that has to be ratified additionally. Article 12 itself does not provide any substantive requirements and does not determine whether liability is to be attached to the State or to civil parties.[617] Similar clauses containing a *pactum de negotiando* can be found in Articles 12 of the Bamako and Waigani Conventions as well as in Article 14 of the Izmir Protocol to the Barcelona Convention. However, it is only in respect of the Basel Convention that a protocol on civil liability and compensation has actually been adopted, albeit still pending its ratification.

1. Evolution of the Protocol

During the negotiations of the Basel Convention the issue of liability and compensation was highly contested, so that in order to promote the adoption of the Basel Convention this issue was put aside for later consideration.[618] Nevertheless,

---

[617] See *supra*, Sect. "Article 12".

[618] See Birnie/Boyle/Redgwell, *International Law and the Environment* (3rd ed., 2009), at 482–483; 232; Gwam, 'Travaux Preparatoires of the Basel Convention', 18 *J. Nat. Resources & Envtl. L.* (2003/2004), at 66–67; Kummer, *The Basel Convention* (1995), at 72.

it is not only the mandate to elaborate and adopt a protocol on liability and compensation as laid down in Article 12 of the Convention that reflects the sense of urgency of this issue. It was also emphasised by Resolution 3 of the 1989 Basel Conference and by Resolution 44/226 of the UN General Assembly that requests the Executive Director of the UNEP to establish an *Ad Hoc* Working Group of legal and technical experts to develop elements that might be included in a protocol on liability and compensation.[619]

The first *Ad Hoc* Working Group convened two sessions in 1990 and 1991 and prepared, in collaboration with the Interim Secretariat to the Basel Convention, a draft set of articles on liability and compensation, which was submitted to the First Conference of the Parties (COP1) held in December 1992.[620] Already at this early stage the view was taken that emergency response should be separated from the compensation topic and be dealt with by an international fund for immediate response action.[621] COP1 then decided to establish a second *Ad Hoc* Working Group to continue and conclude the drafting process on the liability provisions and to possibly include the establishment of an international compensation fund.[622] This second *Ad Hoc* Working Group met ten times from 1993 to 1999 and submitted a final draft protocol on liability and compensation to the Fifth Conference of the Parties (COP5) in December 1999.[623] This draft, however, left open some essential interrelated issues,[624] pending final negotiation and decision by the Conference of the Parties. COP5 convened a legal working group for the consideration and finalisation of these issues which was able to finally achieve a compromise. Although this compromise was criticised as hardly meeting the interests of the respective factions,[625] it was adopted by COP5 as the Basel Protocol on Liability and Compensation for Damage Resulting from the Transboundary Movement of Hazardous Wastes and their Disposal on 10 December 1999.[626] Despite the fact that 13 States signed the Protocol in 2000, it quickly turned out that States remained reluctant to ratify the Protocol.

---

[619] UNGA Res. A/44/226 of 22 December 1989.

[620] Note of the Secretariat of 7 July 1992 (COP1 Doc. UNEP/CHW.1/5).

[621] Soares/Vargas, 'The Basel Liability Protocol', 12 *Yb. Int'l Env. L.* (2001), at 71.

[622] Decision I/5 of COP1, 5 December 1992 (Doc. UNEP/CHW.1/24). This mandate was extended by Decision II/1 of COP2, 25 March 1994 (Doc. UNEP/CHW.2/30), Decision III/2 of COP3, 28 November 1995 (Doc. UNEP/CHW.3/35), and Decision IV/19 of COP4, 18 March 1998 (Doc. UNEP/CHW.4/35).

[623] Report of the 10th Session of the Ad Hoc Working Group of Legal and Technical Experts of 20 September 1999 (Doc. UNEP/CHW.1/WG.1/10/2).

[624] This included the scope of application, strict liability, conflicts with other agreements, financial limits, insurance and other financial guarantees and compensation mechanisms. See in detail Report of the 10th Session of the Ad Hoc Working Group of Legal and Technical Experts of 20 September 1999 (Doc. UNEP/CHW.1/WG.1/10/2); 'No Agreement on Draft Protocol', 29 *Envtl. Pol'y & L.* (1999), at 154.

[625] Krueger, 'The Basel Convention', *YBICED* (2001/2002), at 46; Sharma, 'The Basel Protocol', 26 *Delhi L. Rev.* (2004), at 184–185.

[626] Decision V/29 of COP5, 10 December 1999 (Doc. UNEP/CHW.5/29).

Since at the opening of the Sixth Conference of the Parties (COP6) in December 2002 not a single State had ratified the Protocol, COP6 called on the Parties to expedite the process of ratification in order to enable the Protocol to enter into force. In addition, a questionnaire was developed by which the Parties were invited to submit their comments regarding possible disincentives and obstacles for ratification; furthermore, the Basel Secretariat was requested to organise regional workshops for addressing various aspects and obstacles for ratification as well as to prepare a detailed instruction manual for the implementation of the Protocol.[627] At the Seventh Conference of the Parties in October 2005 (COP7), a first set of reports on the regional workshops were presented, which revealed that possible obstacles to ratification included, *inter alia*, a lack of technical and legal expertise, the non-availability of environmental insurance and an absent awareness of the decision makers regarding the Protocol's advantages and disadvantages.[628] Furthermore, COP7 agreed to convey a draft instruction manual for the implementation of the Protocol to the Open-ended Working Group for consideration and approval and requested the Secretariat to update the final manual on a regular basis.[629] The revised text of the Instruction Manual was finally published in May 2005.[630] At the same time the pace of ratifications of the Protocol increased with seven ratifications occurring between 2004 and 2005,[631] before the acceptance of the Protocol again dropped, with only three additional ratifications occurring by 2007.[632]

At COP8 in November/December 2006, further reports on the regional workshops were presented.[633] The Parties were invited to submit comments on Annexes A and B of the Protocol, and the Secretariat was requested to proceed with its work on the options that may be available with respect to the requirement of insurance or other financial guarantees pursuant to Article 14 of the Protocol.[634] At COP9, which was held in June 2008, no tangible progress regarding the ratification of the Basel

---

[627] Decision VI/15 of COP6, 10 February 2003 (Doc. UNEP/CHW.6/40).

[628] See Reports on the workshops held in Addis Ababa, Ethiopia, 16 September 2004 (Doc. UNEP/CHW.7/INF/11), in Buenos Aires, Argentina, 30 September 2004 (UNEP/CHW.7/INF/11/Add.1), and in San Salvador, El Salvador, 28 October 2004 (UNEP/CHW.7/INF/11/Add.2). See also Summary by the Secretariat of the obstacles and difficulties faced by the Parties in the process of ratification or accession to the Protocol, 26 May 2005 (Doc. UNEP/CHW/OEWG/4/INF/4).

[629] Decision VII/28 of COP7, 25 January 2005 (Doc. UNEP/CHW.7/33).

[630] See Open-ended Working Group Doc. UNEP/CHW/OEWG/4/8.

[631] Botswana, Democratic Republic of the Congo, Ethiopia, Ghana, Liberia, Syrian Arab Republic and Togo.

[632] Colombia, Republic of the Congo and Yemen.

[633] See Reports on the workshops held in Cairo, Egypt, 6 November 2006 (Doc. UNEP/CHW.8/INF/16) and in Yogyakarta, Indonesia, 16 June 2006 (Doc. UNEP/CHW.8/INF/16/Add.1). See as to a further workshop held in Warsaw, Poland, Report of the Open-ended Working Group, 6 February 2006 (Doc. UNEP/CHW/OEWG/5/INF/6).

[634] Decision VIII/25 of COP8, 5 January 2007 (Doc. UNEP/CHW.8/16). To the latter see also Open-ended Working Group, 2 March 2006 (Doc. UNEP/CHW/OEWG/5/2/Add.7).

Protocol could be observed. The Secretariat reported on its elaboration regarding the financial instruments available with respect to the requirement of Article 14 of the Protocol, but it also emphasised that, apart from this, further efforts had to remain uninitiated since the Secretariat had not received any financial contributions by the Parties that would have enabled it to undertake its tasks.[635] COP9 thus could only appeal to Parties "to expedite the process of ratifying the Protocol [...] to facilitate its entry into force at the earliest opportunity" and urge Parties to determine possible means of overcoming perceived obstacles to ratification.[636] The 10th Conference of the Parties (COP10) was held in October 2011 and the 11th Conference of the Parties (COP11) was held in April and May 2013. Both Conferences did not even put the topic of liability and compensation on their agenda.

Since 2007 only one further State has ratified the Protocol.[637] It has today been ratified by a total of 11 States, and it is not possible to predict whether or not the required number of 20 ratifications will be reached in future.

2. Legal Objectives and Main Content

The legal objective of the Basel Protocol is laid down in its Preamble and Article 1, and it consists of the establishment of a comprehensive regime for liability and for adequate and prompt compensation of the victims of pollution and other environmental damage resulting from the transboundary movement of hazardous wastes including illegal traffic.[638] The Protocol aims at supplementing and completing the legal framework established by the Basel Convention and, thus, also functions as a means to ensure compliance with the provisions of the Basel Convention.[639] However, besides this "official" goal, the Basel Protocol is intended by the States to fulfil some further expectations. Initially, the Protocol was intended to respond to and remedy concerns raised especially by developing countries regarding their lack of financial funds and technological expertise for coping with illegal dumping or accidental damage resulting from hazardous waste movements.[640] Another expectation in the Liability Protocol is the creation of a deterrence effect that compels the private actors not to infringe the substantive rules of the Basel Convention in order to not incur liability for the resulting

---

[635] Note by the Secretariat, 17 April 2008 (Doc. UNEP/CHW.9/29). See also Open-ended Working Group, 29 June 2007 (Doc. UNEP/CHW/OEWG/6/14).

[636] Decision IX/24 of COP9, 27 June 2008 (Doc. UNEP/CHW.9/39).

[637] In 2013, Saudi Arabia ratified the Protocol.

[638] Basel Protocol, Article 1 and Preamble 1.

[639] Kummer, *The Basel Convention* (1995), at 210–211; Wolfrum, 'Means of Ensuring Compliance and Enforcement', 272 *RdC* (1998), at 77–81.

[640] Basel Secretariat, *Implementation Manual* (2005), at I. Introduction; 'No Agreement on Draft Protocol', 29 *Envtl. Pol'y & L.* (1999), at 154; de la Fayette, *Compensation for Environmental Damage*, in: Kirchner (ed.) (2003), at 247; Widawsky, 'In My Backyard', 38 *Envtl. L.* (2008), at 609.

damage. The desired effect of the Liability Protocol, thus, is the internalisation of external costs, particularly those relating to environmental damage.[641] Finally, the Protocol is also intended to facilitate legal certainty and prevent forum shopping by establishing international uniform law which includes rules on competent courts, the mutual recognition of judgments and enforcement provisions.[642]

In order to achieve these objectives the Basel Protocol provides for the following basic features: The Protocol is a pure civil liability regime. Unlike the obligation to take back or, alternatively, to dispose of hazardous wastes in case of illegal traffic pursuant to Article 9(2) and (3) of the Basel Convention, the Protocol does not attach any second-tier liability directly to the States. It is, however, also made clear that the general rules of international law with respect to State responsibility are not affected by the Protocol.[643] The Basel Protocol establishes a comprehensive system of interdependent liabilities of the private actors involved. It ensures that in any phase of the transport financial responsibility is attached to at least one private party. To this end, it provides for strict liability that is imposed on the notifier (which may be the generator or exporter[644]) or on the disposer of the wastes, depending on whether the disposer has taken possession of the wastes.[645] The liable person, however, is relieved from strict liability in uncontrollable situations like armed conflicts or exceptional natural phenomenon.[646] In addition to strict liability, the Protocol imposes fault-based liability on any person responsible for a lack of compliance with the provisions of the Basel Convention or in the event of wrongful intentional, reckless or negligent acts or omissions.[647] While fault-based liability is necessarily unlimited in amount, strict liability may be limited by the respective domestic laws of the Contracting States, provided that those limits do not fall below the minimum amounts of liability determined in Annex B of the Protocol.[648] Damage covered by the Protocol includes not only personal damage and damage to property, but also loss of income and costs for reinstatement and preventive measures.[649] Furthermore, the Basel Protocol requires that persons strictly liable under the Protocol[650] be covered by insurance,

---

[641] See Lawrence, 'Negotiation of a Protocol on Liability and Compensation', 7 *RECIEL* (1998), at 250; Murphy, 'Prospective Liability Regimes', 88 *AJIL* (1994), at 62–63.
[642] Lawrence, 'Negotiation of a Protocol on Liability and Compensation', 7 *RECIEL* (1998), at 250.
[643] Basel Protocol, Article 16.
[644] See *ibid.*, Article 6(1).
[645] *Ibid.*, Article 4(1) and (2).
[646] *Ibid.*, Article 4(5).
[647] *Ibid.*, Article 5.
[648] *Ibid.*, Article 12(2) and (1) in connection with Annex B.
[649] *Ibid.*, Article 2(c).
[650] This may be the generator, exporter, importer or disposer of the wastes.

bonds or other financial guarantees for an amount not less than that set out in Annex B to the Protocol.[651] A separate trust fund, however, is not established by the Basel Protocol.[652]

II. 1996/2010 HNS Convention

The International Convention on Liability and Compensation for Damage in Connection with the Carriage of Hazardous and Noxious Substances by Sea (HNS Convention) was elaborated under the auspices of the International Maritime Organization (IMO). It forms part of the triad of maritime civil liability conventions channelling liability to the shipowners or cargo receivers in case of environmental pollution damage resulting from regular shipping activities, particularly from the transportation of persistent oils or hazardous and noxious substances (HNS) or from bunker oil.[653]

1. The Evolution of the 1996 HNS Convention and Its Main Content

Triggered by the *Torrey Canyon* oil pollution disaster, the then existing IMCO[654] decided, as far back as 1967, to advance its work on the issue of liability and compensation for maritime pollution damage caused by persistent oils and HNS.[655] To this end, IMCO installed the permanent Legal Committee and instructed it to prepare a draft convention on civil liability for damage caused by HNS, this draft convention to be considered and adopted at a Diplomatic Conference in 1984. The first draft convention already contained some of the key elements that have been adhered to until today. The draft convention provided for a two-tier system of compensation mechanisms, consisting of limited liability of shipowners as well as compulsory insurance on the first tier and a compensation fund on the second tier that applies if liability fails to attach to the shipowners or if it is exceeded in amount by the actual damage.[656] However, a consensus on this draft convention could not be reached at the 1984 Conference. The main points of conflict turned out to be the application only to bulk cargo as well as uncertain

---

[651] Basel Protocol, Article 14.
[652] See *ibid.*, Article 15.
[653] As to the CLC/Fund Conventions and the Bunker Oil Convention see *infra*, Sect. "Liability for Oil Pollution from Ships".
[654] Until 1982 the IMO was known as the Inter-Governmental Maritime Consultative Organization (IMCO).
[655] IMCO Doc. C/ES.III/5 of 8 May 1967.
[656] Göransson, 'The HNS Convention', 2 *Unif. L. Rev.* (1997), at 250. For the text of the Draft Convention see Annex 1 to the IMO Circular Letter No. 958.

provisions on the limitation of shipowners' liability.[657] As a consequence, this matter was referred back to the Legal Committee for further elaboration and for negotiating a common ground of the essential elements of the proposed regime. After more than 10 years of negotiation the revised draft convention was presented at a Diplomatic Conference held in 1996 under the auspices of IMO and finally was adopted as the 1996 HNS Convention.[658]

The 1996 HNS Convention aims at ensuring adequate, prompt and effective compensation for persons who suffer damage caused by incidents in connection with the carriage by sea of HNS.[659] It is based on the rationale that the economic consequences of such incidents should be borne jointly by the shipping industry and the cargo interests.[660] The Convention applies to HNS that are shipped on board a ship as cargo as well as to residues thereof. HNS are defined by the Convention neither according to an abstract definition nor by means of a comprehensive enumeration of single substances, but rather by reference to existing lists of HNS contained in other IMO instruments, like in the MARPOL 73/78 Convention or in the IMDG Code.[661] The Convention basically covers HNS regardless of whether they are shipped in bulk or in packaged form. It also applies to hazardous wastes, provided those wastes also come within the definition of HNS under the Convention.[662] In contrast, damage caused by radioactive substances and pollution damage as defined in the CLC are excluded from the scope of the HNS Convention.[663] The notion of damage underlying the HNS Convention resembles to a large extent the definitions in comparable civil liability conventions.[664]

Regarding the regime of liability, the HNS Convention adheres to the two-tier system of compensation mechanisms, providing for shipowners' civil liability on the first tier and a supplemental HNS compensation fund on the second tier. At the first layer strict liability is imposed on the shipowner provided that he is not in a position to invoke the exclusion of liability due to specific circumstances that are beyond his control.[665] The shipowner is, furthermore, entitled to limit liability to

---

[657] Drel, *Liability for Damage Resulting from the Transport of Hazardous Cargoes by Sea*, in: Couper/Gold (ed.) (1993), at 360–362; Ganten, 'HNS and Oil Pollution', 27 *Envtl. Pol'y & L.* (1997), at 313; Göransson, 'The HNS Convention', 2 *Unif. L. Rev.* (1997), at 250.

[658] See HNS Conf. Doc. LEG/CONF.10/8/3; in addition to the adoption of the HNS Convention, the 1996 Conference also adopted the Resolution on Setting up the HNS Fund and the Resolution on the Relationship between the HNS Convention and a Prospective Regime on Liability for Damage in Connection with the Transboundary Movement of Hazardous Wastes, see HNS Conf. Doc. LEG/CONF.10/8/3.

[659] 1996 HNS Convention, Preamble 2.

[660] *Ibid.*, Preamble 4; Rengifo, 'The HNS Convention', 6 *RECIEL* (1997), at 192.

[661] 1996 HNS Convention, Article 1(5). See also Ganten, 'HNS and Oil Pollution', 27 *Envtl. Pol'y & L.* (1997), at 312–313.

[662] Göransson, 'The HNS Convention', 2 *Unif. L. Rev.* (1997), at 252–256.

[663] 1996 HNS Convention, Article 4(3).

[664] Wetterstein, 'The HNS Convention', 26 *Ga. J. Int'l .& Comp. L.* (1996/1997), at 601.

[665] 1996 HNS Convention, Article 7.

certain amounts calculated according to the ship's tonnage and by means of constituting a limitation fund with the competent court.[666] As an offset, he is obliged to maintain insurance or other financial security covering the amount of limited liability.[667] As a second layer of liability the HNS Convention establishes the HNS Fund, this being intended to supplement the availability of compensation under the first tier. Compensation from the HNS Fund can be sought if compensation from the shipowner is inadequate or not available, namely if (1) no liability is attached to the shipowner, (2) the shipowner is financially incapable and no financial security is available, or (3) the actual damage exceeds the limit of liability. The aggregate amount of compensation to be claimed from the HNS Fund is limited to 250 million SDR.[668] The Fund is financed by post-event annual contributions of the receivers of cargo consisting of HNS.[669] It is split into a general account and three separate accounts for oil (oil account), liquefied natural gases (LNG account) and liquefied petroleum gases (LPG account), with the latter separate accounts only covering damage caused by the respective specific substance.[670] The HNS Fund will be governed by an Assembly and administered by a Secretariat.[671] In order to ensure that the HNS Fund can operate properly when the Convention enters into force, the 1996 IMO Conference requested the 1992 IOPC Fund to carry out the administrative and organisational tasks necessary for setting up the HNS Fund.[672] The HNS Convention finally provides for rules on claims and actions.[673] It enters into force 18 months after (1) twelve States have ratified the Convention, including four States each with not less than 2 million units of gross tonnage, and (2) those persons liable to contribute to the HNS Fund have received during the preceding calendar year a total quantity of at least 40 million tonnes of cargo contributing to the general account.[674]

2. Perceived Deficiencies and Obstacles to Ratification

The HNS Convention and, in particular, also the HNS Fund are substantially modelled on the oil pollution compensation scheme established by the 1969 CLC

---

[666] *Ibid.*, Article 9.

[667] *Ibid.*, Article 11.

[668] *Ibid.*, Article 14.

[669] For the detailed consideration of the issue of "who should pay the levy?" see IMO Doc. LEG71/3/4, at 2–5.

[670] 1996 HNS Convention, Articles 16 to 21.

[671] *Ibid.*, Article 24.

[672] IMO Resolution on Setting up the HNS Fund, see HNS Conf. Doc. LEG/CONF.10/8/3.

[673] 1996 HNS Convention, Chapter IV.

[674] *Ibid.*, Article 46.

and 1971 Fund Conventions, including their 1992 Protocols.[675] However, there are also basic differences between both regimes as well as between the respective underlying circumstances. First of all, in the HNS Convention both tiers of liability are consolidated into one convention, whereas with respect to oil pollution damage each tier is addressed by an individual convention. Furthermore, the receivers of HNS forming the group of persons contributing to the HNS Fund turned out to be much more numerous, discordant and less cooperative than the oil cargo receivers contributing to the IOPC Fund. In addition, the substances covered by the HNS Convention are much less homogeneous than the "persistent oil" covered by the oil pollution conventions, and they bear a potential threat to the marine environment that is by far more severe than the risk potential of persistent oils.[676]

Three issues in particular were highly contested even after the adoption of the Convention and turned out to be the main obstacles to ratification. These include (1) the contribution of packaged HNS cargo to the HNS Fund, (2) contributions to the LNG account, and (3) the non-submission of contributing cargo reports.[677]

(a) Contribution of Packaged Cargo to the HNS Fund

The first issue concerns the contribution of packaged HNS cargo to the HNS Fund and originates in the Convention's definition of "receiver" in connection with the inclusion of packaged goods into the definition of HNS. The term "receiver" as defined by the Convention means any person who physically receives HNS, including ports and terminals. This, however, will not apply in those cases where the physical receiver is able to disclose the identity of his principal and this principal is also subject to the jurisdiction of any State Party.[678] As far as the carriage of HNS in bulk is concerned, this definition does not cause any major difficulties with regard to either the reporting obligations under the Convention or the calculation of contributions to the HNS Fund.[679] If HNS are carried in bulk, this mostly involves large amounts of cargo discharged and transhipped in the port, making it possible for the port operator to inspect the transport documents and to identify and report the transport particulars including the identity of the principal. This is different with respect to packaged cargo. In practical terms, packaged cargo means that the HNS are packed on pallets or in bales or are filled in IBCs, which are in turn stuffed in a container (that may be a consolidated container with a

---

[675] Rengifo, 'The HNS Convention', 6 *RECIEL* (1997), at 192; Wetterstein, 'The HNS Convention', 26 *Ga. J. Int'l .& Comp. L.* (1996/1997), at 599.

[676] Ganten, 'HNS and Oil Pollution', 27 *Envtl. Pol'y & L.* (1997), at 314 and also at 312; Rengifo, 'The HNS Convention', 6 *RECIEL* (1997), at 195; de la Rue/Anderson, *Shipping and the Environment* (2nd ed., 2009), at 269–270.

[677] See e.g. IMO Doc. LEG 93/13, at 8; IMO Doc. LEG 94/12, at 6–18; IOPC Fund Doc. 92FUND/A/ES.13/5, at 6–11.

[678] 1996 HNS Convention, Article 1(4).

[679] See *ibid.*, Articles 21 and 17.

smorgasbord of different cargoes). Under such conditions it is practically impossible and would only lead to administrative barriers if the port or terminal, which merely provides the logical infrastructure, were obligated to identify and report the particular principal regarding each single package of HNS stuffed in a container.[680] This applies especially since such information cannot always be drawn from the transport documents accompanying the cargo, and since the port or terminal is in general not entitled to open the container for the purpose of inspection.

(b) Contributions of the LNG Account

The second controversial issue concerns the person liable for contributions to the LNG account. Different from the usual contribution scheme involving the general and other separate accounts of the HNS Fund as well as the IOPC Fund, the person liable for contributions to the LNG account is not the receiver of cargo, but the person who "immediately prior to its discharge, held title to an LNG cargo" that is discharged in a port or terminal of a State Party.[681] This deviating approach goes back to the circumstances prevailing in the LNG industry at the time the HNS Convention was drafted. It was intended to take account of the fact that only few States and companies were involved in the LNG trade, so that it was considered reasonable to impose obligations for levies to the LNG Account only on those companies. Since that time, however, the LNG industry has developed tremendously, and this initial approach is no longer considered to reflect a reasonable economic balance between exporters and importers of LNG. Particularly developing countries pointed to the fact that most LNG producers or exporters are situated in developing countries and they thus suspected that the initial arrangement of contribution liability would impose a serious economic burden especially on them. A further objection against this approach was that in most cases the titleholder of HNS cargo may not be subject to the jurisdiction of a State Party to the HNS Convention. Finally, some States also expressed their concerns that it will prove hard to identify the titleholder of LNG immediately prior to discharge given that LNG is traded on the spot market, with the result that the titleholder may change several times during the voyage.[682]

---

[680] See also 1992 IOPC FUND Doc. 92FUND/WGR.5/3, at 2; 1992 IOPC Fund Doc. 92FUND/A/ES.13/5, at 6.

[681] 1996 HNS Convention, Article 19(1)(b).

[682] 1992 IOPC Fund Doc. 92FUND/A/ES.13/5, at 8; 1992 IOPC Fund Doc. 92/FUND/WGR.5/2; Griggs/Shaw, 'IMO Considers Adoption of the HNS Protocol', 111 *Dir. Marit.* (2009), at 278–279; see also Røsæg, 'Non-Collectable Contributions to the LNG Account', 13 *JIML* (2007), at 94 *et seq.;* Weems/Keenan, 'Is the LNG Industry Ready for Strict Liability?', Nov/Dec 2003 *LNG Journal* (2003), at 16.

(c) Non-submission of Contributing Cargo Reports

The last main obstacle for the entry into force of the 1996 HNS Convention was seen in the high reporting requirements of the HNS Convention regarding the quantity and composition of imported HNS cargo. The submission of contributing cargo reports by States is vital for the functioning of the HNS Convention in two regards. First, the entry into force of the Convention depends on the representation of a certain amount of contributing cargo by the ratifying States and, second, the levies to be paid by a State Party to the HNS Fund are calculated according the respective amount of contributing cargo.[683] Strong concerns have been voiced regarding the States' readiness to submit comprehensive and complete HNS cargo reports on a regular basis. Reference was made to the same problems faced with regard to the IOPC Fund. With respect to the HNS Convention the underlying circumstances are considered to be even more complex. This involves not only the number of substances covered by the HNS Convention, but also the lower threshold of amounts to be reported under the HNS Convention as well as the more diverse types of companies involved. As a result, it was feared that the administrative and bureaucratic barriers imposed by the reporting obligation would act as a major obstacle and disincentive for ratification.[684]

(d) Conclusion

In addition to these three main deficiencies and obstacles regarded as preventing the ratification of the 1996 HNS Convention, there are also some minor issues that may have served as a deterrent to Convention ratification for at least some States. However, if one tries to draw a general conclusion from this section it becomes apparent that none of these perceived obstacles for ratification are primarily related to the first tier of liability imposed on the shipowners, instead being almost exclusively related to the second tier consisting of the HNS Fund. The HNS Fund, or to be more precise, the arrangement of contributions to the HNS Fund, therefore, must be regarded as the "elephant in the room" of the HNS liability regime.

Within the oil pollution liability regime the IOPC Fund has been set up by an independent convention[685] adopted subsequent to a civil liability convention imposing limited strict liability on the shipowner.[686] Regarding the HNS compensation regime both elements are combined into one legal instrument, which all

---

[683] 1996 HNS Convention, Article 46(1)(b) and Article 21(2).

[684] 1992 IOPC Fund Doc. 92FUND/A/ES.13/5, at 10; 1992 IOPC Fund Doc. 92FUND/WGR.5/4, of 18 January 2008; Shaw, 'HNS—Is the End in Sight?', 3 *LMCLQ* (2009), at 282–283.

[685] 1971 Convention on the Establishment of an International Fund for Compensation for Oil Pollution Damage as amended by the 1992 Protocol (Fund Convention).

[686] 1969 Convention on Civil Liability for Oil Pollution Damage as amended by the 1992 Protocol (CLC).

the more represents a package deal. This combination of both elements in one legal instrument is surely an advantage since sufficient financial compensation for HNS spills can only be provided by a combination of both elements, and only the inclusion of exporters and importers in the compensation regime may ensure economic balance between the respective industrial interests involved. However, making the adoption of one tier of liability dependent on the adoption of also the other tier involves the increased risk that in the end none of these elements will come into force.

In this context, also the reference to the successful implementation of the IOPC Fund may not be used as a proper argument for a necessary combination of both tiers of liability. First, the oil compensation regime by itself does not consist of only one instrument and, second, there are fundamental differences between the economic background of the transportation of oil, on the one hand, and of HNS on the other. The scope of substances covered by the HNS Convention is by far more extensive than the scope of substances covered by the oil pollution compensation regime. This applies in respect of both its composition and its economic value, the latter ranging from waste substances with a negative economic value to LNGs and LPGs having a tremendous economic value. Furthermore, there are not only a handful of global players as in the oil industry; rather, various types of industries and companies are involved on both ends of HNS trades.[687]

The commercial interests regarding transportation of HNS thus may not be comparable to the commercial interests involved in the oil transportation industry. The perceived deficiencies of the HNS Convention as outlined above and, in particular, the combination of both tiers of liability in one legal instrument may, therefore, function as a strong disincentive for States to subject themselves to a comprehensive international liability regime in the shape of the 1996 HNS Convention. This assessment may well apply at least until that point reliable data on the prospective amounts of liability becomes available. Although 14 States have ratified the 1996 HNS Convention,[688] which is more than the required number of 12 ratifications, only 3 of these States have a merchant fleet of more than 2 million units of gross tonnage, and 13 of these States did not provide any data regarding cargo contributing to the HNS Fund.[689] The Convention, therefore, has not yet entered into force.

Regardless of the attempts to overcome the perceived deficiencies of the HNS Convention by amending particular Articles, one could also raise the question why the Convention rigidly adheres to the initial approach of making both tiers of liability dependant on each other and strictly requires a simultaneous entry into

---

[687] Boga, 'How to Promote the Acceptance of the HNS Convention? (II)', 10 *LHD* (2012), at 130; de la Rue/Anderson, *Shipping and the Environment* (2nd ed., 2009), at 269–270.

[688] Angola, Cyprus, Ethiopia, Hungary, Liberia, Lithuania, Morocco, Russian Federation, Saint Kitts and Nevis, Samoa, Sierra Leone, Slovenia, Syrian Arab Republic, Tonga.

[689] Cyprus, Liberia and the Russian Federation; see Røsæg, *The Rebirth of the HNS Convention*, in: Berlingieri/Boglione/Carbone/Siccardi (ed.) (2010), at 858; and the documentation at http://folk.uio.no/erikro/WWW/HNS/hns.html.

force of both elements. Would one give up too much if, for instance, the Convention in a first step established only civil liability of shipowners and also provided for some reporting obligations regarding the quantities of HNS actually shipped without however the consequence that these amounts would be used to calculate levies to be paid into a trust fund? Then, only in a second step, and after sufficient data regarding the quantities of shipped HNS have been obtained, thus allowing for a realistic evaluation of the amount of contributions to be paid to the trust fund, an HNS Fund could be set up. This second step could be achieved by an individual convention, but it could also be realised within the same convention, provided consultations on the establishment of the fund were scheduled in advance and the entry into force of this second part were made contingent on its further adoption by the States Parties.

3. The 2010 Protocol to the HNS Convention and Further Development

In response to the 1996 IMO Conference Resolution on seeking assistance of the IOPC Fund to prepare the setting up of the HNS Fund, the IOPC Fund Assembly in 1998 decided to carry out all necessary tasks and to give all necessary assistance in this regard.[690] However, the actual contribution of the IOPC Fund was limited at first. By contrast, the IMO Legal Committee held two Special Consultative Meetings in 1998 and 1999 to discuss issues related to the entry into force of the HNS Convention.[691] Additionally, in 1999, IMO established the Correspondence Group on the Implementation of the HNS Convention, which was assigned with the task of monitoring and further encouraging the implementation of the HNS Convention.[692] After an additional informal meeting of the Correspondence Group in 2001 and an IMO Assembly Resolution urging the States to ratify the Convention,[693] a final Special Consultative Meeting of the HNS Correspondence Group was held in Ottawa in June 2003 that proposed to the IMO Legal Committee which action should be taken in order to facilitate the entry into force of the HNS Convention.[694] In the following years, also the IOPC Fund began to play a more important role. This started in 2005 when it published a Guide to the Implementation of the HNS Convention[695] and continued with the extensive discussion at its 12th Session in 2007 of the main causes for non-ratification of the Convention, particularly involving the three key issues outlined above. This

---

[690] 1992 IOPC Fund Doc. 92FUND/A.1/34, para. 33.1.1–33.1.3.
[691] As to the outcome of the second meeting see IMO Doc. LEG/80/10/2.
[692] IMO Doc. LEG/80/11; see also IMO Doc. LEG/80/10/3.
[693] IMO Res. A 22/Res. 932 of 29 November 2001.
[694] See IMO Doc. LEG/87/11 and IMO Doc. LEG/87/11/1. See also Røsæg, *The Rebirth of the HNS Convention*, in: Berlingieri/Boglione/Carbone/Siccardi (ed.) (2010), at 855–856.
[695] This Guide is published e.g. at www.hnsconvention.org.

Session resulted in the establishment of the HNS Focus Group that aimed at developing a Protocol to the Convention in order to overcome the perceived deficiencies of the 1996 HNS Convention.[696]

The HNS Focus Group held two meetings in 2008, at which it elaborated a draft text of a Protocol to the HNS Convention.[697] It invited the IOPC Fund Assembly to submit this draft to the IMO Legal Committee with a view to convening a Diplomatic Conference for the consideration of this Protocol.[698] The draft Protocol and the main issues were considered in detail at the 94th and 95th Sessions of the IMO Legal Committee in 2008/2009, which resulted in the approval of a revised draft Protocol to be considered by a Diplomatic Conference.[699] The Diplomatic Conference on the Revision of the HNS Convention was convened in April 2010 in London. In its Final Act[700] it adopted the final text of the 2010 Protocol to the HNS Convention[701] as well further resolutions including the Resolution on Setting up the HNS Fund, which, reproducing the 1996 Resolution virtually verbatim, sought assistance of the IOPC Fund to carry out the necessary tasks and to provide all necessary assistance for setting up the HNS Fund.[702]

The 2010 Protocol to the HNS Convention states that the Protocol and the 1996 HNS Convention are to be read and interpreted together as one single instrument.[703] It provides for the following main improvements: Regarding the issue of the contribution of packaged cargo to the HNS Fund, one proposal finally gained acceptance at the Diplomatic Conference.[704] According to this, packaged HNS cargo is no longer covered by the definition of contributing cargo, resulting in the exclusion of the receiver of packaged HNS from the obligations to contribute to the HNS Fund and to report the amounts of shipped cargo.[705] For damage caused by packaged HNS, compensation can still be sought from the HNS Fund at the second tier of liability. However, in order to avoid an unequal treatment of receivers of bulk and packaged HNS, the limitation of shipowners' liability on the

---

[696] See IOPC Fund Doc. 1992FUND/A.12/28. See also IOPC Fund Docs. 92FUND/A.12/25/1, 92FUND/A.12/25/2 and 92FUND/A.12/25/3.

[697] The consolidated text of the draft Protocol can be found at IOPC Doc. 92FUND/A/ES.13/5/2.

[698] See IOPC Fund Doc. 92FUND/A.13/22.

[699] IMO Doc. LEG/94/12 and IMO Doc. LEG/95/10. Annex III of LEG/95/10 contains the revised draft Protocol to the HNS Convention.

[700] IMO Doc. LEG/CONF.17/12.

[701] The text of the Protocol of 2010 to the International Convention on Liability and Compensation for Damage in Connection with the Carriage of Hazardous and Noxious Substances by Sea, 1996, is contained in IMO Doc. LEG/CONF.17/10.

[702] See IMO Doc. LEG/CONF.17/11.

[703] 2010 Protocol to the HNS Convention, Article 18(1). The amended Convention, thus, shall be denoted in the subsequent text as the 1996/2010 HNS Convention.

[704] See as to the different proposals put forward regarding this issue: Boga, 'How to Promote the Acceptance of the HNS Convention? (II)', 10 *LHD* (2012), at 123–124.

[705] 1996/2010 HNS Convention, Article 1(10).

first tier is increased by 15 % for HNS carried in packaged form.[706] Regarding the obligation to contribute to the LNG Account the Protocol aligns these rules with the contributing scheme of the other accounts. Unlike in the previous regulation, the person fundamentally liable to pay levies to the LNG Account is the receiver of LNG. In addition, the Convention provides for the possibility of opting for liability of the titleholder of the LNG, provided the titleholder has entered into an agreement with the receiver and the receiver has informed its government about this agreement. In this case, however, the receiver remains a guarantor for the payment of the contributions.[707] Finally, and with regard to the non-submission of contributing cargo reports, the Protocol to the HNS Convention introduces sanctions, i.e. temporary suspension of the status as being a Contracting State if no reports are submitted in the context of ratification of the Convention, and the withholding of any compensation payments from the Fund until the injured State has complied with its reporting obligations.[708]

Even though the amendments provided by the 2010 Protocol to the HNS Convention must be considered a major step towards the creation of a well-balanced and effective legal regime for HNS pollution damage and also towards overcoming the perceived deficiencies and obstacles to ratification, the revised 2006/2010 HNS Convention still contains some highly controversial issues. Therefore, it still cannot be foreseen whether this outcome will meet the needs of the respective interests involved. To date, no State has ratified the 2010 Protocol and it is difficult to foresee when the Convention will enter into force. However, one conclusion in certain: If the States do not perceive an increased incentive in ratifying the Convention in the very near future, the Convention must be considered dead in the water. A repeated attempt at amending the Convention, whereby such attempt could also take into account a possible delayed entry into force of the HNS Fund as outlined above,[709] will not be possible in the foreseeable future. It is to be hoped that no HNS disaster will prove necessary to finally trigger the entry into force of the Convention.

III. Further Civil Liability Conventions

There are additional civil liability conventions that may be partially relevant in the context of transboundary movements of hazardous wastes by sea.

---

[706] *Ibid.*, Article 9(1)(b). See also Shaw, 'IMO Diplomatic Conference Adopts HNS Protocol', January/March 2010 *CMI Newsletter* (2010), at 9.

[707] 1996/2010 HNS Convention, Article 19(1*bis*). See also Boga, 'How to Promote the Acceptance of the HNS Convention? (II)', 10 *LHD* (2012), at 126–127.

[708] 1996/2010 HNS Convention, Articles 45(7) and 21*bis*. See also Boga, 'How to Promote the Acceptance of the HNS Convention? (II)', 10 *LHD* (2012), at 129; Shaw, *The 1996 HNS Convention*, in: Berlingieri/Boglione/Carbone/Siccardi (ed.) (2010), at 910.

[709] See *supra*, Sect. "Conclusion" (p. 179).

1. Liability for Oil Pollution from Ships

Strict and limited liability of shipowners for pollution caused by oil carried as cargo on board or in the bunkers of tankers arises from the 1992 Civil Liability Convention (CLC).[710] The second tier of compensation for such damage is formed by the 1992 Fund Convention, which establishes the 1992 IOPC Fund to provide additional compensation in case liability of shipowners on the first tier is insufficient or unavailable.[711] An optional third tier of liability was finally created by a 2003 Protocol to the CLC/Fund Convention,[712] this establishing a Supplementary Fund providing for higher limits of compensation to those States that have ratified this Protocol.[713] The oil pollution compensation regime applies to any pollution damage caused by persistent oil which has escaped or has been discharged from the ship as a result of an incident, provided this pollution damage has been caused in the territory, including the territorial sea, in the exclusive economic zone (EEZ) or in a 200 nautical mile zone of a Contracting State.[714] Since waste oils are not excluded from the scope of application, the oil pollution compensation regime may basically also apply to constellations that come under the notion of transboundary movements of hazardous wastes by sea.

Regarding oil pollution damage caused by bunker or lubrication oil originating from any ship except oil tankers, and provided this pollution damage has been caused in the territory, including the territorial sea, in the exclusive economic zone (EEZ) or in a 200 nautical mile zone of a Contracting State, the 2001 Bunker Oil Convention establishes strict and limited liability of the shipowner.[715] An additional compensation fund forming a second tier of liability has not been set up. The Bunker Oil Convention is of no significance for the transboundary movement of hazardous wastes and, therefore, can be left out of consideration.

---

[710] The 1969 International Convention on Civil Liability for Oil Pollution Damage entered into force on 19 June 1975. It was amended first by a 1976 Protocol and later by the 1992 Protocol that entered into force on 30 May 1996.

[711] The 1971 International Convention on the Establishment of an International Fund for Compensation for Oil Pollution Damage was amended by a 1976 Protocol and subsequently superseded by the 1992 Protocol which entered into force on 30 May 2002. A further Protocol of 2000, which came to force on 27 June 2001, terminated the original 1971 Fund Convention.

[712] The 2003 Protocol on the Establishment of a Supplementary Fund for Oil Pollution Damage entered into force on 3 March 2005.

[713] See for a detailed discussion of the oil pollution compensation regime Tsimplis, 'Marine Pollution from Shipping Activities', 14 *JIML* (2008), at 105–122; Tan, *Vessel-Source Marine Pollution* (2006), at 286–344.

[714] 1992 CLC, Articles II, III.

[715] The 2001 International Convention on Civil Liability for Bunker Oil Pollution Damage (Bunker Oil Convention) entered into force on 21 November 2008.

2. Liability for Nuclear Damage

As regards the maritime transport of nuclear material, the 1971 NUCLEAR Convention[716] provides for some specific rules on the liability of the private parties involved. This Convention basically refers to the existing civil liability regimes regarding the peaceful use of nuclear energy,[717] i.e. the 1960 Paris Convention[718] and the 1963 Vienna Convention.[719] In conformance with these conventions the 1971 NUCLEAR Convention channels liability to the operator of a nuclear installation and provides a specific exoneration clause for any party otherwise liable under an international convention in the field of maritime transport. It stipulates that if nuclear damage was caused by an incident during a maritime transport of nuclear material, and provided that the operator of a nuclear installation is liable for that damage under either the Paris or Vienna Conventions or by virtue of national law, the liability of shipowners and any other persons involved in this transport arising under any other international convention is excluded.[720] The 1971 NUCLEAR Convention, thus, may also provide for relevant rules regarding the transboundary movement of hazardous nuclear wastes by sea. However, since nuclear wastes are excluded from the scope of the present work, this issue shall not be addressed in any further detail.

3. Liability for the Carriage of Dangerous Goods by Land

If damage was caused by dangerous goods during their carriage by road, rail or inland navigation vessel, the 1989 CRTD Convention[721] imposes strict and limited liability on the person in control of the road vehicle or inland navigation vessel, or,

---

[716] The 1971 Convention Relating to Civil Liability in the Field of Maritime Carriage of Nuclear Material entered into force on 15 July 1975.

[717] The legal regime on liability for nuclear damage is outlined e.g. by Churchill, 'Facilitating Civil Liability Litigation', 12 *Yb. Int'l Env. L.* (2001), at 7–15; de la Fayette, 'Addressing Damage to the Marine Environment', 20 *IJMCL* (2005), at 193–198; Lammers, 'New Developments Concerning International Responsibility', 19 *Hague Y. B. Int'l L.* (2006), at 96–99.

[718] 1960 Paris Convention on Third Party Liability in the Field of Nuclear Energy as amended by the Protocols of 1964, 1982 and 2004. This Convention is supplemented by the 1963 Brussels Supplementary Convention to the Paris Convention as amended by the Protocols of 1964, 1982 and 2004, which provides for additional compensation by means of public funds.

[719] 1963 Vienna Convention on Civil Liability for Nuclear Damage as amended by the Protocol of 1997. This Convention is supplemented by the 1997 Vienna Supplementary Compensation Convention, which provides for an additional tier of compensation by means of contributions to be paid by States Parties. Conflicts arising from the simultaneous application of the 1960 Paris Convention and the 1963 Vienna Convention were resolved by means of the 1988 Joint Protocol Relating to the Application of the Vienna Convention and the Paris Convention.

[720] 1971 NUCLEAR Convention, Article 1.

[721] 1989 Convention on Civil Liability for Damage Caused During Carriage of Dangerous Goods by Road, Rail and Inland Navigation Vessels (CRTD Convention).

alternatively, on the person operating the railway line. The CRTD Convention, thus, represents the counterpart of the HNS Convention regarding all transports carried out by land. It applies to transports exclusively carried out by land, but also to the pre-carriage and on-carriage in respect of multimodal transports involving a sea passage.[722] The CRTD Convention in particular also covers hazardous wastes that come under the definition of dangerous goods of the Convention, so that it basically provides for relevant rules regarding the liability for the transboundary movement of hazardous wastes. However, this only applies to the pre-carriage and on-carriage of shipments by sea. Since the CRTD Convention has, moreover, not yet entered into force, it is not in the main focus of the present consideration.

4. Other Civil Liability Conventions

There are three more global civil liability conventions that should be mentioned at this point, although these conventions do not provide for relevant rules regarding the transboundary movement of hazardous wastes by sea. These conventions establish liability for commercial activities and the operation of industrial installations which are considered to pose in general an increased risk for human health and the environment. Thus, the 1962 Nuclear Ships Convention,[723] which has not yet entered into force, imposes absolute civil liability on the operator of nuclear ships for any nuclear damage. The 1976 Mineral Resources Convention[724] concerns pollution damage caused by the operation of an installation for the exploration or exploitation of crude oil from the seabed or its subsoil within the territory of a State Party. Upon its entry into force, this Convention will establish strict liability for the operators of such installations. Finally, the 2003 Kiev Liability Protocol[725] to the 1992 Helsinki Water Convention[726] and the 1992 Convention on the Transboundary Effects of Industrial Accidents[727] should be mentioned. This Protocol imposes strict liability on the operator for damage caused by the transboundary effects of an industrial accident on transboundary waters. It is, however, similarly not in force.

---

[722] For a detailed discussion see de Boer, 'The New Draft CRTD', 9 *Unif. L. Rev.* (2004), at 51 *et seq.;* Haak, *Presence and Prospects of the CRTD Convention*, in: Basedow/Magnus/Wolfrum (ed.) (2010), at 13 *et seq.*

[723] 1962 Convention on the Liability of Operators of Nuclear Ships.

[724] 1976 Convention on Civil Liability for Oil Pollution Damage Resulting from Exploration and Exploitation of Seabed Mineral Resources.

[725] 2003 Kiev Protocol on Civil Liability and Compensation for Damage Caused by the Transboundary Effects of Industrial Accidents on Transboundary Waters to the Convention on the Protection and Use of Transboundary Watercourses and International Lakes and to the 1992 Convention on the Transboundary Effects of Industrial Accidents.

[726] 1972 Helsinki Water Convention on the Protection and Use of Transboundary Watercourses and International Lakes.

[727] 1992 Helsinki Convention on the Transboundary Effects of Industrial Accidents.

IV. Regional Civil Liability Regulations

1. 1993 Lugano Convention

The 1993 Lugano Convention[728] was elaborated under the auspices of the Council of Europe. It is, however, not limited in its regional scope and also accepts ratifications by non-members to the Council of Europe. The Convention imposes strict and unlimited liability on the operator in respect of a dangerous activity for any damage caused by that activity and on the operator of a site for the permanent deposit of waste for damage caused by that deposited waste.[729] The term "dangerous activity" as defined by the Convention includes, on the one hand, the production, handling, storage, use or discharge of dangerous substances, genetically modified organisms or micro-organisms and, on the other hand, the operation of an installation or site for the incineration, treatment, handling, recycling or permanent deposit of waste.[730] Thus, the Lugano Convention does not apply to damages caused during the transportation of hazardous wastes, but to damages caused during the treatment or final disposal of hazardous wastes.[731]

The Convention has not yet entered into force and it is doubtful whether it will ever come into force. The reason for that is seen in the unlimited liability of the operator and the resulting difficulties in obtaining insurance coverage. In addition, due to the adoption of the European Directive on Environmental Liability[732] in 2004, the EU Member States lack any further incentive to expedite the entry into force of the 1993 Lugano Convention.[733]

2. EU Environmental Liability Directive

In April 2004, the European Parliament and the EU Council adopted Directive 2004/35/CE[734] on environmental liability.[735] This Liability Directive is actually not composed as a civil liability instrument, but rather imposes liability under

---

[728] 1993 Convention on Civil Liability for Damage Resulting from Activities Dangerous to the Environment (Lugano Convention).

[729] Lugano Convention, Articles 6, 7.

[730] *Ibid.*, Article 2.

[731] de la Fayette, *Compensation for Environmental Damage*, in: Kirchner (ed.) (2003), at 245.

[732] Directive 2004/35/CE of the European Parliament and the Council. See the following section.

[733] Churchill, 'Facilitating Civil Liability Litigation', 12 *Yb. Int'l Env. L.* (2001), at 28; Kiss, *Strict Liability in International Environmental Law*, in: Führ/Wahl/von Wilmowsky (ed.) (2007), at 219–220; Lammers, 'Responsibility and Liability for Environmental Damage', 31 *Envtl. Pol'y & L.* (2001), at 96.

[734] Directive 2004/35/CE of the European Parliament and of the Council of 21 April 2004 on Environmental Liability with Regard to the Prevention and Remedying of Environmental Damage.

[735] For a detailed discussion see Beyer, 'Eine neue Dimension der Umwelthaftung in Europa?', 16 *ZUR* (2004), at 257 *et seq.*; Winter/Jans/Macrory/Krämer, 'Weighing up the EC Environmental Liability Directive', 20 *J. Envtl. L.* (2008), at 163 *et seq.*

public law on operators of covered activities vis-à-vis the Member States.[736] Its scope of regulation is limited in two regards, namely in respect of the covered damages and in respect of the covered activities. The Directive applies to "environmental damage" which comprises damage to protected species and natural habitats, damage to the ecological, chemical or quantitative status of waters and contaminations of land.[737] Thus, the Directive only covers supra-individual interests of the community as a whole and excludes personal damage, damage to personal property and economic losses incurred by private persons.[738] In addition, the application of the Directive requires that the environmental damage has been caused by an occupational activity which, in respect of water and land damage, is enumerated in Annex III to the Directive.[739] Liability under the present Directive, therefore, is primarily conceived as liability for industrial installations and operations. In consequence of environmental damage or an imminent threat thereof caused by an occupational activity, the operator of an industrial installation is obliged to take preventive and remedial action and is liable to bear the costs for preventive and remedial action taken by a competent authority.[740]

The EU Directive on environmental liability also applies to any operations related to the management of hazardous wastes, including their treatment, transport and final disposal as well as to any occupational activities related to dangerous substances.[741] However, as long as those activities already fall under the regime of the CLC/Fund Convention, the HNS Convention or the CRTD Convention, and provided the corresponding regime is in force, these activities are excluded from the scope of the EU Directive 2004/35/CE.[742] The Directive, in contrast, remains applicable if the operation in question is also governed by the Protocol to the Basel Convention.

V. The LLMC Convention

It is not a legal basis for claims of compensation but a global right to limit liabilities imposed by maritime civil liability conventions that is provided by the 1976 LLMC Convention.[743] This Convention allows shipowners and persons in

---

[736] Lammers, 'New Developments Concerning International Responsibility', 19 *Hague Y. B. Int'l L.* (2006), at 93; Meßerschmidt, *Europäisches Umweltrecht* (2011), at 641.

[737] Directive 2004/35/CE, Article 2(1).

[738] *Ibid.*, Article 3(3); Meßerschmidt, *Europäisches Umweltrecht* (2011), at 641.

[739] Directive 2004/35/CE, Article 3(1).

[740] *Ibid.*, Articles 5, 6, 8.

[741] *Ibid.*, Annex III (2), (7).

[742] *Ibid.*, Article 4(2) in connection with Annex IV. See also Tsimplis, 'Marine Pollution from Shipping Activities', 14 *JIML* (2008), at 130–132.

[743] The 1976 London Convention on Limitation of Liability for Maritime Claims (LLMC Convention), as amended.

similar positions to limit their liability by establishing a limitation fund with the court at which a claim subject to limitation has been brought,[744] namely claims in respect of loss life, personal injury, loss of or damage to property and consequential loss occurring in direct connection with the operation of the ship as well as for claims resulting from delay in the carriage of cargo, passengers and their luggage.[745] In contrast, claims for oil pollution damage governed by the CLC as well as claims for nuclear damage are excluded from the ambit of the LLMC Convention.[746] From this it follows that claims for damage resulting from the transboundary movement of hazardous wastes by sea are basically eligible for limitation under the 1976 LLMC Convention.[747] This obviously leads to an overlap and possible conflicts with the specific limitation regimes, particularly with those of the Protocol to the Basel Convention and the HNS Convention. This issue, however, shall be addressed later.[748]

VI. Summary

There are several civil liability conventions that may apply to damage caused in the context of transboundary movements of hazardous wastes by sea. However, many of these conventions are not in force so that the legal framework governing such transports is still fragmentary. Liability of private persons involved in hazardous waste transports may arise under conventions originating in three distinct legal areas. The first and most important instrument is the 1999 Protocol to the Basel Convention that applies to hazardous waste shipments as such and covers the entire movement from the moment of its commencement until the completion of the final disposal. The second branch of relevant conventions addresses the issue of the carriage of dangerous substances and comprises the CRTD Convention, applicable, to transports of dangerous substances by land, and the HNS Convention, applicable to such transports by sea. None of these three Conventions, however, have come into force. At a third level, finally, one needs to mention civil liability conventions which apply only to certain kinds of hazardous wastes, like to waste oils and nuclear wastes, or which are restricted to certain regions, such as to the EU area. Although the conventions in this latter cluster are in force, due to their limited scope of application they are only of subordinate importance. In

---

[744] LLMC Convention, Article 11. A detailed description of the limitation procedure can be found at Griggs/Williams/Farr, *Limitation of Liability for Maritime Claims* (4th ed., 2005), at 1–94; de la Rue/Anderson, *Shipping and the Environment* (2nd ed., 2009), at 794–804; Schoenbaum, *Admiralty and Maritime Law* (5th ed., 2011), at 169–205.

[745] LLMC Convention, Article 2.

[746] *Ibid.*, Article 3.

[747] Tsimplis, 'The 1999 Protocol to the Basel Convention', 16 *IJMCL* (2001), at 317.

[748] See *infra*, Sect. "Relationship to the LLMC Convention" in Chap. 5.

conclusion, it must be said, therefore, that the present state of uniform international civil liability in the context of transboundary movements of hazardous wastes by sea is more than unsatisfactory.

E. Relationship Between Civil Liability Conventions and the Principle of State Responsibility

It has been shown in this section that claims for compensation of environmental and other damage caused in the context of transboundary movements of hazardous wastes by sea may basically be raised both under the principle of State responsibility and under civil liability conventions. This leads to the question of the relationship between the two legal concepts.

As a basic rule it can be said that both concepts provide for potentially complementary regimes that are not mutually exclusive.[749] This, for instance, is expressly stated in Article 16 of the Protocol to the Basel Convention.[750] But it also follows from the different purpose and content of both concepts. Whereas the principle of State responsibility applies in consequence of an international wrongful act of the liable State, civil liability conventions achieve a distribution of liabilities among private persons that may in particular also include strict liability for activities that are considered to be generally dangerous or hazardous. Both regimes, therefore, concern different constellations of liability and overlap only partially.

A restriction of this basic rule could result from the local remedies rule. According to this, local remedies, which include civil liability conventions implemented into national laws, need to be exhausted before a claim under the principle of State responsibility may be brought before an international court.[751] It is acknowledged that this rule applies at least concerning claims for diplomatic protection.[752] Regarding transboundary environmental harm, however, the underlying conditions are different. In such cases the injured Party has not voluntarily assumed the risk of being subject to the jurisdiction of a foreign State, and there is also no other relevant connection between the injured and the responsible State. Being reliant as a foreigner on the domestic relief of the polluting State has,

---

[749] Boyle, 'Globalising Environmental Liability', 17 *J. Envtl. L.* (2005), at 23–24; Scovazzi, 'State Responsibility for Environmental Harm', 12 *Yb. Int'l Env. L.* (2001), at 58–59.

[750] Basel Protocol, Article 16, reads: "The Protocol shall not affect the rights and obligations of the Contracting Parties under the rules of general international law with respect to State responsibility."

[751] See Article 44(b) of the ILC Draft Articles on State Responsibility and the Commentary thereof, Article 44, para. 3–5. See also Birnie/Boyle/Redgwell, *International Law and the Environment* (3rd ed., 2009), at 224–225.

[752] See Dugard, *Diplomatic Protection*, in: Crawford/Pellet/Olleson (ed.) (2010), at 1061–1067; Epping/Gloria, in: Ipsen (ed.), *Völkerrecht* (5th ed., 2004), at 340.

therefore, been seen as unreasonable and unfair for the victims of transboundary pollution.[753] Hence, it is generally accepted that the local remedies rule does not apply to transboundary environmental damage. By contrast, the rule does apply where the victim is allowed to bring the claim to a competent court within its own jurisdiction, e.g. by virtue of a choice of jurisdiction clause[754] in a civil liability convention.[755]

Finally, it should be mentioned that there are also constellations possible in which claims under the State responsibility principle are inadmissible by virtue of the principle of specialty (*lex specialis derogat legi generali*). This applies if the conduct of a State simultaneously falls under both the principle of State responsibility and a civil liability convention. This may be the case if, for example, the State acts as a private person in its operation of a State-owned oil tanker and the tanker causes an oil spill which can be attributed to required maintenance work having been neglected. In this case the CLC provides for specific liability rues applicable to the State as the owner of the vessel and, thus, supersedes the application of the general principle of State responsibility.

As a result, it may be concluded that as a general rule the principle of State responsibility and a civil liability convention may apply complementarily. Only in exceptional cases may the application of the principle of State responsibility be excluded, namely if the claimant has not exhausted the local remedies available under its own jurisdiction, or if both regimes apply concurrently because the State acted as a private person and the civil liability convention provides for specific rules superseding the general principle of State responsibility.

F. Summary: Responsibility and Liability *de lege lata*

The *status quo* of the international law on liability and compensation for damage resulting from the transboundary movement of hazardous wastes by sea is not satisfactory.

A customary principle of State liability for lawful but injurious acts is, *de lege lata*, not acknowledged in international law. The customary principle of State responsibility for an unlawful act applies only to a very limited number of cases. This is due to the fact that most waste shipments are initiated and conducted by private persons and entities, whereas the relevant international obligations, such as

---

[753] ILC Doc. A/61/10, at 80–81, para. 7; Lefeber, *Transboundary Environmental Interference* (1996), at 122–124; Pinto-Dobernig, 'Liability for Transfrontier Pollution Not Prohibited by International Law', 38 *Österr. Z. öffentl. Recht* (1987), at 98.

[754] See e.g. Article 17 of the Basel Protocol, which allows for instituting legal proceedings at the courts of the States Parties either where damage was suffered, where the incident occurred, or where the defendant has his habitual residence or principal place of business.

[755] Birnie/Boyle/Redgwell, *International Law and the Environment* (3rd ed., 2009), at 225; Boyle, 'Globalising Environmental Liability', 17 *J. Envtl. L.* (2005), at 23–25.

those of the Basel Convention, are directly addressed to States. Some of those obligations, in fact, impose substantive obligations on the States themselves, such as participation in the PIC procedures. However, the main part of these obligations consists of the general obligations to implement the respective PIC, ESM or other procedural requirements into national law, and to subsequently regulate, control and enforce compliance with these rules. Thus, the States may only be "indirectly" responsible for the non-compliance of private parties with the procedural rules imposed by the respective convention. Furthermore, these latter indirect obligations are conceived as due diligence obligations, so that a breach at least pre supposes that the State knew or ought to have known the actual circumstances of the case and the illegal conduct of the respective private party. A further obstacle of this "indirect" responsibility of the State is the requirement that in each case a causal link between the unlawful act and the actual damage must be established. For example, even if it can be established that a State failed to monitor compliance with the PIC requirements in a particular case, compensation for the resulting damage can only be claimed from the State if it is additionally established that this damage would not have occurred if the State had fulfilled its obligation. Finally, claims under the principle of State responsibility involve the burden of proving that the respondent State has not complied with the due diligence standard, which is hardly possible from an external position. Furthermore, for diplomatic and political reasons States are reluctant to bring environmental claims before the international courts. All these difficulties clearly show that the customary principle of State responsibility cannot be considered an effective instrument for the regulation and compensation of damage occurring in the course of transboundary movements of hazardous wastes by sea.

The existing civil liability conventions also do not represent an effective remedy for damage resulting from hazardous waste shipments. This is due to the simple fact that none of the relevant conventions, such as the Protocol to the Basel Convention or the HNS Convention, have come into force yet.

As a general result, it must be concluded that as far as the international level is concerned there is *de lege lata* no legal instrument that imposes an effective regime of responsibility and liability on the parties involved in transboundary movements of hazardous wastes by sea.

Chapter 4
Attempting an Interim Conclusion: Preconditions for an Effective Legal Regime on Liability and Compensation

The application of the customary principle of State responsibility to damage resulting from the transboundary movement of hazardous wastes by sea as well as the application of civil liability conventions, whether in force or not, to such movements have been outlined in the previous chapter. As an overall result, it has been ascertained that the current legal situation at the international level regarding liability and compensation for damage resulting from hazardous waste movements is insufficient and unsatisfactory. The question now must be whether the Protocol to the Basel Convention can be regarded as an appropriate and effective solution and what can be done to encourage ratification of this convention. To this end, a first step entails examining in this chapter the basic preconditions for a feasible liability regime.

A. Necessity of a Regime of Liability and Compensation

I. Insufficiency of Non-financially Oriented Treaty Compliance Mechanisms

One could argue that there is no actual need for additional rules on liability and compensation if the objectives of damage compensation and treaty compliance can be achieved by other mechanisms that are either already existing or that can be established without considerable efforts. Such alternative mechanisms could be seen in the treaty compliance mechanisms available in respect of the Basel Convention.

The practical benefit of an international convention, particularly in the field of international environmental law, depends on whether the Contracting States act in compliance with the respective substantive obligations. If non-compliance with the treaty obligations remains without consequences for the responsible State, the convention may eventually end up as a "sleeping treaty".[1] Traditional mechanisms to induce treaty compliance involve confrontational and mostly bilateral methods and strategies, such as countermeasures, the invocation of State responsibility and procedures of dispute settlement.[2] These confrontational measures, however, have proven inadequate in the end, particularly as regards environmental matters. This is due to the fact that, in a large number of cases, primarily interests of private parties are involved, for which diplomatic and intergovernmental channels do not seem to provide an adequate forum. In addition, States are reluctant to adopt drastic and unfriendly measures for the sake of environmental concerns and rather prefer avoiding any adverse effects on their relations with other States. The more effective approach to ensure compliance, hence, is seen in pursuing preventive and co-operative mechanisms that take effect prior to the wrongful conduct in the form of collective supervision.[3]

One of those co-operative treaty compliance mechanisms is seen in reporting systems established under the respective conventions.[4] Accordingly, the Parties of the Basel Convention are under the obligation to transmit annual reports, *inter alia*, on the status of implementation of the Convention through the Secretariat to the Conference of the Parties (COP).[5] By means of this requirement the actual status of implementation of the respective States is made transparent to the other Parties and the treaty organs, which are thus in a position to evaluate the measures taken by each State. The reporting State, in turn, is urged to undertake the required implementation in order to avoid having to give a public justification for the non-compliance.[6] This basic reporting obligation of the Basel Convention is accompanied by the function of the Secretariat to prepare and transmit reports based on the information received,[7] as well as by the mandate of the COP to continuously review and evaluate the effective implementation of the Convention, including the taking of appropriate measures to support this effort.[8] Despite the availability of these legal means, in practice the reporting system of the Basel Convention has been poorly utilised by the COP and the Secretariat for the purpose of achieving

---

[1] Birnie/Boyle/Redgwell, *International Law and the Environment* (3rd ed., 2009), at 240.

[2] Rules on dispute settlement are contained in Article 20 of the Basel Convention.

[3] Beyerlin/Marauhn, *International Environmental Law* (2011), at 318; Birnie/Boyle/Redgwell, *International Law and the Environment* (3rd ed., 2009), at 211–212, 238–239; Daniel, 'Civil Liability Regimes as a Complement to MEAs', 12 *RECIEL* (2003), at 238; Handl, 'Compliance Control Mechanisms', 5 *Tul. J. Int'l & Comp. L.* (1997), at 32–37.

[4] For instance, as to the reporting obligations under Article 22 of the OSPAR Convention, which form part of the compliance mechanism of this Convention, see Lagoni, *Monitoring Compliance and Enforcement of Compliance*, in: Ehlers/Mann-Borgese/Wolfrum (ed.) (2002), at 158–160.

[5] Basel Convention, Article 13(3)(c).

[6] Shibata, *Ensuring Compliance with the Basel Convention*, in: Beyerlin/Stoll/Wolfrum (ed.) (2006), at 70–72.

[7] Basel Convention, Article 16(1)(b).

[8] These measures are specified in Article 15(5) of the Basel Convention and, *inter alia*, include the possibility of establishing subsidiary bodies.

treaty compliance.[9] Furthermore, this mechanism aims at ensuring compliance via the implementation requirements rather than the compliance of the acting private parties vis-à-vis the substantive obligations of the Basel Convention. Therefore, it cannot be considered an effective instrument for directly preventing illegal trade and ensuring compensation.

Another compliance feature of the Basel Convention is the right of every Party to request verification in case there is reason to believe that another Party has acted in breach of its obligations under the Convention.[10] This provision, however, suffers from the substantial weakness that it neither empowers the Secretariat to undertake its own investigations nor imposes any other specific legal consequences.[11] Thus, this feature can similarly not be considered an effective treaty compliance mechanism.

The Basel Convention, however, provides for another specific mechanism to ensure treaty compliance. In 2002, at COP6, the Mechanism for Promoting Implementation and Compliance was established as a subsidiary device of the COP.[12] The objectives of this mechanism are to facilitate, promote, monitor and aim to secure the implementation of and compliance with the Convention's obligations; additionally, the mechanism is designed to be non-confrontational, transparent and preventive in nature.[13] Under the Mechanism, a Committee is established that is assigned with different functions. First, it is assigned with a "specific submission" task according to which the Committee, pursuant to a submission made by the Secretariat or a Party that failed to resolve the matter through consultations, determines whether an act of non-compliance actually exists. The non-confrontational nature of this function is evidenced by the fact that the submitting State is only allowed to participate in the proceedings if this is permitted by the accused State.[14] The measures that may be taken by the Committee are non-binding and include the provision of the purportedly non-complying Party with advice, recommendations and information. If these measures are not considered sufficient, the Committee may present the issue to the COP to consider further action.[15] Finally, the Committee is also assigned with an auxiliary "general review" task, according to which the Committee, as directed by the COP, reviews

---

[9] Shibata, *Ensuring Compliance with the Basel Convention*, in: Beyerlin/Stoll/Wolfrum (ed.) (2006), at 71–72.

[10] Basel Convention, Article 19.

[11] Kummer, *The Basel Convention* (1995), at 233; Shibata, *Ensuring Compliance with the Basel Convention*, in: Beyerlin/Stoll/Wolfrum (ed.) (2006), at 73–74.

[12] COP6 Decision VI/12 on the Establishment of a Mechanism for Promoting Implementation and Compliance (Doc. UNEP/CHW.6/40, at 45). This mechanism is based on Article 15(5)(e) of the Convention. The terms of reference (ToR) of this mechanism are annexed to this Decision.

[13] ToR, Paragraphs 1–2.

[14] *Ibid.*, Paragraphs 9–18. See also Shibata, *Ensuring Compliance with the Basel Convention*, in: Beyerlin/Stoll/Wolfrum (ed.) (2006), at 80–81.

[15] ToR, Paragraphs 19–20. This involves, *inter alia*, the issuing of a cautionary statement by the COP. See also Shibata, 'The Basel Compliance Mechanism', 12 *RECIEL* (2003), at 193–194.

general issues of compliance and implementation under the Basel Convention.[16] The Basel Compliance Mechanism, despite its rather modern approach of resorting to co-operative instead of confrontational measures, has been criticised, in the end, exactly because of these specific features. It is argued that the Committee lacks sufficient authority and means to conduct investigations and is not vested with sufficient powers to impose and enforce effective measures which might ensure compliance. The success of this Mechanism, therefore, largely depends on the good-will of the Party allegedly in non-compliance.[17] It could be encountered that exactly the absence of confrontational means enhances the incentive of States to find an amicable solution by means of the Basel Compliance Mechanism in order to avoid further confrontational measures. However, the same effect could be achieved by bilateral consultations. It must be concluded, therefore, that the Basel Compliance Mechanism cannot replace the existing, general mechanisms of treaty compliance, but may only function as a supplementary device.

In summary, it can be concluded that there are traditional confrontational as well as rather modern co-operative compliance mechanisms available in respect of the Basel Convention. These mechanisms may provide certain reasonable approaches to encourage treaty compliance. However, these mechanisms fail to ensure compensation of the victims of pollution, and, most importantly, they are addressed to States Parties only and therefore do not have any direct effect on the private parties involved in the shipment of hazardous wastes. The objectives of liability and compensation, thus, cannot be entirely achieved through the existing mechanisms of treaty compliance.

II. Liability Rules as a Remedy for Environmental Damage

Rules on liability and compensation serve three main objectives: The first is a preventive one. The threat of incurring liabilities functions as a deterrent and creates an incentive to prevent damage; thus, liability aims at promoting compliance with the respective substantive rules. The second objective concerns the repressive function of liability and aims at denouncing the wrongfulness of the conduct. In doing so, liability contributes to an *ex post* enforcement of the substantive obligations, as well as to legal certainty and predictability. The third objective, finally, covers the compensatory aspect of liability and aims at shifting

---

[16] ToR, Paragraph 21. See also Shibata, 'The Basel Compliance Mechanism', 12 *RECIEL* (2003), at 194–195.

[17] Shibata, *Ensuring Compliance with the Basel Convention*, in: Beyerlin/Stoll/Wolfrum (ed.) (2006), at 83–84; and with regard to the compliance mechanisms of the Basel Convention in general Kummer, *The Basel Convention* (1995), at 236–238.

the injurious consequences from the injured party to the wrongdoer or the source of harm. It, thus, intends to restore the *status quo ante* as far as possible.[18]

In the absence of an effective international legal regime on civil liability, private parties involved in transboundary hazardous waste movements, from an economic point of view, have no direct interest in acting in conformity with the procedural obligations of the Basel Convention. In order to avoid the restrictive regulations and the incurrence of further costs for preventive measures, the legal *status quo* rather provides an economic incentive to ship hazardous wastes illegally or to circumvent the substantive provisions by transferring the hazardous wastes to a different company, which does not fully comply with the export and notification procedures and simply disappears from the scene in case of damage. Rules imposing financial responsibilities on private parties may provide an opposite economic incentive that may urge that acting party to comply with the procedural and other rules of the Basel Convention in order to avert the imposition of liability. Rules on civil liability, thus, are also part of the treaty compliance mechanisms.[19] They may, moreover, be combined with additional financial mechanisms, such as a trust fund for ensuring immediate response action, or the requirement of compulsory insurance. By establishing international uniform law, both at the substantive and procedural level, non-discriminatory access to national remedies is guaranteed, and it is also ensured that judgments based on this uniform law are mutually recognised and enforced among the Signatory States. This offers the opportunity for the victim of pollution to obtain a more favourable legal position by choosing one of several possible jurisdictions (so-called "forum shopping").[20]

B. The Appropriate Form of Liability

To the extent it is now established that a regime of liability and compensation is the only sufficient remedy regarding damage caused in the context of hazardous waste shipments, some further questions arise in respect of the basic conditions of this envisaged liability regime. This particularly involves the issue of whether an international approach is necessary as well as the issue of whether a regime on State liability or civil liability should be favoured.

---

[18] Dahm/Delbrück/Wolfrum, *Völkerrecht*, vol. I/3 (2nd ed., 2002), at 867; Gaines, 'International Principles for Transnational Environmental Liability', 30 *Harv. Int'l L. J.* (1989), at 324–328; Kummer, *The Basel Convention* (1995), at 210–211; Lefeber, *Transboundary Environmental Interference* (1996), at 1; Wolfrum/Langenfeld/Minnerop, *Environmental Liability in International Law* (2005), at 496–498.

[19] Wolfrum, 'Means of Ensuring Compliance and Enforcement', 272 *RdC* (1998), at 77, 79.

[20] Birnie/Boyle/Redgwell, *International Law and the Environment* (3rd ed., 2009), at 304–315, 316–318; Boyle, 'Globalising Environmental Liability', 17 *J. Envtl. L.* (2005), at 9–12; Daniel, 'Civil Liability Regimes as a Complement to MEAs', 12 *RECIEL* (2003), at 240.

I. National, Regional or Global Approach?

Rules on liability and compensation can be established on a national, regional or global scale. Each of these approaches is connected with different advantages and disadvantages, so that it seems appropriate to evaluate which approach suits best the needs of liability in the context of transboundary hazardous waste shipments, which is by nature an international activity involving several private and States Parties situated in different countries.

It has been outline before that with regard to the procedural requirements of hazardous waste shipments, the Basel Convention provides for a basically suitable and effective legal regime on a global scale. This effectiveness particularly concerns the interaction with potential import bans established unilaterally by States or by means of regional Conventions, such as the Bamako or Waigani Conventions. In consequence of such a unilateral or regional ban the Basel Convention ensures that any export of hazardous wastes to this area is considered "illegal trade" and comes under the obligation of "re-importation".[21] By means of this interrelation the particular needs and autonomous decisions of Parties to either ban or admit hazardous waste trade are taken into account at the international level.

In more general terms it can be said that autonomous national regulations can be better adapted to the particular needs of the respective countries. This is possible only to a limited extent as far as the domestic implementation of international uniform law is concerned. In these cases the State is bound by a pre-defined international standard, which may be the result of a compromise among the involved States.[22] National laws, therefore, may be regarded the more suitable solution for the regulation of activities that are taking place within the borders of that country, or which concern the prohibition or cessation of a certain activity.

However, every legal regime can only be as effective as its enforceability. The same consideration applies to the issue of liability and compensation. Since the transboundary movement of hazardous wastes involves several parties based in different States, any isolated national approach faces considerable obstacles when it comes to the issue of enforcing domestic judgments in foreign countries. A necessary condition for an effective regime on liability and compensation, therefore, is the mutual recognition and enforcement of judgments. This can be ensured by means of international law. A further advantage of an international regime on liability is that a certain set of minimum standards is guaranteed in regards to substantive aspects. The injured party, thus, is not required to call upon the courts of a foreign and possibly unfavourable jurisdiction, but may choose a more convenient court having jurisdiction. Hence, the international approach also strengthens legal protection and certainty. In summary, it must therefore be stated that in respect of

---

[21] *Supra*, at Sect. "Background of the Convention" in Chap. 3.

[22] In some cases, however, it is possible to impose stricter rules domestically, provided the international rules are to be considered minimum standards and do not, at the same time, guarantee a certain maximum standard.

commercial activities taking place on a global scale, a global international regime on liability and compensation is necessary. Such an approach, however, also involves certain drawbacks. Negotiating such a regime of liability may last for a long time, resulting in a rather general and weak consensus. The entry into force of such a convention is clearly uncertain and depends on an additional act of ratification by each State Party. A remedy to these obstacles could be seen, for example, in including accessory or framework provisions that either leave space for a detailed adjustment on the regional or national level, or that refer to existing regional or national rules and establish international enforcement mechanisms.

Regional international approaches, finally, incorporate some advantages and disadvantages of both national and international regimes. On the one hand, it might be easier to achieve consensus on specific rules, better meeting the expectations of the respective States. On the other hand, regional regimes do not ensure mutual recognition and enforcement on a global level. In consequence, they can only be considered an additional instrument rather than a real alternative to a global regime on liability and compensation.

II. State Liability or Civil Liability?

A further basic issue regarding the prospective regime on liability and compensation is whether this regime should impose liabilities on States or on private parties.

It should first be kept in mind that unlike the civil liability conventions applicable to the trade in and transport of hazardous wastes, which have not yet entered into force, the customary principle of State responsibility represents valid international law. The principle of State responsibility has, nevertheless, proven to be a rather blunt sword as regards the largest portion of transboundary pollution. This is to be explained by two sets of reasons: The first set concerns the requirement of the claimant State to produce evidence of the existence of a breach of an international obligation by the respondent State. This requirement faces considerable difficulties due to the fact that shipping hazardous wastes is an activity mainly conducted by private parties, so that the State's obligations with regard to this activity are largely confined to an obligation of control and enforcement. Moreover, these obligations are mostly conceived as obligations of due diligence. The burden of proof that lies with the claimant State—taken in combination with the fact that this State will hardly be able to gather sufficient information on the internal circumstances leading to the pollution—thus acts as a formidable obstacle for the successful invocation of State responsibility.[23] The second set of reasons for the insufficiency of the customary principle of State

---

[23] Boyle, 'Globalising Environmental Liability', 17 *J. Envtl. L.* (2005), at 6–8; Daniel, 'Civil Liability Regimes as a Complement to MEAs', 12 *RECIEL* (2003), at 238.

responsibility is to be seen in the fact that this legal instrument is simply not intended to redress damage that is caused and in most cases also sustained by private parties. Only States and other subjects of international law are entitled to bring claims under the customary principle of State responsibility. The absence of any direct civil liability also means that there is no direct economic incentive for private parties to avoid the occurrence of damage. States could, moreover, decide to countenance transnational (fault-based and strict) liabilities in order to subsidise their industries.[24] Furthermore, on the interstate level political and diplomatic aspects play a major role and may eventually lead to the reluctance of States to strain diplomatic relations for the sake of environmental claims. Consequently, only few cases have been brought before international or arbitral courts that seek compensation for environmental damage.[25]

Some of these obstacles and disadvantages of State responsibility could be overcome by establishing a convention on the liability of States that expressly determines in more detail the legal prerequisites of liability for damage resulting from the transboundary movement of hazardous wastes. This particularly concerns the first set of reasons. A regime of liability could, for instance, impose strict liability with regard to specific incidents, or it could determine and modify the heavy onus of proof, possibly by establishing a lighter burden of proof.[26] However, the systemic obstacles comprised by the second set of reasons can hardly be overcome even by an explicit regime of liability imposed on States. This concerns the issue that the interstate level is not the appropriate forum for ensuring liability and compensation for harmful activities conducted by private parties.

In conclusion, it must be stated, therefore, that the imposition of responsibilities and liabilities on States in general may represent a sufficient and effective legal remedy for transboundary damage.[27] In respect of damage caused by the activity of private parties, however, the States are simply not the correct respondents for realising the purposes of liability and compensation.[28] Liability for damage

---

[24] Boyle, 'Globalising Environmental Liability', 17 *J. Envtl. L.* (2005), at 8.

[25] Beyerlin/Marauhn, *International Environmental Law* (2011), at 380; Boyle, 'Globalising Environmental Liability', 17 *J. Envtl. L.* (2005), at 8; Kummer, *The Basel Convention* (1995), at 224–225; see also Birnie/Boyle/Redgwell, *International Law and the Environment* (3rd ed., 2009), at 303.

[26] A possible solution could be to establish a "secondary burden of proof", according to which in a first step the claimant State has to prove that the damage was most likely caused by the respondent State, which in turn has to produce evidence of compliance with pertinent rules by disclosing its internal documents, findings and data records.

[27] See also Birnie/Boyle/Redgwell, *International Law and the Environment* (3rd ed., 2009), at 431.

[28] The contrary view is held by Bergkamp, 'Proposals for International Environmental Liability', 8 *Eur. Envtl. L. Rev.* (1999), at 326–327.

resulting from the transboundary movement of hazardous wastes thus should be primarily imposed on the private parties actually involved in these activities. This outcome also reflects recent developments in the work of the ILC.[29]

C. Limitations Set by Other Areas of Law

The scope and extent of rules of liability and compensation regarding damage caused in the context of hazardous wastes shipments by sea is also affected and can be limited by other areas of law. This concerns most of all the fact that the provisions imposed, for example, the Basel Convention and any related regime on civil liability represent trade restrictions that must be judged against the applicable trading systems, such as the General Agreement on Tariffs and Trade (GATT), which today forms part of the World Trade Organization (WTO). However, this conflict arises primarily with regard to the substantive trade restrictions established by the Basel Convention and other regional conventions. It does not represent a conflict that is specifically related to the issue of liability and compensation and, therefore, cannot be addressed in more detail at this point.[30]

D. Summary

In this chapter some fundamental ideas have been identified which form the basic conditions for a prospective efficient regime on liability and compensation. It has been shown that liability pursues preventive, repressive and compensatory functions. From the perspective of the victim of pollution, the compensatory function might often be at the centre of attention; whereas from a global standpoint characterised by anxiousness over a conserving and sustainable use of the environment and concerns about the protection of human health and biodiversity, the preventative function of liability, which aims at ensuring treaty compliance, represents the main focus. Against this background, the economic ramifications of liability have been depicted. It has been shown that environmental liability represents the only efficient strategy for the internalisation of the external costs of pollution. By means of civil liability an inherent economic interest of the private parties actually

---

[29] See *supra*, sect. "State Liability" in Chap. 3.

[30] For an analysis of the relationship between the Basel Convention and the GATT see Alam, 'Trade Restrictions Pursuant to MEAs', 41 *J.W.T.* (2007), at 1005–1010; Friedrich-Ebert-Stiftung (ed.), *Zehn Jahre Basler Übereinkommen* (1999), at 41–43; Krueger, *International Trade and the Basel Convention* (1999), at 64–81; Wirth, 'Trade Implications of the Basel Ban', 7 *RECIEL* (1998), at 237 *et seq*. As to the relationship to the EC rules on the free movement of goods see Frenz, 'Grenzüberschreitende Abfallverbringungen und gemeinschaftliche Warenverkehrsfreiheit', 20 *UPR* (2000), at 210 *et seq*.

involved in the initialisation and conduct of hazardous waste shipments is created. It has also been shown that other mechanisms of ensuring treaty compliance which are available with regard to the Basel Convention are not sufficient to achieve the legal aims associated with an instrument of liability. An effective regime on liability and compensation for damage resulting from the transboundary movement of hazardous wastes by sea must inevitably be established on a global scale. Although rules imposing responsibilities and liabilities on States generally provide for suitable solutions with regard to transboundary damage, as regards transboundary shipments of hazardous wastes they cannot be considered appropriate. This is due to the fact that only rules of civil liability can have a direct economic effect on the conduct of private parties, the latter being the main actors in the context of transboundary hazardous waste movements.

Chapter 5
The 1999 Basel Protocol on Liability and Compensation

In this chapter the provisions of the 1999 Protocol to the Basel Convention on Liability and Compensation for Damage Resulting from the Transboundary Movement of Hazardous Wastes and their Disposal (Basel Protocol or Protocol) shall be addressed and analysed in more detail. For this purpose, also a comparison between the provisions of the Basel Protocol and the respective provisions of comparable civil liability regimes is provided at the relevant points.

A. The Regulatory Content of the Basel Protocol

I. Scope of Application

The scope of application of the Basel Protocol is determined in Article 3. This Article represents a very detailed and sophisticated provision that consists of 1,095 words, which makes up roughly 20 % of all words of the Convention.[1]

As a general rule it can be said that the Basel Protocol applies to damage due to an incident occurring during a transboundary movement of hazardous wastes and other wastes and their disposal, including illegal traffic.[2] Article 3 regulates in a detailed fashion the question to which movements, incidents and damage the Basel Protocol applies, and it moreover defines the temporal and geographical scope of application. Finally, it also prescribes under which conditions the States Parties may exclude the application of the Basel Protocol.

---

[1] Article 3 of the Basel Protocol was subject to controversy and long-lasting discussions during the negotiation process of the Protocol. Initially it consisted of two paragraphs and 173 words. See AHWG Doc. UNEP/CHW.1/WG.1/1/5.
[2] Basel Protocol, Article 3(1).

1. Transboundary Movements of Hazardous Wastes and Other Wastes

The Basel Protocol applies to transboundary movements of hazardous wastes and other wastes and their disposal, including illegal traffic.

(a) Wastes Subject to the Basel Protocol

(aa) Hazardous Wastes

The term "hazardous wastes and other wastes" is not defined independently within the Basel Protocol, but with reference to the Basel Convention.[3] According to this, "wastes" are defined as substances or objects which are disposed of or are intended to be disposed of or are required to be disposed of by provisions of national law.[4] Wastes are considered to be "hazardous wastes" in two cases:

(1) Autonomous Definition

The first case concerns wastes that come under the autonomous definition of hazardous wastes established within the Basel Convention framework. According to this, wastes are considered hazardous if they fall under any category contained in Annex I of the Basel Convention, unless they do not possess any of the characteristics contained in Annex III of the Convention.[5] Annex I of the Basel Convention enumerates specific waste streams and constituents of wastes that are deemed to be generally or potentially hazardous, whereas Annex III contains a detailed list of particular hazardous characteristics. The definition of hazardous wastes, thus, requires two elements cumulatively. First, the waste in question must be attributed to a category of wastes that is considered to be generally or potentially hazardous, and, second, the waste must demonstrate specific hazardous characteristics in the individual case. The present inquiry does not allow for an in-depth analysis of the particular waste streams, constituents and hazardous characteristics of hazardous wastes covered by the Basel Convention. However, as considered in comparison with the relevant definitions found in other conventions and regulations, some brief conclusions can be drawn.

The definitions of hazardous wastes contained in other conventions and regulations dealing in particular with the transboundary movement of hazardous wastes resemble to a large extent the definition of the Basel Convention. This applies to the Bamako and Waigani Conventions,[6] to the Izmir Protocol to the Barcelona

---

[3] Basel Protocol, Article 2(2)(b), refers to Basel Convention, Article 1, which in turn refers to Basel Convention, Article 2(1) in respect of the term "wastes".

[4] Basel Convention, Article 2(1).

[5] *Ibid.*, Article 1(1)(a).

[6] Bamako Convention, Article 1(1), (2), Article 2(1) and Annexes I and II; Waigani Convention, Article 1, Article 2(1) and Annexes I and II.

Convention and the Tehran Protocol to the Kuwait Convention,[7] as well as to the OECD and EU regulations.[8] The only substantial difference concerns the fact that some conventions and regulations, as in the case of the Basel Convention, require that both elements of the definition simultaneously apply to the wastes in question.[9] Other conventions or regulations, by contrast, only require one or the other of the two elements,[10] or focus only on whether they display certain hazardous characteristics.[11] The practical implication of these differences, however, is limited as they only have consequence in respect of atypical wastes, i.e. those which although generally considered hazardous are not considered as such in the individual case or those wastes which are generally non-hazardous display hazardous characteristics in the particular case.[12]

In contrast, the definitions of hazardous and noxious substances (HNS) as used by the HNS Convention and the definition of dangerous goods as used by the CRTD Convention differ considerably from the pattern of the Basel Convention, even disregarding the fact that the Basel Convention applies exclusively to waste substances and items. Both the HNS Convention and the CRTD Convention largely refer to substances and articles enumerated or referred to in existing conventions and regulations, like in Annexes I and II of MARPOL 73/78, in the IMDG and ADR Codes or in other IMO regulations. Due to the fact that such enumerated lists of covered substances may contain gaps, the definition of hazardous wastes based on abstract categories as used by the Basel Convention is more comprehensive. However, the HNS and CRTD Conventions also apply to substances, materials and articles that are not covered by the combined definition of hazardous wastes of the Basel Convention. To this extent, the scope of substances covered by the Basel Convention's definition is narrower.

(2) National Definitions

Wastes are also considered hazardous under the Basel Convention if they are defined as, or are considered to be, hazardous wastes pursuant to the domestic legislation of the State of export, import or transit.[13] In order for national definitions to be considered valid under the Basel Convention they must be reported by

---

[7] Izmir Protocol to the Barcelona Convention, Article 1(c), (d), Article 3(1) and Annexes I and II; Tehran Protocol to the Kuwait Convention, Article 2(1), Article 1(1)(a) and Annexes I and III.

[8] OECD Council Decision C(2001)107/FINAL, A(1), (2)(i) and Appendixes 1 and 2; EU Directive 2008/98/EG, Article 3(1), (2) and Annex III.

[9] This applies to the Waigani Convention, the OECD Council Decision C(2001)107/FINAL and the Tehran Protocol to the Kuwait Convention.

[10] This applies to the Bamako Convention and the Izmir Protocol to the Barcelona Convention.

[11] Directive 2008/98/EG.

[12] In addition, such wastes often fall under the national definition of hazardous wastes as outlined in the next section.

[13] Basel Convention, Article 1(1)(b).

the respective State to the Secretariat of the Basel Convention.[14] As of today, 24 States have notified the Secretariat with regard to wastes that are defined as hazardous pursuant to their domestic legislation.[15] However, as regards the Basel Protocol—as opposed to the Basel Convention—further requirements must be fulfilled in order that the provisions apply to wastes considered hazardous pursuant to the national definition of Contracting States. This includes, first, that the national definition of hazardous wastes has been notified to the Basel Secretariat, particularly by the State of export or import, or both. By contrast, notification made only by a State of transit is not sufficient. The second requirement is that the wastes must in particular also be defined as hazardous (and so duly notified) by that State of export, import or transit in whose area of national jurisdiction the damage actually occurs.[16]

(bb) "Other Wastes"

The Basel Protocol also applies to "other wastes". This term is defined by the Basel Convention as wastes belonging to any of the categories contained in Annex II of the Convention. This annex comprises exclusively household wastes and residues arising from the incineration of household wastes.[17] Thus, despite the fact that the term "other wastes" suggests an understanding of "non-hazardous" wastes, the real meaning is more comparable to "hazardous-like" wastes.

Household wastes and the residual ash from their incineration cannot per se be considered hazardous according to the definition of "hazardous wastes" under the Convention. The inclusion of household wastes, therefore, represents a considerable expansion of the scope of the Basel Convention and, thus, has been criticised as eliminating a clear-cut distinction between covered hazardous and non-covered non-hazardous wastes.[18] Nonetheless, household wastes, and particularly the residual ash from their incineration, contain as a general rule considerable amounts of matter such as heavy metals and other harmful substances. Therefore, they can also be considered generally or potentially hazardous. In view of this and also in the aftermath of the then recent *M/V "Khian Sea"* case, the States represented at the 5th Session of the Conference of Plenipotentiaries to adopt the Basel Convention introduced the definition of "other wastes" into the convention text so as to include household wastes and the residual ash from their incineration.[19]

---

[14] *Ibid.*, Article 3.

[15] A list of these States and their respective national definitions of hazardous wastes can be obtained at www.basel.int.

[16] Basel Protocol, Article 3(6)(b).

[17] Basel Convention, Article 1(2) and Annex II.

[18] Abrams, 'Regulating the International Hazardous Waste Trade', 28 *Colum. J. Transnat'l L.* (1990), at 820.

[19] Kummer, *The Basel Convention* (1995), at 50.

Although it seems reasonable to include household wastes in the scope of the Convention, it must also be stressed that the choice of the term "other wastes" is unfortunate. The use of this term is the result of a compromise that was reached during the drafting process in response to the controversy of whether household wastes should be covered by the Convention or not. In the end, it was decided to include such wastes in the scope of the Convention but to abstain from defining them as hazardous.[20] Thus, this second category of wastes covered by the Convention was introduced. However, since there is no difference in the treatment of both kinds of wastes throughout the Basel Convention and Protocol, the distinction between hazardous and "other" wastes is purely terminological.[21] This artificial, terminological distinction gives much occasion for confusion and suggests coverage of the Convention which is far beyond the actual scope of application. But most of all, there is no need to refrain from defining household wastes as being hazardous under the Convention. An enumeration of household wastes and their incineration residues in Annex I of the Convention would allow for an even more differentiated classification. In that case household wastes would only be considered hazardous under the Convention if, in addition, they displayed in each individual case at least one of the hazardous characteristics laid down in Annex III of the Convention.

Other conventions relevant to the transboundary movement of hazardous wastes follow the example of the Basel Convention and include household wastes and their incineration residuals in their scope of application.[22] However, only one convention introduced a similar distinction between hazardous and "other" wastes,[23] which further shows that there is no need to introduce a second category of "other wastes".

For the sake of simplicity, the term "hazardous wastes" shall in the following discussion be used to cover both hazardous and "other" wastes as defined by the Basel Convention.

(cc) Wastes Excluded from the Scope of the Protocol

The Basel Protocol does not apply to radioactive wastes, provided they are subject to other international control systems applying specifically to radioactive materials.[24] This means that in the end most radioactive wastes will be excluded from the scope of the Basel Convention, since most radioactive wastes are subject to the

---

[20] Kummer, *The Basel Convention* (1995), at 50.

[21] Kummer, *The Basel Convention* (1995), at 50.

[22] This applies to the Bamako and Waigani Conventions as well as to the Izmir Protocol to the Barcelona Convention. The EU Regulation 2008/98/EG provides a more detailed regulation of household wastes. Household wastes are not covered by the HNS and CRTD Conventions or by OECD Decision C(2001)107/FINAL.

[23] Tehran Protocol to the Kuwait Convention.

[24] Basel Convention, Article 1(3).

relevant conventions and regulations of the IAEA.[25] This applies in the first instance to radioactive wastes falling under the 1979 Physical Protection Convention.[26] Other non-binding instruments include the 2009 IAEA Regulations for the Safe Transport of Radioactive Material[27] and the 1990 Code of Practice on the International Transboundary Movement of Radioactive Waste.[28] The Basel Convention, however, remains applicable to radioactive wastes which do not fall under the IAEA instruments. This particularly applies where the level of radioactivity remains below the limits of the IAEA instruments.[29]

In addition to the exclusion of radioactive wastes, the Basel Convention also excludes wastes which derive from the normal operation of a ship, provided that their discharge is covered by another international instrument.[30] This exemption mainly applies to wastes falling under the MARPOL 73/78 Convention.[31] The wording "normal operation of a ship" has been criticised as being ambiguous and potentially including routine cleaning operations on board a ship, and excluding such operations has not been seen as one of the aims of the Basel Convention.[32] With regard to hazardous wastes generated on board a ship, the Secretariat of the Basel Convention, in the wake of the *M/V "Probo Koala"* case, has been concerned with the question under which conditions the movement of such wastes on board a ship to another country falls within the scope of the Basel Convention. The Secretariat of the Basel Convention, in a revised legal analysis prepared in October 2011, found that wastes generated on board a ship which conform to the definition of hazardous wastes under the Basel Convention basically fall within the scope of the Basel Convention and may basically come under the definition of "transboundary movement".[33] The Secretariat, furthermore, found that "specific industrial processes or activities on board a ship (such as refining oil products [...]) might be considered distinct from the 'normal operation of ships'",[34] so that those wastes do not fall under the exclusion clause for wastes that come under the MARPOL 73/78 Convention. However, in the particular case of the *M/V "Probo*

---

[25] See for an overview on the international regulatory regime on shipments of radioactive substances Nadelson, 'The Contemporary Shipment of Radioactive Substances in the Law of the Sea', 15 *IJMCL* (2000), at 215–220.

[26] 1979 Convention the Physical Protection of Nuclear Material. It entered into force on 8 February 1987.

[27] Published on 19.05.2009 (IAEA Safety Standards Series, No. TS-R-1, Vienna 2009).

[28] The Code of Practice (INFCIRC/386) was adopted on 21.09.1990 by the IAEA.

[29] Kummer, *The Basel Convention* (1995), at 51.

[30] Basel Convention, Article 1(4).

[31] Kummer, *The Basel Convention* (1995), at 52.

[32] Kummer, *The Basel Convention* (1995), at 52; Tsimplis, 'The 1999 Protocol to the Basel Convention', 16 *IJMCL* (2001), at 299.

[33] Revised legal analysis of the application of the Basel Convention to hazardous wastes and other wastes generated on board a ship of the Secretariat of the Basel Convention of 7 October 2011, COP10 Doc. UNEP/CHW.10/INF/16, at 3–4.

[34] *Ibid.*, at 7.

Koala" the Secretariat was, due to a lack of information, unable to establish that the hazardous wastes causing pollution damage were distinct from wastes which derive from the normal operation of a ship.[35]

Similar exclusions of radioactive wastes and wastes covered by the MARPOL 73/78 Convention can also be found in other relevant conventions and regulations.[36]

(b) Transboundary Movements and Other Activities Covered by the Protocol

The Basel Protocol applies to the transboundary movement and the disposal of hazardous wastes, including illegal traffic.[37]

(aa) Transboundary Movement

The definition of the term "transboundary movement" requires that at least two States are involved in the particular movement. The starting point of the transport must be within an area under the national jurisdiction of one State, and it must either end in or cross an area under the national jurisdiction of another State. As long as this requirement is fulfilled, also those transports are covered which end in or cross an area not under the national jurisdiction of a State.[38]

The term "area under the national jurisdiction of a State" is defined as any land, marine area or airspace within which a State exercises administrative and regulatory responsibility in accordance with international law in regard to the protection of human health or the environment.[39] It is clear that this term does not only cover any land territory, but also extends to the territorial sea of a coastal State.[40] In contrast, it is questionable whether this definition furthermore comprises the EEZ of a coastal State. Practical importance is attached to this question, however, only in cases where the hazardous wastes are generated on board a vessel anchoring in the EEZ of a coastal State prior to the commencement of the

---

[35] The Report of the Basel Convention Secretariat's technical assistance mission to Ivory Coast can be found at Doc. UNEP/SBC/BUREAU/8/1/INF/2, and annexed to OEWG Doc. UNEP/CHW/OEWG/6/2.

[36] Bamako Convention, Article 2; Waigani Convention, Article 2; EU Regulation 2008/98/EG, Article 2; an exclusion for only radioactive wastes is provided for by the Tehran Protocol to the Kuwait Convention, Article 1; HNS Convention, Article 4; CRTD Convention, Article 4; an exclusion for only wastes covered by the MARPOL 73/78 Convention is provided for by the Izmir Protocol to the Barcelona Convention, Article 3 and the 1972/1996 London Dumping Convention, Article 1.

[37] Basel Protocol, Article 3(1).

[38] Basel Convention, Article 2(3).

[39] *Ibid.*, Article 2(9).

[40] See UNCLOS, Article 2(1), and *supra*, Sect. "3rd Tier: Prior Informed Consent Principle" in Chap. 3.

transboundary movement. Since regular movements of hazardous wastes by sea are conducted at least from port to port, the territorial sea of the coastal State would be part of the movement anyway. If, however, the wastes are generated on board a vessel anchoring in the EEZ, then it should first be noted that, different to the term "State of transit", the term "transboundary movement" is not restricted to movements "through" a State, but rather explicitly includes further marine areas within which a State exercises administrative and regulatory responsibility in regard to the protection of the environment.[41] Such marine must be seen in the EEZ, since in the EEZ the coastal State exercises limited jurisdiction with regard to the prevention, reduction and control of pollution from vessels.[42] The fact that this jurisdiction is limited to certain measures as provided for by the UNCLOS does not mean that the EEZ is not covered by this definition. This definition, firstly, does not require the competence of the coastal State to take specific measures. And, secondly, it can well be argued that according to the provisions of the UNCLOS the coastal State is given the competence to adopt laws and regulations as well as to enforce these rules in regard to the generation of hazardous wastes on board a vessel anchoring in its EEZ. Unlike the mere transit of a ship carrying hazardous wastes through the EEZ of a coastal State the generation of hazardous wastes on board a vessel anchoring in the EEZ may in general pose a significant threat for the marine environment.[43] It should be concluded, therefore, that the term "area under the national jurisdiction of a State" also covers the EEZ of a coastal State.[44]

The Basel Protocol, finally, does not apply to transports of hazardous wastes from the high seas to a State, or to transports from a State directly to the high seas for disposal.[45] It should also be stressed that the Protocol applies only to transboundary movements that have commenced after the entry into force of the Protocol for the respective Contracting Party.[46]

(bb) Other Activities

The scope of application of the Basel Protocol also comprises the subsequent disposal of hazardous wastes. The term "disposal" includes all operations

---

[41] The question of whether the term "State of transit" also covers the EEZ of a coastal State is discussed *supra*, Sect. "3rd Tier: Prior Informed Consent Principle" in Chap. 3.

[42] UNCLOS, Article 56(1)(b)(iii), provides that the coastal State has jurisdiction as provided for in the relevant provisions of the UNCLOS with regard to the protection and preservation of the marine environment.

[43] See *ibid*, Articles 211(5) and 220 (3), (5) and (6).

[44] See also Kummer, *The Basel Convention* (1995), at 21. Not entirely clear on this issue is Rummel-Bulska, *The Basel Convention and the UNCLOS*, in: Ringbom (ed.) (1997), at 87–88.

[45] Kummer, *The Basel Convention* (1995), at 52; Tsimplis, 'The 1999 Protocol to the Basel Convention', 16 *IJMCL* (2001), at 299.

[46] Basel Protocol, Article 3(6)(a).

specified in Annex IV to the Basel Convention.[47] Finally, the Basel Protocol extends its application not only to cases of illegal traffic, but to all cases of re-importation according to Article 8 and Article 9(2)(a), (4) of the Basel Convention.[48]

2. Incidents Covered by the Basel Protocol

The application of the Basel Protocol presupposes the existence of an incident. The term "incident" represents the central term determining the application of the Basel Protocol in temporal and geographical regard.

(a) Terminological Scope of "Incidents"

An "incident" is defined by the Basel Protocol as any occurrence, or series of occurrences having the same origin that causes damage or creates a grave and imminent threat of causing damage.[49] Based on this wording, it becomes apparent that the term "incident" has a broad meaning. First, it is not restricted to single events, but also covers series of events in case it cannot be established which single event has actually caused the damage. This applies even if neither of the single events taken by itself was capable of causing the damage. Furthermore, this term covers not only events actually leading to damage but also encompasses events that have not yet caused damage but create a grave and imminent threat of causing damage. By means of this extension it is ensured that incidents are covered even if it cannot, or cannot yet, be established whether later damage will actually occur or not. This shows that the existence of an incident must be determined from an *ex ante* point of view. The term "grave and imminent threat" is not defined in more detail. However, it seems appropriate to give it a wide meaning excluding only incidents where the damage is in any way unlikely to occur in the short term.[50] This is due to the fact that the term "incident" fulfils two functions within the framework of the Basel Protocol. On the one hand, it determines the scope of application of the Protocol pursuant to its Article 3. On the other hand, it entitles and obliges the person in operational control of the hazardous wastes to take preventive measures in order to mitigate damage as of the time of an incident.[51] To this end, however, it is rigidly required that the preventive measures are taken prior

---

[47] Basel Convention, Article 2(4).

[48] Basel Protocol, Article 3(4). The reference made to "illegal traffic" in Article 3(1) of the Protocol does not sufficiently describe the covered activities. "Illegal traffic" is defined by Article 2(21) and Article 9(1) of the Basel Convention as not including cases of re-importation according to Article 8 of the Convention.

[49] Basel Protocol, Article 2(2)(h).

[50] See Tsimplis, 'The 1999 Protocol to the Basel Convention', 16 *IJMCL* (2001), at 307.

[51] Basel Protocol, Article 6(1).

to the actual occurrence of damage, namely as soon as a grave and imminent threat of causing damage is assessed from an *ex ante* perspective. Finally, the term "incident" also has a broad meaning as far as the nature of the event is concerned. It is not limited to unexpected and unintentional events, as is the case with the term "accident". It rather includes all events that may result in damage, including, for example, nautical measures of the ship's master, usual handling operations during a transboundary movement, inappropriate packing and deliberate pollution.[52]

The definition of "incidents" under the Basel Protocol is identical to the definitions of the same term as adopted by other civil liability conventions.[53]

(b) Geographical and Temporal Coverage of Incidents

The application of the Basel Protocol to incidents is not unlimited, instead being restricted in geographical and temporal regard. In this context the particular difficulties associated with the application of the Basel Protocol to the international transport of hazardous wastes become apparent.

Transboundary movements of hazardous wastes involve at least one State of export and either a State of transit or a State of import. Sometimes even more States of transit are involved. As far as the hazardous wastes are shipped by sea, it is not only the territorial seas and the EEZ of the coastal States which are crossed but in most cases the high seas are crossed, too. In instances of a multimodal transport of hazardous wastes, also the legs of pre-carriage and on-carriage need to be taken into account. Finally, the application of the Basel Protocol is determined by the fact of whether or not the respective States are contracting Parties to the Protocol. It becomes apparent that several constellations of involved States and areas are conceivable, all of which the Basel Protocol attempts to deal with.

The starting point of the application of the Basel Protocol is defined in geographical terms. Where the State of export is a Party to the Basel Protocol the application of the Protocol begins at that point where the wastes are loaded on the means of transport in an area under the national jurisdiction of a State of export.[54] The State of export is nevertheless given the right to redefine the point where the application of the Protocol begins to the point where the wastes leave the area of its national jurisdiction.[55] This may be any land border or the point where the

---

[52] Basel Secretariat, *Implementation Manual* (2005), at 4; these examples are given by *ibid.*, at 4–5.

[53] HNS Convention, Article 1(8); Bunker Oil Convention, Article 1(8); 1969/1992 Civil Liability Convention, Article 1(8); CRTD Convention, Article 1(12).

[54] Basel Protocol, Article 3(1).

[55] *Ibid.*, Article 3(1). This option represents a compromise that was reached during the negotiating process of the Basel Protocol; see AHWG Docs. UNEP/CHW.1/WG.1/1/5, at 5; UNEP/CHW.1/WG.1/2/4, at 5; UNEP/CHW.1/WG.1/3/2, at 5–7; UNEP/CHW.1/WG.1/7/2, at 4. The exercise of this option requires that prior notification be given to the Secretary-General of the United Nations, Article 32.

territorial sea of that State ends. The exercise of this option, however, requires that both the incident and the resulting damage occur within the area of national jurisdiction of the State of export; otherwise the point where the wastes are loaded on the means of transport remains decisive. If the State of export is not a Party to the Basel Protocol, but the State of import is a Party, the starting point of the application of the Protocol is defined in temporal terms. The Protocol then applies to incidents which take place after the moment at which the disposer[56] has taken possession of the hazardous wastes.[57]

The endpoint of the application of the Basel Protocol is defined in temporal terms. Where the State of import is a Party to the Protocol, the application of the Protocol ends with the completion of the disposal process, i.e. with the notification of completion of disposal or, alternatively, with the notification of completion of the subsequent disposal operations, as far as movements destined for interim storage or processing operations are concerned.[58] Where the State of import is not a Party to the Basel Protocol but the State of export is, the Protocol applies to incidents which occur prior to the moment at which the disposer takes possession of the hazardous wastes.[59] If neither the State of export nor the State of import is a Contracting Party, the Basel Protocol does not apply, even if a State of transit is involved that is a Contracting Party to the Basel Protocol.[60] In cases of re-importation according to Article 8 or 9 of the Basel Convention, the application of the Protocol ends when the hazardous wastes reach the original State of export.[61]

In summary, it can be concluded that the application of the Basel Protocol to incidents is restricted in geographical and temporal regard. The main criterion is the moment when the incident occurs. The Basel Protocol basically applies to incidents occurring up until the completion of the final disposal process. The commencement of the Protocol's application, however, is defined in terms of location. This does not ultimately cause any difficulties, since the point "where" the wastes are loaded on the means of transport also defines the point "when" the wastes are loaded. The geographical description, thus, also contains a temporal aspect. The fact that the beginning of the application is defined in geographical terms is due to a dispute regarding the issue of whether damage in the State of export should be covered by the Protocol.[62] A compromise was found by allowing the State of export to redefine the point "where" the application of the Protocol begins. This seems to be a politically reasonable solution. However, it would have

---

[56] According to Article 2(19) of the Basel Convention, the term "disposer" means any person to whom hazardous wastes are shipped and who carries out the disposal.

[57] Basel Protocol, Article 3(3)(b).

[58] *Ibid.*, Article 3(2)(a) or, as concerns movements destined for the operations specified in D13, D14, D15, R12 or R13 of Annex IV of the Basel Convention, Article 3(2)(b).

[59] *Ibid.*, Article 3(3)(b).

[60] *Ibid.*, Article 3(3)(b).

[61] *Ibid.*, Article 3(4).

[62] See AHWG Docs. UNEP/CHW.1/WG.1/1/5, at 5; UNEP/CHW.1/WG.1/2/4, at 5; UNEP/CHW.1/WG.1/3/2, at 5–7; UNEP/CHW.1/WG.1/7/2, at 4.

been equally effective and resulted in increased clarity if both the endpoint and the starting point were defined in temporal terms. Furthermore, it should be emphasised that in cases where not all States involved in the transboundary movement of hazardous wastes are Parties to the Basel Protocol, the moment at which the disposer takes possession of the wastes constitutes the key threshold for the application of the Basel Protocol. Incidents prior to this moment are attributed to the sphere of the State of export; incidents after this moment are attributed to the sphere of the State of import.

3. Damage Covered by the Protocol

The Basel Protocol, finally, requires the occurrence of damage. The application of the Protocol to damage is, however, limited to certain types of damage as well as in respect of the place where the damage occurs.

(a) Terminological Scope of "Damage"

The Basel Protocol defines certain types of damage which are eligible for compensation under the Protocol.

(aa) Personal Damage and Damage to Property

Damage covered by the Protocol includes, first, loss of life or personal injury and loss of or damage to property other than property held by the person liable under the Protocol.[63]

Personal damage and damage to property are covered by the Basel Protocol to a different extent. Whereas loss of life and personal injury is covered per se,[64] loss of or damage to property is eligible for compensation only as far as property is concerned that is not held by the person liable under the Protocol.[65] The different coverage of these two types of damage corresponds to the legal position under other international regimes of civil liability for damage resulting from hazardous

---

[63] Basel Protocol, Article 2(2)(c)(i) and (ii).

[64] The Basel Secretariat, *Implementation Manual* (2005), at 4, gives the following example: Hazardous wastes are discharged from a truck due to a traffic accident. The persons present at the scene of the accident were poisoned by the emitted fumes and suffered personal injury within the scope of the Basel Protocol.

[65] The following example is provided in Basel Secretariat, *Implementation Manual* (2005), at 4: Corrosive wastes are packed in inappropriate containers and this results in leakage and damage to the truck. The damage to the truck is covered by the Basel Protocol, whereas the damage to the containers, if among the property of the liable person, is not covered by the Protocol.

substances.[66] It differs, by contrast, from the cover provided by the international regimes of civil liability for oil pollution damage. Those conventions require "pollution damage", which presupposes damage caused outside the ship[67] by contamination[68] resulting from the escape or discharge of oil or bunker oil from the ship.[69] Thus, "pollution damage" basically also comprises personal damage, however, only to the extent it occurred outside the ship due to a contamination from oil or bunker oil.[70] The coverage of personal damage under the Basel Protocol is, therefore, broader to the extent that personal damage caused *on* the ship is covered. As regards both types of damage, coverage under the Basel Protocol is broader to the extent that such damage is covered irrespective of the actual cause of damage.

The fact that personal damage, unlike damage to property, is covered by the Basel Protocol even if the injured person is the person liable under the Protocol has an effect only in some specific constellations. The person liable would have to bear the damage to his own property anyway; however, in those cases where the person strictly liable under the Protocol is entitled to take recourse against a person at fault under the Protocol, this recourse action also covers the damage to property of the person strictly liable. A further constellation concerns the case where a claim for compensation against a person liable under the Protocol is limited in amount. If damage to one's own property were also covered and considered in calculating the maximum amount of liability, the *pro rata* amount of compensation to be paid to third parties would be lower. And, finally, if there is a compensation fund active, then the person liable would be entitled to obtain immediate payments and response action from the fund, albeit pending a final legal assessment. Bearing in mind these particular advantages it seems to be an appropriate solution to privilege personal damage and to let it benefit from these advantages.

(bb) Loss of Income

The Basel Protocol covers loss of income to the extent that it is derived from an economic interest in any use of the environment and was incurred as a result of an impairment of the environment, taking into account savings and costs.[71]

---

[66] HNS Convention, Article 1(6)(a) and (b); CRTD Convention, Article 1(10)(a) and (b). See de la Rue/Anderson, *Shipping and the Environment* (2nd ed., 2009), at 276–277.

[67] The restriction "caused outside the ship" is tantamount to the restriction "property other than property held by the person liable" according to the Basel Protocol.

[68] The inclusion of this term makes clear that damage resulting from accidental occurrences like fire or explosions are not covered. See Altfuldisch, *Haftung und Entschädigung nach Tankerunfällen* (2007), at 23; Gunasekera, *Civil Liability for Bunker Oil Pollution Damage* (2010), at 71–74.

[69] CLC 69/92, Article I(6); Bunker Oil Convention, Article 1(9).

[70] Altfuldisch, *Haftung und Entschädigung nach Tankerunfällen* (2007), at 26–27.

[71] Basel Protocol, Article 2(2)(c)(iii).

According to this definition the coverage of economic losses under the Basel Protocol is restricted in three respects. First, the person affected must have an economic interest in any use of the environment; second, the loss must have been caused in consequence of an impairment of the environment; and third, the person affected must suffer a loss of income, taking into account savings and costs. Since this explicit wording includes "any" use of the environment, it is made clear that both direct[72] and indirect[73] uses of the environment are considered sufficient.[74] By further requiring an economic interest it is also made clear that non-economic interests, such as the protection of endangered species and plants by means of a non-commercially managed marine conservation park,[75] are not eligible for compensation under the Basel Protocol. However, those non-economic interests are partially taken into account within the scope of preventive or reinstatement measures and the related costs for prevention or reinstatement may be recoverable in this context.[76]

Under the HNS and the CRTD Convention, loss of profit is covered as well; the same applies to the CLC and the Bunker Oil Convention.[77] However, the coverage of loss of income under the Basel Protocol is defined in a more detailed manner than is the case in respect of the other international civil liability conventions.

(cc) Measures of Reinstatement and Preventive Measures

(1) Definitions

Finally, the Basel Protocol covers costs of reinstatement measures as well as costs incurred for preventive measures.

---

[72] There can be no doubt that economic interests in the direct use of the environment are covered by the Protocol, such as fishery, whale watching, deep sea mining or other economic offshore activities.

[73] The Basel Secretariat, *Implementation Manual* (2005), at 4, gives the following example of an economic interest in an indirect use of the environment: In a tourist region, close to a restaurant that is known for its panoramic view and clean air, a lorry carrying hazardous wastes overturns and guests are now staying away due to a bad smell emanating from the wastes. Although no physical damage has been incurred by the restaurant and it is not physically prevented from offering its services, the loss of income deriving from the economic interest in the direct use of the environment is to be considered an economic loss covered by the Basel Protocol.

[74] The contrary view is voiced by Tsimplis, 'The 1999 Protocol to the Basel Convention', 16 *IJMCL* (2001), at 310.

[75] This example is given by Tsimplis, 'The 1999 Protocol to the Basel Convention', 16 *IJMCL* (2001), at 310.

[76] Measures of reinstatement and preventive measures may be taken in respect of any impact on the environment; hence, they are not limited to the protection of economic losses. See Basel Protocol, Article 2(2)(d) and (e).

[77] HNS Convention, Article 1(6)(c); CRTD Convention, Article 1(10)(c); CLC 69/92, Article 1(6)(a); Bunker Oil Convention, Article 1(9)(a).

"Measures of reinstatement" are defined as any reasonable measures aiming to assess, reinstate or restore damaged or destroyed components of the environment.[78] Costs incurred for such measures are eligible for compensation under the Protocol, albeit limited to the costs of measures actually taken or to be undertaken.[79] The Protocol, furthermore, provides that the domestic law may indicate who will be entitled to take measures of reinstatement.[80]

"Preventive measures" means any reasonable measures taken by any person in response to an incident, to prevent, minimise, or mitigate loss or damage, or to effect environmental clean-up.[81] The Basel Protocol recognises costs for preventive measures, including any loss or damage caused by such measures, to be eligible for compensation, to the extent that the damage arises out of or results from hazardous properties of the wastes involved.[82]

From these definitions it follows that preventive measures can be undertaken by any person in response to an incident, whereas measures of reinstatement can be taken only by persons entitled to do so by the respective domestic law. This distinction in the entitlement to claim compensation for expenses is to be explained by the fact that the Basel Protocol intends to provide different incentives to the respective persons. Thus, measures of reinstatement are to be undertaken only by competent personnel who possess the necessary knowledge, equipment and capacities to restore the impaired environment in an appropriate and sufficient manner. By contrast, all individuals should be encouraged to take preventive measures and, hence, to prevent and mitigate damage. An incentive for everyone to take preventive measures can be established only by allowing them to claim compensation for such costs from the person liable under the Basel Protocol. Apart from that, the same consideration applies to the fact that measures of reinstatement are covered only in an area under the national jurisdiction of a Contracting Party, whereas preventive measures are eligible for compensation even when they are taken on the high seas.[83]

As regards the definition of "measures of reinstatement", some problems may arise from the inclusion of the reference to domestic law for the purpose of determining who will be entitled to take such measures. This particularly applies in international constellations in which a person entitled to take reinstatement measures under the domestic law of one Contracting Party may not be considered

---

[78] Basel Protocol, Article 2(2)(d). The Basel Secretariat, *Implementation Manual* (2005), at 4, mentions as an example, the costs for growing new plants in response to the destruction of plants caused by a spillage of obsolete pesticides during a transboundary movement.

[79] Basel Protocol, Article 2(2)(c)(iv).

[80] *Ibid.*, Article 2(2)(d).

[81] *Ibid.*, Article 2(2)(e). The Basel Secretariat, *Implementation Manual* (2005), at 4, mentions the example of costs incurred for the removal of hazardous wastes that have been illegally dumped in a river, threatening the fresh water supply of a village.

[82] Basel Protocol, Article 2(2)(c)(v).

[83] *Ibid.*, Article 3(3)(c). As to the geographical coverage see also *infra*, Sect. "Geographical Scope of Covered Damage".

authorised to do so under the domestic law of another State.[84] In the end, this situation will have a number of impacts on the availability of reinstatement capacities. Persons or companies specialised in salvage and reinstatement measures will provide their services only if they can be certain of being rewarded for their services. If this is not the case in certain areas, they will refrain from providing their services in those areas, or they will require the conclusion of an explicit service contract in advance.[85] However, unlike usual salvage cases where the shipowner has a strong economic interest in salvage measures being conducted immediately in order to salve the vessel and the goods, there is no such incentive with regard to measures of reinstatement after an incident involving hazardous wastes. Therefore, no party has an urgent interest in concluding such an agreement. In consequence, reinstatement measures will be only minimally available in areas under the national jurisdiction of a State Party that does not in its domestic law designate parties that are entitled to take reinstatement measures.

Other international civil liability regimes provide cover for costs of measures of reinstatement and preventive measures to a similar extent.[86]

(2) Distinction from Salvage Remuneration

The coverage of costs of preventive measures under the Basel Protocol may lead to difficulties regarding their distinction from remunerations for salvage services under the 1989 Salvage Convention.[87]

The 1989 Salvage Convention gives salvors the right to claim remuneration for services carried out which result in the maritime salvage of a ship or cargo. The Convention basically pursues the "no cure-no pay" approach, which means that the salvor is given the right to reward only if the salvage operations have been successful. In this case, the reward is fixed with a view to encouraging salvage operations. By contrast, no payment is due in case the salvage operations have had no useful result.[88] The Convention, furthermore, contains some specific regulations regarding environmental damage and in this respect breaks with the basic "no cure-no pay" rule.[89] According to this, the salvor is entitled to special compensation, on the condition that he has carried out salvage operations in respect of a vessel which threatened damage to the environment, and provided that the

---

[84] See Tsimplis, 'The 1999 Protocol to the Basel Convention', 16 *IJMCL* (2001), at 311.

[85] This is the usual procedure with regard to salvage contracts. See e.g. the LOF 2000 Agreement and the supplementary SCOPIC clause.

[86] HNS Convention, Article 1(6)(c), (d), (7); CRTD Convention, Article 1(10)(c), (d), (11); CLC 69/92, Article 1(6), (7); Bunker Oil Convention, Article 1(9), (10).

[87] 1989 International Convention on Salvage.

[88] 1989 Salvage Convention, Article 12(1), (2) and Article 13(1).

[89] In this respect the 1989 Salvage Convention provides for an innovative approach compared to the former 1910 Convention for the Unification of Certain Rules of Law respecting Assistance and Salvage at Sea including the 1967 Protocol thereto. For the historical background see Herber, *Seehandelsrecht* (1999), at 390–393; Rabe, *Seehandelsrecht* (4th ed., 2000), at 1016–1018.

compensation rewarded on the basis of the usual "no cure-no pay" rule is insufficient to cover his expenses.[90] If, in addition, such salvage operations have been successful in preventing or minimising damage to the environment, the salvor is rewarded an increased remuneration of up to 30 % of the expenses incurred.[91] The right to claim salvage remuneration under the 1989 Salvage Convention results by operation of law, irrespective of any explicit salvage contract.[92] In practice, however, the salvor will mostly be able to push through the conclusion of a salvage contract under the Lloyd's Standard Form of Salvage Agreement (LOF 2000/ 2011).[93]

Difficulties may arise in constellations where a maritime salvor undertakes salvage operations aiming at salving the ship and cargo and, at the same time, intends to prevent or minimise environmental damage caused by hazardous wastes. In such constellations of dual-purpose operations a distinction must be made between the remuneration for salvage operations and the compensation of costs incurred for preventive measures. Whereas costs for preventive measures are recoverable under the Basel Protocol and, to a different extent, also by means of a salvage reward, costs for salving the ship and cargo are eligible for compensation only under the "no cure-no pay" rule of the Salvage Convention. The first question, therefore, must be how to differentiate between salvage operations and prevention measures. A similar situation applies with regard to oil spills after tanker accidents and with regard to the question which preventive measures are covered by the legal regime of the CLC/Fund Convention. In this respect, international judgments and academic consensus have established the "primary purpose" rule, according to which the primary purpose of an operation is exclusively decisive for the classification of the entire operation, irrespective of whether any secondary intention exists.[94] The determination of the primary purpose by means of a subjective criterion, however, involves difficulties and in some cases it simply fails. Therefore, it is considered to be a slight indication that pollution prevention is the primary intention if the expenses incurred by the salvor exceed the salved values of the ship and cargo.[95] However, this consideration cannot sufficiently take

---

[90] 1989 Salvage Convention, Article 14(1).

[91] *Ibid.*, Article 14(2).

[92] *Ibid.*, Article 6(1).

[93] The LOF is approved and published by the Council of Lloyd's. It is reprinted, along with supplementary clauses, such as the SCOPIC Clause, in Rose, *Law of Salvage* (8th ed., 2013), at 747 *et seq.*

[94] IOPC Fund Doc. 71FUND/EXC.52/9, at para 2.2; Altfuldisch, *Haftung und Entschädigung nach Tankerunfällen* (2007), at 33–34; Ganten, 'Die Regulierungspraxis des internationalen Ölschadensfonds', 40 *VersR* (1989), at 333; Jacobsson, 'Entwicklung des Schadensbegriff im Recht der Haftung für Ölverschmutzungsschäden', 70 *DVIS, Reihe A* (1990), at 12; Jacobsson, 'Schadensersatzrecht für Ölverschmutzungsschäden', 90 *DVIS, Reihe A* (1998), at 14; Kappet, *Tankerunfälle und der Ersatz ökologischer Schäden* (2006), at 111; Wolfrum/Langenfeld/ Minnerop, *Environmental Liability in International Law* (2005), at 15.

[95] Altfuldisch, *Haftung und Entschädigung nach Tankerunfällen* (2007), at 34; Ganten, 'Die Regulierungspraxis des internationalen Ölschadensfonds', 40 *VersR* (1989), at 333–334.

into account the diverse circumstances of the individual case. Thus, if the primary purpose of an operation cannot be determined, it is acknowledged in international State practice that the costs for dual-purpose operations are divided into costs for pollution prevention and salvage costs.[96] This rather Solomonic approach seems to provide a solution which is appropriate and is the most practicable under the circumstances. It is, therefore, consistent to also apply these criteria to marine accidents involving hazardous wastes as covered by the Basel Protocol.[97]

The second issue that arises in this context concerns the relationship between claims for compensation of preventive measures under the Basel Protocol and the right to claim salvage remuneration for preventive measures under Article 14 of the Salvage Convention. The content of both claims is different. Whereas under the Basel Protocol only the costs for reasonable preventive measures are recoverable,[98] the remuneration awarded to the salvor must also take into account a fair rate for equipment and personnel, depending on the promptness of services and the availability, efficiency and value of the used equipment.[99] Where the salvage operations have successfully prevented or minimised pollution damage the salvor is even awarded a remuneration increase of up to 30 %.[100] Furthermore, and different from the CLC/Fund Convention regime, the person liable under the Basel Protocol will in most cases not be the shipowner whereas under the 1989 Salvage Convention, liability for special compensation rests with the owner of the vessel.[101] As regards the relationship of both claims, one first need to bear in mind that a salvor is entitled to special compensation under Article 14 of the 1989 Salvage Convention only if he has carried out salvage operations in respect of the relevant vessel, regardless of whether this was the primary purpose.[102] Thus, persons taking preventive measures under the Basel Protocol are not in all cases entitled to salvage reward; instead, it is only where a maritime salvor undertakes supplementary measures of damage prevention or minimisation that the claim for compensation of expenses is subject to the regimes of both the Basel Protocol and the Salvage Convention. In respect of the relationship between claims under the CLC/Fund Convention and the Salvage Convention, it is argued that such

---

[96] IOPC Fund Annual Reports of 1991, at 51; Annual Report 1992, at 57; Annual Report 1999, at 76–77; Altfuldisch, *Haftung und Entschädigung nach Tankerunfällen* (2007), at 34; Jacobsson, 'Schadensersatzrecht für Ölverschmutzungsschäden', 90 *DVIS, Reihe A* (1998), at 14; Kappet, *Tankerunfälle und der Ersatz ökologischer Schäden* (2006), at 111.

[97] Tsimplis, 'The 1999 Protocol to the Basel Convention', 16 *IJMCL* (2001), at 311 argues that in the context of the Basel Protocol no managerial or legal basis for such a distinction exists. However, the Basel Protocol does not determine under which conditions an intention of the salvor to prevent damage exists. This is to be determined by legal interpretation.

[98] Basel Protocol, Article 2(2)(c)(v) and (e).

[99] 1989 Salvage Convention, Article 14(1), (3), Article 13(1)(h), (i) and (j).

[100] *Ibid.*, Article 14(2).

[101] *Ibid.*, Article 14(1).

[102] Reeder (ed.), *Brice on Maritime Law of Salvage* (4th ed., 2003), at 6-88–6-89; Rose, *Law of Salvage* (8th ed., 2013), at 6.024.

pollution measures are either "purely incidental" and recoverable under the Salvage Convention alone,[103] or that they are eligible for compensation under the respective civil liability regime alone.[104] However, both views fail to provide convincing arguments. This applies at least with regard to claims raised under the Basel Protocol. Distinct from the legal position under the CLC/Fund Convention, the Basel Protocol does not generally impose liability on the shipowner, this corresponding to the approach of the Salvage Convention. The legal position of the salvor under the Basel Protocol, therefore, differs significantly from his legal position under the 1989 Salvage Convention. Since there is no legal justification for restricting the application of one of these conventions, the only appropriate approach seems to be to give the salvor the right to choose on which legal basis to claim compensation and against which liable person.

(dd) Purely Ecological Damage

Ecological or environmental damage is not covered as such by the Basel Protocol. Ecological damage may consist, for example, in the impairment of the marine environment and in damage to natural resources due to the contamination by hazardous and noxious substances. The particular feature of ecological damage consists in the fact that the consequences of such damage, e.g. the impairment or destruction of natural resources or entire ecosystems, cannot be undone in a short time. In many cases ecological damage is of a permanent nature or the natural environment will need a long period to recover. This particular feature of ecological damage also shows the particular difficulties related to the compensation of ecological damage. Since there is no allocation of exclusive rights of use for the marine environment, there is no State or private party that can invoke the infringement of individual rights with regard to purely ecological damage. In addition, since ecological damage cannot be undone in most cases, the determination of the exact amount of damage is not possible. An assessment of loss by means of abstract or theoretical models does not seem to sufficiently take into account the unique circumstances of the individual case.[105] Particular difficulties may arise with regard to the issue of causation in cases where ecological damage may be attributed to different causes.

---

[103] Reeder (ed.), *Brice on Maritime Law of Salvage* (4th ed., 2003), at 6-180.

[104] Altfuldisch, *Haftung und Entschädigung nach Tankerunfällen* (2007), at 34; Jacobsson, 'Schadensersatzrecht für Ölverschmutzungsschäden', 90 *DVIS, Reihe A* (1998), at 14.

[105] In this context the IOPC Fund adopted in 1980 a resolution, according to which "the assessment of compensation to be paid by the International Oil Pollution Compensation Fund is not to be made on the basis of an abstract quantification of damage calculated in accordance with theoretical models". This resolution was adopted in response to a claim submitted by the Government of the USSR to the IOPC Fund in 1979, concerning ecological damage due to an oil spill that was caused by the first *M/V "Antonio Gramsci"* incident in 1979. See IOPC Fund Annual Report 1988, at 61–64; Altfuldisch, *Haftung und Entschädigung nach Tankerunfällen* (2007), at 34–35; de la Rue/Anderson, *Shipping and the Environment* (2nd ed., 2009), at 481–482.

Under the Basel Protocol ecological damage is covered, therefore, only to a limited extent. The Protocol acknowledges the existence of ecological damage and its eligibility for compensation within the definitions of loss of income, costs of measures of reinstatement of the impaired environment and costs of preventive measures. Particularly the explicit clarification of the Protocol that reinstatement measures are covered only to the extent that such measures are actually taken or to be undertaken[106] underlines the fact that purely ecological damage, to the extent it goes beyond the actual costs of reinstatement and is not covered by the other types of damage, is not recoverable under the Basel Protocol.

This outcome is, furthermore, in line with the coverage of ecological damage under other international civil liability regimes, such as under the HNS and CRTD Conventions[107] and under the CLC/Fund Convention[108] and the Bunker Oil Convention.[109]

(b) Geographical Scope of Covered Damage

The scope of application of the Basel Protocol is not only restricted by the time and place an incident takes place, but also with regard to the place where the consequent damage occurs.

The place where an incident takes place and the place where the actual damage occurs are not necessarily the same. For example, it is well possible that as a result of an incident taking place in an area under the jurisdiction of State A, hazardous wastes will be released in the marine environment and ultimately drift along with the ocean current to the coastal regions of State B, where they case pollution damage.

The Basel Protocol itself determines its scope of application with regard to such constellations. It basically applies to damage that occurs in an area under the national jurisdiction of a Contracting Party, regardless of whether a State of export, import or transit is concerned.[110] It even applies if damage is suffered in an area under the national jurisdiction of a State that is not at all involved in the transboundary movement of hazardous wastes, provided this State is a Party to the

---

[106] Basel Protocol, Article 2(2)(c)(iv).

[107] See Lawrence, 'Negotiation of a Protocol on Liability and Compensation', 7 *RECIEL* (1998), at 251; de la Rue/Anderson, *Shipping and the Environment* (2nd ed., 2009), at 490.

[108] Unlike the legal position under CLC 69, CLC 69/92 in Article I(6)(a) contains an explicit exclusion of purely ecological damage apart from the costs of reasonable measures of reinstatement actually undertaken or to be undertaken. See also Altfuldisch, *Haftung und Entschädigung nach Tankerunfällen* (2007), at 34–40; Kappet, *Tankerunfälle und der Ersatz ökologischer Schäden* (2006), at 73–79; de la Rue/Anderson, *Shipping and the Environment* (2nd ed., 2009), at 482–490.

[109] Gunasekera, *Civil Liability for Bunker Oil Pollution Damage* (2010), at 175; de la Rue/Anderson, *Shipping and the Environment* (2nd ed., 2009), at 490.

[110] Basel Protocol, Article 3(3)(a).

Protocol. Moreover, the Basel Protocol also applies to damage suffered in an area under the national jurisdiction of a State of transit which is not a Party to the Protocol, but which is listed in Annex A of the Protocol and has acceded to an operative multilateral or regional agreement concerning transboundary movements of hazardous wastes.[111] In contrast, the Basel Protocol does not basically apply to damage suffered in areas beyond any national jurisdiction, in particular on the high seas.[112] An exemption from this rule is made for certain types of damage, such as for loss of life or personal injury, loss of or damage to property, and costs for preventive measures. Those damages are covered by the Protocol even if they occur on the high seas. This, however, does not include damage suffered in an area under the jurisdiction of State that is not a Party to the Protocol.[113] In contrast, loss of income and costs for measures of reinstatement of the impaired environment are not covered by the Basel Protocol when they are suffered on the high seas.

This geographic restriction of covered damages under the Basel Protocol leads to appropriate results. On the high seas measures of reinstatement are mostly ineffective and must fail due to practical reasons. Since there is no allocation of exclusive rights of use on the high seas, there is also no economic incentive for an individual State or party to invest in the reinstatement of the marine environment. But what is even more significant is that in the absence of exclusive individual rights of use, a reinstatement of the impaired environment cannot be claimed since no State or party has seen its individual rights infringed such that a right of this nature could be enforced. Reinstatement measures conducted on the high seas, therefore, are of no practical relevance. This is different with regard to preventive measures. Such measures are taken in response to an incident and aim at preventing or minimising further damage, which may spread over great distances. Therefore, the Basel Protocol creates an incentive to take such measures by giving any person the right to claim compensation for such measures and by allowing recovery even for preventive measures that are taken on the high seas. Thus, the different geographic coverage of costs for reinstatement measures and preventive measures under the Protocol seems appropriate.

4. Summary

In conclusion, it can be summarised that the scope of application of the Basel Protocol is regulated in a very detailed and not always very clear manner. The application of the Protocol is determined by means of the definitions of transboundary movements of hazardous wastes, incidents and damages. The application to incidents and damages, furthermore, is restricted in temporal and geographical regard.

---

[111] *Ibid.*, Article 3(3)(d). In this case Article 3(3)(b) applies *mutatis mutandis*.
[112] *Ibid.*, Article 3(3)(a).
[113] *Ibid.*, Article 3(3)(c).

As regards the definition of hazardous wastes the Basel Protocol refers to the definitions of the Basel Convention. Wastes are covered by the Protocol if they belong to any category contained in Annex I of the Convention, unless they do not possess any of the characteristics contained in Annex III of the Convention. In addition, the Basel Protocol also applies to wastes that are defined as hazardous by the domestic legislation of the respective Contracting States. The explicit coverage of "other wastes" by the Protocol is somewhat confusing since this term does not mean "non-hazardous" wastes, as this wording would suggest, but only household wastes and residues from their incineration, which are in most cases hazardous as well. The conceptual distinction between hazardous and household wastes and the terminology of "other wastes", thus, must be considered a shortcoming of the Basel Convention framework. The Basel Protocol does not apply to radioactive wastes subject to the control system of the IAEA and to wastes deriving from the normal operation of a ship as covered by the MARPOL 73/78 Convention. The term "transboundary movement" requires the involvement of at least two States. Thus, the Protocol does not apply, for example, to transports of hazardous wastes to the high seas for dumping or incineration.

The application of the Basel Protocol, further, requires an incident. This term has a broad meaning and also covers long-term events as well as events that have not yet caused damage but which are likely to cause damage in the future. In other words, all those events are covered that create a grave and imminent threat of causing damage. The coverage of incidents under the Basel Protocol is also restricted in geographical and temporal regard. The application of the Protocol basically begins at that point and at that moment the wastes are loaded on the means of transport in the State of export. The State of export, however, may postpone the beginning of the application to the moment the wastes leave the area of its jurisdiction. If the State of export is not a Party to the Basel Protocol, the application begins at that moment the disposer takes possession of the wastes. The application of the Protocol ends with the notification of completion of disposal or, in case the State of import is not a Party to the Protocol, when the wastes are delivered to the disposer.

Finally, the Basel Protocol only applies to certain types of damage, and this is, moreover, restricted in geographical regard. The Basel Protocol covers loss of life and personal injury as well as damage to property other than the property held by the person liable, irrespective of whether this damage occurs in an area under the national jurisdiction of a Contracting Party or on the high seas. Loss of income is covered by the Protocol only if this loss derives from an economic interest in any use of the environment and if it results from an impairment of the environment. Such losses, however, are only covered if they are suffered in an area under the national jurisdiction of a Contracting Party. Finally, the Basel Protocol also covers costs for measures of reinstatement of the impaired environment and costs for preventive measures taken in response to an incident. Whereas reinstatement measures can be taken only by persons designated by the domestic law and only within an area under the national jurisdiction of a Contracting Party, preventive measures are covered that are taken by any person and even if they are conducted on the high seas. This is to be explained by the intention of the Basel Protocol to encourage any potential person to undertake preventive measures, while only certain designated persons are to be encouraged to undertake measures of reinstatement.

II. Relationship to Other Civil Liability Regimes

The relationship of the Basel Protocol to other international regimes of civil liability and compensation is explicitly regulated in Article 3(7) and in Article 11 of the Protocol. These provisions provide a kind of a general subsidiarity of the Basel Protocol in relation to other civil liability regimes, on the condition that those regimes meet certain regulative standards as defined by Article 3(7) and Article 11 of the Protocol.

1. Requirements in General

There is one major difference between the two scenarios dealt with by Article 3(7) and Article 11 of the Basel Protocol. Whereas Article 3(7) of the Protocol concerns the relationship of the Protocol to other liability agreements regarding particularly the *transboundary movement of hazardous wastes*, Article 11 of the Protocol regulates the relationship to other *general* agreements on liability and compensation. Since both provisions impose different legal standards which must be met by the respective agreement of liability and compensation in order to override the application of the Basel Protocol, it is necessary to make a clear distinction between these two scenarios.

(a) Article 3(7) of the Protocol

Article 3(7) concerns the relationship of the Basel Protocol to other liability agreements regarding, particularly, the transboundary movement of hazardous wastes that are concluded and notified in accordance with Article 11 of the Basel Convention. Article 3(7) provides that an agreement concluded and notified in accordance with Article 11 of the Basel Convention overrides the application of the Basel Protocol provided that certain requirements are fulfilled by that agreement.

(aa) Legal Requirements

According to Article 3(7) of the Protocol, the alternative agreement must first come under the ambit of Article 11 of the Basel Convention. Article 11 of the Basel Convention addresses the relationship of the Basel Convention to other bilateral, multilateral or regional agreements or arrangements regarding the transboundary movement of hazardous wastes. Thus, those agreements must not necessarily contain rules of liability and compensation. Within Article 11 of the Basel Convention, a fundamental distinction[114] is made between agreements the States accede to after the Basel Convention enters into force for them[115] and agreements the States enter into before the Basel Convention enters into force for them.[116] As regards the former kind of agreements, the Basel Convention stipulates that the Parties to the Basel Convention may enter into such "post-Basel" agreements provided these agreements do not derogate from the environmentally sound management of hazardous wastes as required by the Basel Convention and on the condition that the provisions stipulated by such agreements are not less environmentally sound than those of the Basel Convention.[117] In addition, the States that accede to such agreement must notify the Secretariat of the Basel Convention of the respective agreement.[118] As regards agreements the Parties to the Basel Convention entered into before the Basel Convention became applicable to them, the Convention stipulates that such "pre-Basel" agreements will not be affected by the provisions of the Basel Convention provided those agreements are compatible with the environmentally sound management of hazardous wastes as required by the Basel Convention and on the condition that those agreements have been notified to the Secretariat of the Basel Convention.[119] A further requirement in order for both "post-Basel" and "pre-Basel" agreements to take precedence over the Basel Convention is that that the agreement in question applies only to hazardous waste movements which take place exclusively among the Parties to this agreement.[120] In other words, in the event that only one of the States involved in a movement of hazardous wastes is a Party to such agreement but at least one of the other States involved is not, the Basel Convention remains applicable to the entire transport. At present, 12 bilateral and 9 multilateral agreements coming under the ambit of Article 11 of the Basel Convention have been notified to the Secretariat of the Basel Convention.[121]

However, not all agreements that come under Article 11 of the Basel Convention are capable of overriding the application of the Basel Protocol. In this respect, Article 3(7) of the Basel Protocol stipulates further requirements. According to this, the Protocol foregoes applicability[122] only on the conditions that

---

[114] See also Kummer, *The Basel Convention* (1995), at 94; Basel Secretariat, *Implementation Manual* (2005), at 7–8.
[115] This scenario is dealt with by Article 11(1) of the Basel Convention.
[116] This scenario is covered by Article 11(2) of the Basel Convention.
[117] Basel Convention, Article 11(1).
[118] *Ibid.*, Article 11(2).
[119] *Ibid.*, Article 11(2).
[120] See Kummer, *The Basel Convention* (1995), at 94–96.
[121] A list of these agreements is available at www.basel.int.
[122] Basel Protocol, Article 3(7)(a), see also Article 3(7)(c).

the damage occurred in an area under the national jurisdiction of one of the Parties to the respective agreement[123] and—of primary importance—that there is a liability and compensation regime which is in force, is applicable and stipulates provisions that fully meet or exceed the objective of the Basel Protocol by providing a high level of protection to persons who have suffered damage.[124] Such a liability and compensation regime must not necessarily be part of the respective agreement or arrangement, but may also be contained in another applicable legal instrument or, for instance, be found in the domestic law that is referred to.[125] In such cases, for the purpose of promoting transparency as to the liability and compensation regime applicable instead of the Basel Protocol, the State Party to such an agreement must notify the Secretariat of the Basel Convention regarding the applicable liability and compensation regime and enclose a description of this regime.[126] Article 3(7) of the Basel Protocol, furthermore, requires that the Secretary-General of the United Nations will have been notified of the non-application of the Protocol by the State in which the damage occurs[127] and that the respective agreement does not declare that the Basel Protocol will be applicable.[128]

Article 3(7) of the Basel Protocol, read in conjunction with Article 11 of the Basel Convention, also determines that agreements within the ambit of Article 3(7) of the Basel Protocol are eligible to override the application of the Basel Protocol only in case of "internal" transports among the Parties to such agreements. Constellations in which the applicable regime of liability and compensation changes during the course of the movement are not possible, to the extent the movement is covered by the Basel Protocol. It has been outlined above that Article 11 of the Basel Convention only covers agreements applying to transports that take place exclusively among the Parties to this agreement. In case only the State of export or the State of import is Party to this agreement, this agreement will not be covered by Article 11 of the Convention. Due to the express wording "entirely among the Parties" in Article 11 of the Convention, this also applies to any State of transit involved in the movement. In addition, according to Article 3(7)(a)(i) of the Protocol, damage must occur in an area under the national jurisdiction of a Party to this agreement. Following this, it becomes apparent that there might not be any constellation where in respect of a particular movement two regimes of liability will apply. Either the Basel Protocol is displaced by the other agreement in respect of the entire transport or the Basel Protocol remains applicable for the entire transport.

---

[123] *Ibid.*, Article 3(7)(a)(i).
[124] *Ibid.*, Article 3(7)(a)(ii).
[125] Basel Secretariat, *Implementation Manual* (2005), at 8.
[126] Basel Protocol, Article 3(7)(b).
[127] *Ibid.*, Article 3(7)(a)(iii).
[128] *Ibid.*, Article 3(7)(a)(iv).

(bb) Objections Raised Against This Provision

Article 3(7) of the Basel Protocol has been subject to broad criticism. Particularly developing countries and NGOs have argued that the primary purpose of this provision is to exclude the application of the Basel Protocol for industrialised countries that fall under the relevant OECD decisions and regulations concerning the transboundary movement of hazardous wastes. Were that the case, the majority of hazardous waste movements would not be covered by the Basel Protocol and, in respect of waste movements between OECD countries and non-OECD countries, only the importing countries would remain liable under the Basel Protocol.[129] Industrialised countries, in turn, have argued that in respect of intra-OECD or, alternatively, intra-EU movements, the provisions of the Basel Protocol would not be necessary since there are already relevant legal systems in force.[130] Neither of these arguments is fully convincing.

The objection that allowing the application of regional agreements and arrangements would undermine the significance of the Basel Protocol cannot be raised in respect of the OECD and EU regulations alone. The same consideration basically applies, for example, to the Bamako and Waigani Conventions as well as to other regional or bilateral conventions and agreements, provided that either the envisaged liability protocols to the respective conventions will enter into force, or that these conventions or agreements will be amended with regard to an explicit reference to an existing liability regime, which may also be contained in the respective domestic laws. What is important, however, is that the Basel Protocol stipulates a qualitative requirement with regard to the agreement at issue. Such agreement may only take precedence over the Basel Protocol if it fully meets or exceeds the protection level of the Protocol. By means of this provision the legal aim of the Basel Protocol becomes apparent. The purpose of the Protocol is not the application of the Protocol as such, but rather to ensure the application of a regime of liability and compensation that provides a specific minimum standard of protection that is defined to be the standard of the Basel Protocol. This purpose, however, can also be achieved by allowing the application of regional or bilateral agreements containing or referring to a regime of liability and compensation that has been previously notified to and evaluated by the Secretariat of the Basel Convention.

Furthermore, the objection that in respect of hazardous waste movements between OECD and non-OECD States, only the importing countries would remain liable under the Basel Protocol is not correct. Irrespective of the question of

---

[129] See 'Compensation and Liability Protocol Adopted', 30 *Envtl. Pol'y & L.* (2000), at 44; Choksi, '1999 Protocol on Liability and Compensation', 28 *Ecology L. Q.* (2001), at 525–526; Long, 'Protocol on Liability and Compensation', 11 *Colo. J. Int'l Envtl. L. & Pol'y* (2000), at 259–260.

[130] See Choksi, '1999 Protocol on Liability and Compensation', 28 *Ecology L. Q.* (2001), at 526; Long, 'Protocol on Liability and Compensation', 11 *Colo. J. Int'l Envtl. L. & Pol'y* (2000), at 259–260; Soares/Vargas, 'The Basel Liability Protocol', 12 *Yb. Int'l Env. L.* (2001), at 89.

whether the OECD regulation may be considered an agreement within the ambit of Article 3(7) of the Basel Protocol, this regulation would not be able to replace the Basel Protocol in respect of such transports anyway. It has been set out above that regional or bilateral agreements may only override the application of the Basel Convention and Protocol if all Parties involved in the transboundary movement are Contracting Parties also to the respective agreement. This does not apply to hazardous waste movements from OECD countries to non-OECD countries.

However, another issue becomes obvious that must actually be considered a major weakness of this provision. The Basel Protocol does not define the procedure how to determine whether the respective regional or bilateral agreement actually meets or exceeds the protection level of the Basel Protocol. In this context it should be considered that any Party to the Basel Protocol that is also a Party to a regional or bilateral agreement is obliged to notify the Secretariat of the Basel Convention about the applicable regime of liability and compensation and to enclose a description of this regime.[131] Therefore, it seems appropriate that the decision of whether the other regime of liability fully meets or exceeds the protection level of the Basel Protocol be prepared by the Secretariat of the Basel Convention or any subsidiary body for later consideration and final decision at the Meeting of the Parties.[132] However, to the extent the Protocol does not explicitly specify this or any other procedure, a certain lack of clarity remains.

In summary, it must be concluded, therefore, that the objections voiced with regard to Article 3(7) of the Basel Protocol and the possibility of Contracting States to exclude the application of the Protocol with regard to movements among Parties to other agreements are not persuasive. Such exclusion only encompasses transports taking place exclusively among Contracting Parties to such an agreement; hence, it is not conceivable that two different regimes of liability could apply in respect of one movement. Moreover, the Basel Protocol requires in a qualitative regard that the respective agreement fully meets or exceeds the level of protection provided for by the Basel Protocol. By this means, the Basel Protocol constitutes a minimum standard of liability and ensures that the respective agreement does not fall below this standard. From a legal perspective there is, therefore, no cause for concern. From a practical perspective, a weakness of the Protocol may consist in the fact that the Protocol does not provide just how to determine whether or not the respective agreement fully meets or exceeds the level of protection of the Basel Protocol. Such rules of procedure, however, may be elaborated in detail at the Meeting of the Parties.[133]

---

[131] Basel Protocol, Article 3(7)(b).

[132] Such a mandate can be derived from Article 25(1)(a) of the Basel Protocol regarding the preparation of decisions by the Secretariat of the Basel Convention, and from Article 24(4)(d) of the Basel Protocol as regards the competence of the Meeting of the Parties to consider and make decisions about those issues.

[133] See Basel Protocol, Article 24(3).

(b) Article 11 of the Protocol

In contrast to Article 3(7), Article 11 of the Basel Protocol determines the relationship of the Protocol to other general agreements on liability and compensation that are not specially focused on the transboundary movement of hazardous wastes. According to this provision, the Basel Protocol does not apply to damage caused by an incident during the same portion of a transboundary movement, provided this damage is covered by an alternative bilateral, multilateral or regional agreement on liability and compensation, which is in force for the Party concerned and had been opened for signature before the Basel Protocol was opened for signature,[134] even if the alternative agreement was amended afterwards.[135]

(aa) Difficulties Related to the Dual Coverage of Movements

It becomes apparent that Article 11 of the Protocol, in contrast to Article 3(7), does not require that the alternative agreement applies to all States involved in the particular hazardous waste movement. Thus, the Basel Protocol abandons the approach of Article 3(7) for avoiding the application of two different liability regimes in respect of one particular movement. Pursuant to Article 11, it may be possible that in respect of one particular transport of hazardous wastes the application of the Basel Protocol is excluded only for certain stages of the transport, whereas in respect of other stages the Basel Protocol remains applicable. This situation has been criticised as causing uncertainty about the applicable regime and as leading to protracted litigation.[136] This objection seems valid.

The possible coverage of one transport by two different regimes of liability and compensation may cause difficulties also with regard to the question when and to which extent the alternative agreement overrides the application of the Basel Protocol. Article 11 stipulates that the alternative agreement overrides the Basel Protocol "whenever the provisions of the Protocol and the provisions of a[n alternative] agreement apply to liability and compensation for damage caused by an incident". This definition is far from clear. In particular, it remains unclear whether the alternative agreement prevails over the Basel Protocol even where it generally applies but does not, for instance, cover the particular type of damage.[137] In abstract terms the question is whether the fact that the alternative agreement does not impose liability in a specific case re-opens the application of the Basel Protocol, or whether the decision of the alternative agreement not to impose

---

[134] The Basel Protocol was opened for signature on 6 March 2000, see Article 26.
[135] Basel Protocol, Article 11.
[136] See Birnie/Boyle/Redgwell, *International Law and the Environment* (3rd ed., 2009), at 483.
[137] Further constellations involve cases in which the alternative agreement does not cover the particular hazardous wastes, attaches liability to different persons, imposes a different standard of liability (fault-based or strict), provides for a deviating statute of limitation, stipulates certain exclusions of liability, excludes, for example, damage on the high seas from its coverage, or simply concerns another aspect of liability, such as liability of the carrier for cargo damage.

liability in a specific case is also binding and precludes the application of the Basel Protocol. In the end, this question can be answered in most cases by a simple reference to the distinction between the legal prerequisites of application and the legal consequences thereof. The scope of application is defined by each convention and agreement in terms of substance as well as in geographical and temporal regard. Where a particular event of damage falls under the definition of application of an alternative agreement, the same event of damage may not be covered at the same time by the Basel Protocol. This means that if the alternative agreement considers itself applicable to that certain kind of waste as well as to that incident and damage in a geographical and temporal regard, then any legal consequences that will follow from this event of damage are governed exclusively by that agreement. This applies even in case the alternative agreement does not impose liability in the specific case, although liability would be established under the Basel Protocol.[138]

Conversely, the Basel Protocol remains applicable in any instance the event of damage does not fall under the scope of application of the alternative agreement. In particular, this also applies to incidents involving different hazardous wastes, only some of which are covered by the alternative liability regime. In that case both the alternative regime as well as the Basel Protocol applies concurrently. This outcome is reaffirmed by the Basel Protocol, which provides that in such situations a person liable under an alternative agreement will be liable under the Protocol only "in proportion to the contribution made by the wastes covered by the Protocol to the damage".[139]

(bb) Meaning of "Portion of a Transboundary Movement"

A further ambiguity of Article 11 arises from the requirement that both the incident and the resulting damage must have occurred during the same portion of a transboundary movement. The term "portion of a transboundary movement" is defined neither in the Basel Protocol nor in the Basel Convention. The extent or duration of such portion is, therefore, unclear. One could think about dividing a transboundary movement into a land portion and a sea portion. Another possible distinction could be to assume one portion for each area that comes under the national jurisdiction of a different State. But still it remains unanswered whether, for example, the passage through the high seas may constitute its own discrete portion of a transboundary movement, or whether a new portion might begin with any transhipment of the cargo. And even more: Could it be possible to assume that

---

[138] This may be the case if the alternative agreement does not attach liability to persons who would be liable under the Basel Protocol, if it does not provide compensation for types of damage that would be deemed recoverable under the Basel Protocol, if it establishes a shorter period of limitation, or if it stipulates exclusions of liability that are not recognised under the Basel Protocol.

[139] Basel Protocol, Article 7(1).

in any instance where a vessel alters her course this must be considered a new portion of a transboundary movement?

In order to give an answer, the legal purpose of this requirement needs to be taken into account. Article 11, by means of this requirement, intends to avoid as far as possible inconsistencies and ambiguities regarding the applicable regime of liability and compensation, which could arise particularly in cases where the incident and the resulting damage occur at different places. In such cases the legal consequence could be that the liability regime applicable to the incident does not cover the damage in geographical or temporal regard, whereas the other regime of liability covers only the damage and not the incident. In order to avoid that none of such liability regimes would ultimately apply, Article 11 of the Basel Protocol allows for a prevailing application of the alternative agreement only if it covers both the incident and the damage. The concept of a "portion of a transboundary movement" must take this into account. Possible alternative agreements, such as the HNS Convention, contain detailed rules regarding the question what kind of damage is covered in which maritime zones.[140] Consequently, it seems appropriate to assume a "portion of a transboundary movement" in accordance with the distinction of national territories and maritime zones and to ensure, by this means, that an event of damage is excluded from the application of the Basel Protocol only where it is in its entirety subject to another regime of liability and compensation. An individual portion of a transboundary movement, thus, should be assumed for each territory including the territorial sea, the EEZ and the high seas. Any further distinction, such as the consistent course of a vessel, is superfluous.

(cc) Insufficiency of the Formal Criterion

Article 11 of the Basel Protocol has been strongly criticised, moreover, for providing far-reaching exclusions from its scope of application. It is objected that Article 11 does not impose any qualitative requirement regarding the alternative agreement which is comparable with the requirement imposed by Article 3(7) of the Protocol. It is, thus, possible that the alternative agreement overrides the application of the Basel Protocol even where it provides a standard of liability that is below the protection level of the Basel Protocol.[141] Instead of imposing a qualitative requirement, Article 11 rather focuses on a formal criterion. The alternative regime will prevail over the Basel Protocol on the condition that this regime had been opened for signature before 6 March 2000, the date when the

---

[140] See, for instance, the definition of the scope of application in Article 3 of the HNS Convention and the distinction between the territory of a State Party, including the territorial sea, the EEZ or a zone of 200 nautical miles, and the high seas.

[141] See Birnie/Boyle/Redgwell, *International Law and the Environment* (3rd ed., 2009), at 483; Tsimplis, 'The 1999 Protocol to the Basel Convention', 16 *IJMCL* (2001), at 315.

Basel Protocol was opened for signature. This provision is said to be specifically designed for the HNS Convention,[142] but other conventions also fall within its scope.[143]

It is obvious that from a formal perspective this objection is true. By virtue of Article 11 the Basel Protocol declares itself inapplicable in case the event of damage is covered by another regime of liability and compensation, irrespective of whether this alternative regime fully meets or exceeds the level of protection established by the Basel Protocol. To this extent Article 11 can indeed be considered a major loophole of the Basel Protocol. From a practical perspective, however, the impact of Article 11 remains limited. By means of the formal requirement of Article 11, according to which the alternative agreement had to be opened for signature before the Basel Protocol was opened for signature on 6 March 2000, it is ensured that any convention or agreement that possibly comes under the ambit of Article 11 was known at the time the Basel Protocol was signed. Therefore, it is not possible to circumvent the provisions of the Basel Protocol by creating a new agreement with a lesser standard of liability. One could surely argue that States could undermine the application of the Basel Protocol by subsequently amending an agreement that comes under Article 11, to the extent that liability under this amended agreement would be considerably lower than under the Basel Protocol.[144] However, such approach appears to be rather improbable.

The formal criterion of Article 11 is, nevertheless, insufficient in another respect. Since the temporal aspect is in the first instance the decisive factor for the application of the alternative regime on liability and compensation, Article 11 does not take into account that a convention or agreement may be established after the Basel Protocol was opened for signature and that such a convention or agreement could provide a more specialised and sophisticated regime applying only to particular aspects of transboundary movements of hazardous wastes. Thus, it might be possible that with regard to certain types of hazardous and noxious substances, such as for PCBs or POPs, a specific regime of liability and compensation might be set up in future which provides a more detailed and more suitable solution as regards liability and compensation for these particular substances. With regard to such cases it would have been the better approach to establish a quantitative criterion in Article 11 which was modelled on Article 3(7) of the Protocol, rather than establishing a rigid and formal criterion by which more suitable, but newly established agreements are simply ignored.

---

[142] 'Compensation and Liability Protocol Adopted', 30 *Envtl. Pol'y & L.* (2000), at 44; Lawrence, 'Negotiation of a Protocol on Liability and Compensation', 7 *RECIEL* (1998), at 253.

[143] The respective conventions are outlined in the following section.

[144] According to Article 11 of the Protocol the alternative agreement prevails over the Basel Protocol if it had been opened for signature when the Basel Protocol was opened for signature, even if the agreement was amended afterwards.

2. Relationship to Single Civil Liability Instruments

(a) Relationship to the HNS Convention

The 1995/2010 HNS Convention, which has not yet entered into force, establishes rules on liability and compensation for damage arising from the carriage of hazardous and noxious substances (HNS) by sea.[145] It applies to HNS that are shipped on board a ship as cargo as well as to residues thereof. As regards the definition of HNS, the Convention refers to existing lists of HNS contained in other IMO instruments, such as in the MARPOL 73/78 Convention or in the IMDG Code.[146] Damage caused by radioactive substances and pollution damage as defined in the CLC are excluded from the scope of application, whereas waste materials are basically covered by the HNS Convention.[147]

Since wastes are not excluded from the scope of the HNS Convention, the HNS Convention and the Basel Protocol potentially overlap. The HNS Convention imposes a regime of liability and compensation without being particularly focused on the transboundary movement of hazardous wastes. It, therefore, falls within the ambit of Article 11 of the Basel Protocol. The HNS Convention, furthermore, was opened for signature before the Basel Protocol,[148] which means that, provided the incident and damage occur during the same portion of the transboundary movement, the HNS Convention will by virtue of Article 11 of the Basel Protocol prevail over the Basel Protocol to the extent the HNS Convention applies.

However, it should be noted that the scope of application of the HNS Convention and the scope of the Basel Protocol are not fully coincident. The scope of the Basel Protocol is broader in three respects: First, the Basel Protocol covers in temporal and geographical regard the entire transboundary movement, also including the pre-carriage and the on-carriage by, for instance, land transport, as well as the subsequent disposal process. The HNS Convention, by contrast, is limited to the carriage by sea from tackle to tackle.[149] Second, the scope of the Basel Protocol may be broader with regard to the covered substances in the individual case. It applies to broad categories of hazardous wastes defined by means of their characteristics, even comprising household wastes, whereas the HNS Convention applies to substances previously defined in enumerated lists. Finally, the Basel Protocol imposes liabilities on several persons involved in hazardous waste movements, as there is the notifier (which may be the generator or

---

[145] The HNS Convention has been outlined in detailed *supra*, Sect. "1996/2010 HNS Convention" in Chap. 3.

[146] 1996/2010 HNS Convention, Article 1(5).

[147] *Ibid.*, Article 4(3). See also *supra*, Sect. "The Evolution of the 1196 HNS Convention and its Main Content" in Chap. 3.

[148] Namely on 1 October 1996, see 1996 HNS Convention, Article 45(1).

[149] 1996/2010 HNS Convention, Article 1(9).

the exporter), the importer and the disposer of the hazardous wastes.[150] Liability under the HNS Convention, by contrast, is channelled to the shipowner.[151]

According to this, even in the event that both conventions enter into force, they are not fully coincident. The Basel Protocol will remain applicable in addition to the HNS Convention as far as concerns aspects of the transboundary movement that are not covered by the HNS Convention.[152] Consequently, it will be possible that two regimes of liability will apply with regard to one transport of hazardous wastes. This, however, only applies in a temporal and geographical regard concerning the different legs of transport. There is an unmistakeable border between the application of the Basel Protocol prior to the loading on and after the discharge of the hazardous wastes from the ship (the tackle-to-tackle period). To this extent, both Conventions are mutually exclusive.

A simultaneous application of both Conventions, moreover, appears to be improbable. Such situation may arise only in those cases where damage was caused by hazardous wastes, only part of which are deemed to be HNS, and where it is possible to distinguish between the portions of damage caused by each kind of wastes. The HNS Convention provides that where it is not reasonably possible to separate damage caused by the HNS from that caused by other factors, all such damage will be deemed to be caused by the HNS.[153] Since it will not be possible in most cases to distinguish between the portion of damage caused by the hazardous wastes deemed to be HNS and the portion of damage caused by hazardous wastes deemed not to be HNS, the HNS Convention, hence, will be applicable in most cases by virtue of this provision. However, if it is reasonably possible to separate between the portions of damage caused by hazardous wastes deemed to be HNS and those deemed not to be HNS, both the HNS Convention and the Basel Protocol remain applicable to the respective portion of damage.[154]

Finally, the issue as to which person is liable for compensation depends on the legal consequences stipulated by the respective applicable regime of liability and compensation, rather than being related to the question of which convention applies. Where it is established that a convention applies to a particular event of damage, then this convention will solely and exclusively determine to which person liability is attached and to which extent.

In conclusion, it should be pointed out that, as a general rule, the HNS Convention prevails over the Basel Protocol to the extent the particular event of damage is covered by the HNS Convention. For the remaining cases the Basel Protocol remains applicable as a supplement to the HNS Convention. This

---

[150] In addition, any person may be liable based on fault. See in detail *infra*, Sect. "Fault-Based Liability According to Article 5".

[151] 1996/2010 HNS Convention, Article 7.

[152] Sharma, 'The Basel Protocol', 26 *Delhi L. Rev.* (2004), at 191.

[153] 1996/2010 HNS Convention, Article 1(6).

[154] See explicitly, Basel Protocol, Article 7(1). See also Tsimplis, 'The 1999 Protocol to the Basel Convention', 16 *IJMCL* (2001), at 317.

particularly concerns damage occurring during the pre-carriage or on-carriage of the hazardous wastes, including the subsequent disposal process. It also applies in the rare cases where the portion of damage caused by hazardous wastes deemed to be HNS can be separated out from the portion of damage caused by hazardous wastes deemed not to be HNS. The basic prevalence of the HNS Convention during the maritime carriage of the hazardous wastes should be viewed with some degree of scepticism. Notwithstanding the indisputable advantages of this regime of liability, which is modelled after the generally approved maritime civil liability pattern and which is equipped with its own HNS Fund, the HNS Convention does not attach liability to a person other than the shipowner and does not establish fault-based liability. The possibility that one transboundary movement could be subject to the application of two different liability regimes with fundamentally divergent distributions of liability does not contribute to legal certainty. In the end, one will have to ask why States should undertake the efforts to implement the sophisticated provisions of the Basel Protocol if the largest part of its scope of application is overridden by the 1996/2010 HNS Convention, anyway. One could counter that most incidents might occur outside the "tackle-to-tackle" period to which the HNS Convention applies. However, in order to increase the importance of the Basel Protocol and to ensure a consistent and uniform arrangement of liability throughout an entire transboundary movement of hazardous wastes, it would be advantageous to exclude waste substances from the scope of application of the HNS Convention.[155]

(b) Relationship to the CRTD Convention

The 1989 CRTD Convention represents the counterpart of the HNS Convention regarding the land transport of dangerous goods.[156] It imposes strict and limited liability on the person in control of the road vehicle or inland navigation vessel or on the person operating the railway line for damage caused by any dangerous goods during their carriage by road, rail or inland navigation vessel.[157] It covers the period from the beginning of the process of loading the goods onto the vehicle for carriage until the end of the process of unloading the goods.[158] Dangerous goods are defined by means of reference to the classes of the ADR,[159] also including waste substances and articles. The Convention also contains a provision, according to which in cases where it is not reasonably possible to separate damage

---

[155] See also Tsimplis, 'The 1999 Protocol to the Basel Convention', 16 *IJMCL* (2001), at 317. However, it should be noted that the purpose of Article 11 of the Basel Protocol is to make the HNS Convention applicable, *supra*, Sect. "Insufficiency of the Formal Criterion".

[156] The CRTD Convention is outlined *supra*, Sect. "Liability for the Carriage of Dangerous Goods by Land" in Chap. 3.

[157] CRTD Convention, Article 5(1).

[158] *Ibid.*, Article 3(3).

[159] *Ibid.*, Article 1(9).

caused by the dangerous goods from that caused by other factors, all such damage is to be deemed as having been caused by the dangerous goods.[160]

Since the CRTD Convention is not particularly focused on the transboundary movement of hazardous wastes, it will fall within the ambit of Article 11 of the Basel Protocol upon its entry into force.[161] It will, therefore, override the application of the Basel Protocol to the extent it applies to the particular event of damage. The findings made above in respect of the relationship of the Basel Protocol to the HNS Convention and the Basel Protocol's remaining scope of application apply *mutatis mutandis* to the relationship of the Basel Protocol to the CRTD Convention. If both, the HNS and the CRTD Convention were to enter into force, the remaining scope of application of the Basel Protocol would be limited to the final disposal process, to the cases in which wastes defined as hazardous under the Basel Protocol would not be considered HNS or dangerous goods, and to cases in which the incident and the resulting damage do not occur during the same portion of a transboundary movement as required by Article 11 of the Basel Protocol.

(c) Relationship to the Civil Liability Convention

Strict and limited liability of the shipowner for pollution damage caused by oil carried as cargo on board or in the bunkers of a tanker is established by the Civil Liability Convention (CLC).[162] The CLC applies to oil that has escaped or has been discharged from a ship as a result of an incident, provided the damage occurred in the territory, including the territorial sea, in the EEZ or in a zone of 200 nautical miles of a Contracting State.[163] Since waste oils are not excluded from the scope of application, the CLC also applies to certain aspects of transboundary movements of hazardous wastes. To this extent it prevails over the provisions of the Basel Protocol by virtue of Article 11 of the Protocol.[164] Particularly on the high seas, however, the application of the Basel Protocol is not affected.

---

[160] *Ibid.*, Article 1(10).
[161] The CRTD Convention was opened for signature on 1 February 1990, Article 22(1).
[162] The CLC is outlined *supra*, Sect. "Liability for Oil Pollution from Ships" in Chap. 3.
[163] 1992 CLC, Articles II, III.
[164] Tsimplis, 'The 1999 Protocol to the Basel Convention', 16 *IJMCL* (2001), at 316.

(d) Relationship to Other Regimes of Liability and Compensation in the Field of Transboundary Movements of Hazardous Wastes

The Bamako Convention and the Waigani Convention do not contain their own regime on liability and compensation, but contain the mandate of the Contracting States to elaborate a protocol setting out relevant rules of liability and compensation.[165] The same applies to the Izmir Protocol to the Barcelona Convention,[166] whereas the Tehran Protocol to the Kuwait Convention does not contain a "*pactum de negotiando*" for developing a supplementary regime on liability and compensation. However, in none of these cases has a liability protocol yet been adopted. Provided that one of these instruments enters into force, it will be considered an agreement falling within the ambit of Article 3(7) of the Basel Protocol. Thus, it would prevail over the provisions of the Basel Protocol on the condition that its standard of liability fully meets or exceeds the level of protection of the Basel Protocol.

(e) Relationship to the OECD and EU Regulations

It has already been outlined above that it is doubtful whether OECD Council Decision C(2001)107/FINAL on the Control of Transboundary Movements of Wastes Destined for Recovery Operations comes under the ambit of Article 11 of the Basel Convention.[167] But even if this were the case, the additional requirements set out in Article 3(7) of the Basel Protocol would not be fulfilled. Neither does this Decision contain a regime of liability and compensation nor does it refer to any other regime already existing in another legal instrument or in domestic law. OECD Council Decision C(2001)107/FINAL, therefore, cannot override the application of the Basel Protocol.

The same outcome applies to the EU legal framework on the transboundary movement of hazardous wastes, which is comprised by the Directive 2008/98/EC on Waste and the Waste Shipment Regulation No 1013/2006/EC.[168] These EU instruments must be considered agreements within the meaning of Article 11 of the Basel Convention.[169] However, in order to override the application of the Basel Protocol, these instruments must contain or refer to an existing liability and compensation regime which is in force, applicable, and fully meets or exceeds the level of protection as provided by the Basel Protocol.[170] Neither the Directive on

---

[165] See Article 12 of both the Bamako and the Waigani Convention.
[166] Izmir Protocol to the Barcelona Convention, Article 14.
[167] See *supra*, Sect. "OECD Council Decisions" in Chap. 3.
[168] See in detail *supra*, Sect. "European Union Legislation" in Chap. 3.
[169] See *ibid*.
[170] Basel Protocol, Article 3(7)(a)(ii).

Waste nor the Waste Shipment Regulation contains or refers to an existing regime of liability and compensation.[171] The 1993 Lugano Convention is not in force and the EU Environmental Liability Directive imposes liability on operators of particular activities under public law, rather than imposing civil liability.[172] The domestic laws of the EU Member States may not be considered sufficient in this respect either. Specifically, it is not ensured that the domestic laws provide for civil liability to an extent that is comparable to the level of protection provided by the Basel Protocol. This particularly applies in respect of the liable persons as well as to the coverage of activities during the course of a transboundary movement.

(f) Relationship to Regimes of Liability for Cargo Damage

The Hague/Visby Rules[173] and the Hamburg Rules[174] impose limited liability for presumed fault on the carrier in case of loss of or damage to cargo.[175] These regimes of liability and compensation, therefore, concern a different aspect of liability and the legal relation among a different circle of persons. Liability for cargo damage under the Hague/Visby and Hamburg Rules is based on a presumed breach of a contractual obligation of the sea carrier of a cargo vis-à-vis the contractual partner in a contract of affreightment or the lawful holder of a bill of lading. It does not concern liability for damage suffered in consequence of an incident by third persons and, thus, does not fall within the meaning of "damage due to an incident" or "damage caused by an incident" as required by Article 3(7) and Article 11 of the Basel Protocol.

(g) Relationship to the LLMC Convention

The LLMC Convention does not provide a legal basis for liability and compensation, but rather establishes the right of shipowners and persons in comparable positions to limit their liability in respect of claims raised by third parties in respect of personal damage, damage to property and consequential losses occurring in direct connection with the operation of a ship.[176] Since claims for compensation

---

[171] The Waste Shipment Regulation No 1013/2006/EC merely provides that "this Article shall be without prejudice to Community and national provisions concerning liability", see Articles 23(2), 24(10) and 25(5).

[172] See *supra*, Sect. "EU Environmental Liability" in Chap. 3.

[173] 1924 International Convention for the Unification of Certain Rules of Law related to Bills of Lading (Hague Rules) as amended by the 1968 Visby Protocol and by the 1979 Protocol.

[174] 1978 United Nations Convention on the Carriage of Goods by Sea (Hamburg Rules).

[175] Under the Hamburg Rules the carrier is liable also for delay in delivery, see Article 5.

[176] For a detailed description see *supra*, Sect. "The LLMC Convention" in Chap. 3.

that are raised as a result of a transboundary movement of hazardous wastes are not excluded from the scope of the LLMC Convention, those claims are basically subject to limitation under the LLMC Convention.[177]

It could be argued that the LLMC Convention falls within the ambit of Article 11 of the Basel Protocol, with the consequence that during the sea leg of a transboundary movement of hazardous wastes no liability at all is imposed on any person involved. This would be due to the fact that the Basel Protocol would hold itself inapplicable by virtue of its Article 11 and the LLMC Convention does not provide a legal basis for liability.[178] This conclusion, however, cannot be seen as convincing in the end. Although it must be admitted that the wording of Article 11 of the Basel Protocol allows such interpretation,[179] the terms and the wording of a convention text are not the sole basis for the construction of treaty provisions. The systematic context and the *telos* of the provision in question need to be taken into consideration in equal measure.[180] Accordingly, it should be noted that Article 11 of the Basel Protocol aims at governing the relationship to other regimes of liability and thus tries to solve possible conflicts with other legal instruments that establish liability in a different manner. Article 11, thus, requires by implication that the alternative agreement itself provides for a legal basis of liability and compensation. Consequently, it may well be argued that the LLMC Convention, which only establishes the right to limit liabilities, does not fall within the scope of Article 11 of the Basel Protocol.[181]

---

[177] *Ibid.*

[178] This issue is raised by Tsimplis, 'The 1999 Protocol to the Basel Convention', 16 *IJMCL* (2001), at 317–318 and 329–331.

[179] Article 11 of the Basel Protocol requires that the provisions of the alternative agreement "apply to liability and compensation for damage caused by an incident".

[180] The methodology of interpreting treaty provisions is laid down in Articles 31–33 of the 1969 Vienna Convention on the Law of Treaties (VCLT). The VCLT basically pursues the objective approach, according to which the treaty text, instead of the historic intention of the Parties, provides the basis for any interpretation and adaption of an ambiguous rule in the individual case. It distinguishes between, first, the general rule of interpretation which is set out in Article 31 ("A treaty shall be interpreted in good faith in accordance with the ordinary meaning to be given to the terms of the treaty in their context and in the light of its object and purpose.") and, second, the supplementary means of interpretation as specified in Article 32. The particular components of the general rule of interpretation, i.e. the interpretation according to the ordinary meaning, the systematic context and the *telos* of the provisions, are of equal importance and have to be applied complementary and in a combined manner. An interpretation according to the ordinary meaning rule has the consequence that the usual meaning of an expression, which may also derive from its technical or terminological context, is explored. The systematic context of a provision comprises the meaning that can be attached to a term or wording by taking into account the entire treaty including the preamble. The teleological interpretation of a treaty provision, finally, considers the objective and purpose of a provision, which are to be taken from the wording of the treaty itself. See to the methodology of the interpretation of treaties Crawford, *Brownlie's Principles of Public International Law* (8th ed., 2013), at 379–384; Heintschel von Heinegg, in: Ipsen (ed.), *Völkerrecht* (5th ed., 2004), at 137–147.

[181] This issue is left open by Tsimplis, 'The 1999 Protocol to the Basel Convention', 16 *IJMCL* (2001), at 317, 330.

A further possible interpretation of Article 11 is that the application of the Basel Protocol is excluded only to the extent the provisions of an alternative agreement apply. With regard to the LLMC Convention this would mean that the Basel Protocol remains applicable as far as the establishment of liability is concerned, whereas the provisions regarding the limitation of liability in the Basel Protocol are overridden by the provisions of the LLMC Convention. Such an interpretation, however, can hardly be brought into accordance with the wording of Article 11, which explicitly states that "the Protocol shall not apply" instead of solely stating that "the provisions of the Protocol shall not apply". Furthermore, the systematic context and the object and purpose of Article 11 do not indicate that a distinction between different provisions or parts of the Basel Protocol as regards its application is allowed. Since the LLMC Convention only concerns the limitation of a shipowner's liability, such a solution would also be of minor expediency concerning the liability of any other persons involved in a transboundary movement of hazardous wastes. It is therefore to be concluded that by virtue of Article 11, the LLMC Convention excludes the application of the Basel Protocol neither in *toto* nor in parts.

In consequence of this interpretation, the relationship of the LLMC Convention to the Basel Protocol independent of Article 11 of the Basel Protocol needs to be clarified. The first possible solution would be that as soon as the Basel Protocol is applicable the operation of the LLMC Convention would be entirely excluded. However, there are no rules or indications in the provisions of the Basel Protocol or in the LLMC Convention that would justify such a conclusion. Furthermore, the Basel Protocol only intends to establish certain minimum standards of liability, leaving room for additional provisions established by other civil liability conventions.[182] Consequently, the more convincing arguments speak in favour of the alternative solution. According to this, the LLMC Convention applies additionally and supplementary to the Basel Protocol. The overlapping areas, however, are limited. Practical significance may emerge only, if at all, in case fault-based liability according to Article 5 of the Basel Protocol is attached to the shipowner. In such a case liability under the Basel Protocol would necessarily be unlimited according to Article 12(2), whereas under the LLMC Convention the shipowner is entitlement to limit his liability to a certain extent.[183]

3. Summary

The Basel Protocol provides for two different rules governing its relationship to other international regimes of civil liability and compensation: both of these rules establish a kind of a general subsidiarity of the Basel Protocol on the condition that certain regulatory standards are met by the respective legal regimes. The Basel

---

[182] Tsimplis, 'The 1999 Protocol to the Basel Convention', 16 *IJMCL* (2001), at 318, 330.

[183] As to this conflict see *infra*, Sect. "Potential Conflicts with the LLMC Convention".

Protocol basically distinguishes between its relationship to other regimes regulating in particular the *transboundary movement of hazardous wastes*, which is governed by Article 3(7) of the Protocol, and its relationship to other *general* regimes of liability and compensation, which is determined by Article 11 of the Protocol.

Article 3(7) stipulates that the alternative agreement must be a bilateral, multilateral or regional agreement regarding the transboundary movement of hazardous wastes within the meaning of Article 11 of the Basel Convention. This agreement must contain or refer to a regime of liability and compensation which is in force, applicable and stipulates provisions that fully meet or exceed the objective of the Basel Protocol by providing a high level of protection to persons who have suffered damage. By stipulating this qualitative requirement it is ensured by the Basel Protocol that only those alternative agreements may take precedence over the Basel Protocol which fully meet or exceed the protection level of the Protocol. No reference is made to formal criteria like the entry into force of the alternative agreement or its geographical scope. A particular weakness of this qualitative requirement, however, is that the Protocol does not define the procedure by which it is to be determined whether the alternative agreement actually meets or exceeds the protection level of the Basel Protocol.

According to Article 11 the Basel Protocol does not apply if another general agreement on liability and compensation is in force for the Party concerned and the agreement had been opened for signature before the Basel Protocol was opened for signature. This provision has encountered broad criticism for different reasons. It has been criticised, first, that it abandons the approach of avoiding the application of two different legal regimes in respect of one movement of hazardous wastes and, furthermore, that it makes reference to the ambiguous term "portion of a transboundary movement". But above all else, the establishment of a formal, temporal requirement for the application of the Basel Protocol—instead of a qualitative requirement—has met with criticism. Even though it is not possible to circumvent the substantive requirements of the Basel Protocol by creating a new agreement with a lesser standard of liability, the strict temporal requirement is insufficient in another regard. Irrespective of any considerations of adequacy, it excludes the later ratification of a convention or agreement even where that subsequent instrument provides a more specialised and sophisticated regime of liability which is established with a view to particular aspects of transboundary movements of hazardous wastes, such as liability for damage caused by PCBs or POPs.

In consideration of these findings it must be concluded that the establishment of the formal criterion in Article 11 of the Basel Protocol cannot be considered an optimal solution (although the practical relevance of this shortcoming is limited). It would have represented a better solution to have abandoned the approach of a formal, temporal criterion and to have established a qualitative criterion modelled on Article 3(7). Under such an approach, the legal distinction between the two scenarios governed by Article 3(7) and Article 11 of the Basel Protocol would be superfluous and both cases could have been regulated within one single provision.

As regards the relationship to the individual international civil liability conventions, it can be summarised that the HNS Convention, upon its entry into force, will prevail over the Basel Protocol by means of Article 11 of the Protocol. This, however, only concerns the sea leg of the transport; the Basel Protocol remains applicable in respect of the pre-carriage and the on-carriage. The application of the Basel Protocol will also be excluded by means of its Article 11 if and to the extent the CRTD Convention or the CLC will apply to the wastes in question. Any future liability Protocol that will be established in connection with the Bamako or Waigani Conventions or the Izmir Protocol to the Barcelona Convention will prevail over the Basel Protocol by means of the latter's Article 3(7), provided that these future instruments contain or refer to a regime of liability and compensation that fully meets or exceeds the protection level of the Basel Protocol. Finally, the LLMC Convention remains applicable in addition to the Basel Protocol, although the LLMC Convention would only be relevant in respect of transboundary movements of hazardous wastes as regards the limitation of a shipowner's fault-based liability.

III. The Liability Regime of the Basel Protocol

There are several features that can typically be found in international civil liability conventions. A common ground of those conventions is that uniform law is established among the Contracting States, by which uniform standards of liability are set and the judicial enforcement of liability is ensured. The scheme of liability of those civil liability conventions also follows a consistent structure. Usually, strict liability is established which is channelled to the person in operational control of the dangerous activity or hazardous substance (such as the shipowner, the carrier or the plant operator). In some cases secondary strict liability might be attached to other persons involved, or an additional fault-based liability might be imposed, which in this instances comes along with the right of any liable person to take recourse against any other persons liable. Strict liability is usually excluded if damage resulted under circumstances that are beyond the control of the person liable. Liability is, moreover, typically limited in amount and time, and it is accompanied by supplementary financial mechanisms like compulsory insurance or the establishment of a compensation fund.[184]

The present section examines the liability regime of the Basel Protocol, meaning the entirety of the provisions establishing civil liability of the persons involved in transboundary movements of hazardous wastes by sea.

---

[184] Boyle, 'Globalising Environmental Liability', 17 *J. Envtl. L.* (2005), at 12; Churchill, 'Facilitating Civil Liability Litigation', 12 *Yb. Int'l Env. L.* (2001), at 33.

1. The Basic Concept of Liability of the Basel Protocol

The Basel Protocol basically follows the usual pattern of civil liability conventions as outlined just above. It establishes, in Article 4, strict liability of the notifier and the exporter or, alternatively, of the disposer of the hazardous wastes. Strict liability is supplemented and adjusted by means of fault-based liability according to Article 5, which is imposed on any person responsible for a lack of compliance with the substantive provisions of the Basel Convention or for a wrongful intentional, reckless or negligent act or omission. Article 8 of the Protocol, finally, provides that any person liable under the Protocol is entitled to take recourse against any other person also liable under the Protocol.

(a) The Concept of Combining Strict and Fault-Based Liability

Strict liability and fault-based liability each possess characteristics which can have positive effects in different situations. By combining both types of liability in one legal instrument the respective positive effects can be consolidated and, thus, as optimal a solution as possible can be achieved.

Strict liability has the advantage that compensation for damage is made available not only if damage was caused by a deliberate or negligent act, but also in case of an accident or misfortune.[185] A further major advantage of strict liability is that it overcomes the obstacle of proving fault or negligence on the part of the defendant party. Particularly if data and information of complex and technical industrial processes or installations are required to which the claimant party has no access, if the observance of appropriate standards of reasonable care by the defendant State is in dispute, or if with regard to environmental damage a multitude of possible causes come into question, the burden of proof that rests with the claimant State in case of fault-based liability may amount to a heavy and insurmountable obstacle that inhibits the successful enforcement of claims.[186] This outcome is considered to be hardly acceptable, especially in constellations where the injured party does not derive any benefit from the harmful activity.[187] Strict liability, which only requires proof of damage and the establishment of a causal link to an activity conducted by the defendant party that comes under the ambit of the respective convention, avoids lengthy disputes about the existence of fault or negligence on the part of the defendant party. A further advantage of strict liability is that in practice an increased incentive is created to amicably settle most claims

---

[185] Blay/Green, 'Liability Annex to the Madrid Protocol', 25 *Envtl. Pol'y & L.* (1995), at 28.

[186] Boyle, 'Globalising Environmental Liability', 17 *J. Envtl. L.* (2005), at 13; Churchill, 'Facilitating Civil Liability Litigation', 12 *Yb. Int'l Env. L.* (2001), at 34–35; See also Basel Secretariat, *Implementation Manual* (2005), at 9.

[187] Boyle, 'Globalising Environmental Liability', 17 *J. Envtl. L.* (2005), at 13.

out of court.[188] Finally, strict liability strengthens the incentive of potentially liable parties to invest in damage precaution, rather than to subsequently dispute the existence of fault on their own part.[189]

The imposition of fault-based liability, by contrast, ensures that if any person contributed to the occurrence of damage by fault or negligence, this person is liable particularly because he failed to meet the required standard of reasonable care. Fault-based liability, thus, is also an expression of the polluter-pays principle.[190] It contributes to legal certainty and imposes financial responsibilities on grounds of the moral aspects of fairness and justice.[191] However, since the determination of fault presupposes the existence of a legal rule that has been neglected, it may lead to appropriate results only in highly regulated areas.[192]

The combination of strict liability and fault-based liability in one legal instrument consolidated the respective positive effects of these elements. Strict liability averts enduring disputes regarding the establishment of fault or negligence on the part of the liable person. By this means it is ensured that in the first instance fast and efficient compensation is made available, which allows an optimal combating of environmental damage and a prompt compensation of victims of pollution. Only on a secondary level is the strictly liable person entitled to take recourse against any other person who is strictly liable or liable based on fault or negligence. At this secondary level the final allocation of financial responsibilities among the liable persons involved, as well as the aspects of who is to be considered the "real" responsible party based on categories of fault and negligence, are prioritised, irrespectively of any predominant concerns of promptness and efficiency. At this stage, the person who takes recourse hence may also decide whether to take action against another person liable on the basis of strict liability, which is limited in amount but which does not impose a high burden of proof, or to take recourse against another person liable based on fault, which includes the burden of proving fault or negligence, but which is unlimited in amount.

(b) No Subsidiary Liability of the State

The Basel Protocol imposes strict liability on any of the persons mentioned in Article 4 and fault-based liability on any person according to Article 5. The term "person" is defined in Article 2(14) of the Basel Convention as any natural or legal

---

[188] Churchill, 'Facilitating Civil Liability Litigation', 12 *Yb. Int'l Env. L.* (2001), at 34; Jacobsson, 'Oil Pollution Liability and Compensation', 1 *Unif. L. Rev.* (1996), at 263.

[189] Blay/Green, 'Liability Annex to the Madrid Protocol', 25 *Envtl. Pol'y & L.* (1995), at 28. In respect of the functions of strict liability see *supra*, Sect. "Liability Rules as a Remedy for Environmental Damage " in Chap. 4.

[190] Wolfrum/Langenfeld/Minnerop, *Environmental Liability in International Law* (2005), at 505.

[191] See also Bergkamp, *Liability and Environment* (2001), at 5 and 119.

[192] Wolfrum/Langenfeld/Minnerop, *Environmental Liability in International Law* (2005), at 504.

persons, and thus basically also includes States, provided the State acts in a private capacity and does not exercise sovereign rights in the particular case. Consequently, States may be exposed to civil liability under the Basel Protocol if they themselves or by means of a State-owned company are to be considered the exporter, importer or disposer within the meaning of Article 4 or to be considered any person within the meaning of Article 5 of the Protocol.

The Basel Protocol, however, attempts to avoid the imposition of civil liability on States. This becomes clearly apparent in the provisions of Article 4(1) and (2), which provide that if the State has notified the transport, it is not the State which will be held liable but the exporter or the importer of the wastes. The Basel Protocol also fails to establish any explicit rule imposing subsidiary liability on the State which would apply in case sufficient compensation cannot be obtained from a liable person. Nevertheless, this approach for avoiding the imposition of civil liability on States cannot be considered a weakness or insufficiency of the Protocol. By means of compulsory insurance and the establishment of a trust fund the risk of insufficient compensation being available from the liable person is minimised. Moreover, subsidiary liability of the State does not seem to be an appropriate approach. It has already been outlined above[193] that with regard to damage caused by the activity of private persons the State usually lacks sufficient information about the conduct of such activities and, consequently, cannot sufficiently control and supervise these activities. The imposition of subsidiary civil liability would thus involve a substantial extension of the States' financial commitment, even though they are not directly involved in those commercial activities. Given the fact that States have shown themselves reluctant to ratify the Basel Protocol in any event, it is unlikely that a consensus among States for an additional financial commitment could be found. It is to be expected that the establishment of a subsidiary civil liability of States would only further hamper the entry into force of the Basel Protocol.

(c) Common but Differentiated Civil Liabilities?

The legal concept of common but differentiated responsibilities has found its way into the international practice of States particularly in the aftermath of the 1972 Stockholm Conference.[194] This concept is characterised by the understanding that the traditional rigid approach of employing identical obligations on all States and

---

[193] See *supra*, Sect. "Summary" (p. 155) in Chap. 3.

[194] An explicit formulation of this concept was later incorporated into Principle 7 of the 1992 Rio Declaration, which reads: "States shall cooperate in a spirit of global partnership to conserve, protect and restore the health and integrity of the Earth's ecosystem. In view of the different contributions to global environmental degradation, States have common but differentiated responsibilities. The developed countries acknowledge the responsibility that they bear in the international pursuit to sustainable development in view of the pressures their societies place on the global environment and of the technologies and financial resources they command."

treating them strictly equally, which is derived from the principle of sovereign equality of all States, does not appropriately take into consideration the different capabilities and stages of development of States. According to this competing conception, an equal treatment of all States rather requires that different obligations be imposed on States depending on their individual capabilities and their stage of development. In other words, this approach tries to avoid the fact that the equal treatment of unequal States ultimately leads to inequity.[195]

The concept of common but differentiated responsibilities has been developed primarily in the context of the protection of the ozone layer and the climate change debate. Recent fields of application involve the international protection of biological diversity and combating desertification. One could now argue that also the transboundary movement of hazardous waste represents an issue requiring a differentiated treatment of developing and developed States, since mainly the latter group of States profited from these activities in the past. One could, furthermore, argue that the concept of common but differentiated responsibilities could be implemented in the context of global waste movements by imposing common but differentiated civil liabilities on the acting persons. This could be achieved by means of a civil liability regime that provides for a differentiated scheme of liability which applies different rules of liability on the acting persons or companies depending on whether they are based in a developing or developed country.

It needs to be considered, however, that the environmental risk underlying the transboundary movement of hazardous waste is not fully comparable to the environmental risks associated with the protection of the ozone layer or the climate change debate. The common feature of those cases is that they involve a steady process of deterioration of the global environment, which has commenced in the past due to an exhaustive use of natural resources and the environment by the industrialised countries. The transboundary movement of hazardous waste is indubitably an issue of global concern, yet it does not per se cause a steady deterioration of the global environment in a manner that affects the global community as a whole. In most cases it rather poses a significant threat to human health and the environment which is limited to a certain area. Moreover, the person or at least the State of origin of this pollution may in most cases be identifiable. Hence, there is basically no actual need to stipulate differentiated responsibilities of States on a global level. The Basel Convention nevertheless establishes elements of a "common but differentiated responsibility" by making the general obligations of States regarding the generation and transportation as well as technology transfer contingent on their respective social, technological and economic capabilities.[196]

---

[195] A detailed description of the concept of common but differentiated responsibilities is provided by Beyerlin/Marauhn, *International Environmental Law* (2011), at 61–71; Birnie/Boyle/Redgwell, *International Law and the Environment* (3rd ed., 2009), at 132–136; French, 'The Importance of Differentiated Responsibilities', 49 *Int'l & Comp. L. Q.* (2000), at 35 *et seq.*

[196] This applies with regard to the general obligation of States to minimise the generation and transportation of hazardous wastes according to Article 4(2)(a)(b) and (d) of the Basel Convention, the obligation to co-operate in the development and implementation of new

A further implementation of this concept, particularly within the scheme of civil liability, would not be appropriate. The purpose of imposing civil liability is not to allocate global responsibilities for the exhaustive use of natural resources in the past, but rather to ensure compliance of the acting private persons with a standard of due precaution as regards the conduct of ultra-hazardous activities.[197] The concept of common but differentiated civil liabilities thus does not represent an appropriate approach for determining civil liabilities in the context of transboundary movements of hazardous waste.

2. Strict Liability According to Article 4

In the transboundary movement of hazardous waste by sea usually several persons are involved. This includes on the outgoing side the generator and the exporter of the wastes and on the incoming side the importer and the disposer. Further persons involved are the person who notifies the transport, the waste brokers and dealers, in most cases a forwarding agent, one or more actual carriers and the person owning, operating or chartering the respective means of transport.[198]

Article 4 of the Basel Convention undertakes the function of allocating strict liability among these persons. In this section, this allocation of strict liability shall be examined in more detail.

(a) The Approach of Channelling Strict Liability According to Spheres of Responsibility

According to the traditional approach of civil liability regimes, strict liability is usually channelled to the person in operational control of the industrial installation or means of transport or to the owner or operator of such installation or transport vehicle.[199] This approach of "channelling liability" through only one person has

---

(Footnote 196 continued)
environmentally sound low-waste technologies (Article 10(2)(c)), the establishment of regional or sub-regional centres for training and technology transfer (Article 14(1)). See also Widawsky, 'In My Backyard', 38 *Envtl. L.* (2008), at 595, and *supra*, Sect. "General Obligation: Minimisation of Generation and Transportation of Hazardous Wastes" in Chap. 3.

[197] See also Birnie/Boyle/Redgwell, *International Law and the Environment* (3rd ed., 2009), at 136.

[198] In addition, there may also be other persons involved in hazardous waste movements, such as insurance underwriters providing coverage for different risks and perils associated with those movements.

[199] Under the HNS Convention strict liability is attached to the person registered as the owner of the ship, or in the absence of registration, to the person owning the ship (Article 4 HNS Convention). A similar regulation is provided for by Article III of the CLC. Article 5 of the CRTD Convention imposes strict liability on the carrier. The term "carrier" is defined by the Convention as the person operating the railway line or, alternatively, the person in operational

the advantage that it simplifies the identification of the person liable in consequence of a particular incident and avoids disputes and uncertainties concerning contributory fault of other persons involved.[200] Furthermore, it facilitates the availability of reasonably priced insurance coverage since it reduces the number of persons required to obtain insurance cover and, thus, also avoids overlapping insurance coverage.[201] By channelling liability particularly through the operator or owner of the industrial installation or means of transport, account is taken of the fact that the operator or owner of an industrial installation or transport vehicle is usually in the best position to exercise effective control over the source of potential danger and, thus, may most effectively prevent the occurrence of damage.[202] There are, however, also drawbacks related to this approach. Channelling of liability to only one person also means that no other potentially liable person may be addressed by the claimant, which may turn out to be unfavourable particularly in cases where the liable party is dissolved, becomes insolvent, lacks sufficient funds or insurance cover, or simply invokes a limitation of liability which he is entitled to claim.[203] Moreover, in the context of transboundary movements of hazardous wastes by sea the actual circumstances differ from the usual pattern underlying civil liability conventions. The process of exporting hazardous wastes is not limited to the transportation of such wastes alone. Rather, it also involves the entire set of procedures related to the export including the final disposal process of the hazardous wastes. Hence, there is not only one primary player involved in the transboundary movement, but rather different persons exercise operational control over the hazardous wastes, depending on which stage of the movement is

---

(Footnote 199 continued)
control of the road vehicle or inland navigation vessel, which is presumed to be the person in whose name the vehicle is registered in a public register or, in the absence of such registration, the owner of the vehicle (see Article 1(8) CRTD Convention). The operator of a nuclear installation is strictly liable according to Articles 3 and 4 of the Paris Convention or is, alternatively, absolutely liable according to Article II of the Vienna Convention. Absolute liability is also imposed on and channelled to the operator of a nuclear ship by Article II of the 1962 Nuclear Ships Convention. Under the 2003 Kiev Liability Protocol strict liability for damage caused by an industrial accident is channelled to the operator, meaning the person or authority in charge of an activity (Art. 1 of the 1992 Helsinki Convention). Finally, strict liability is imposed on the operator of an installation according to Article 3 of the 1976 Mineral Resources Convention.

[200] Boyle, 'Globalising Environmental Liability', 17 *J. Envtl. L.* (2005), at 14; Churchill, 'Facilitating Civil Liability Litigation', 12 *Yb. Int'l Env. L.* (2001), at 37; Kummer, *The Basel Convention* (1995), at 241; Murphy, 'Prospective Liability Regimes', 88 *AJIL* (1994), at 51–52.

[201] Churchill, 'Facilitating Civil Liability Litigation', 12 *Yb. Int'l Env. L.* (2001), at 37; Murphy, 'Prospective Liability Regimes', 88 *AJIL* (1994), at 51.

[202] Boyle, 'Globalising Environmental Liability', 17 *J. Envtl. L.* (2005), at 14; Wolfrum/Langenfeld/Minnerop, *Environmental Liability in International Law* (2005), at 505.

[203] Kummer, *The Basel Convention* (1995), at 241.

concerned.²⁰⁴ Channelling liability to only one person would not sufficiently take into account these underlying conditions of transboundary waste movements. It would rather create a disincentive in the other persons involved to exercise the best possible care in order to prevent the occurrence of damage.²⁰⁵

There are two further general approaches for the allocation of strict liability that might come into consideration in the context of transboundary movements of hazardous wastes by sea. First, a combination of primary strict liability of the person in operational control of the wastes could be supplemented by a secondary strict liability of any other person involved in case sufficient compensation cannot be sought from the person primarily liable.²⁰⁶ Another approach is to establish joint and several liability, which would allow the claimant to bring claims against any of the persons involved in the transboundary movement in full amount, even if this person contributed in only small proportion to the occurrence of damage. The person who compensated the claimant in full would, in turn, be entitled to claim compensation payment from the other liable persons involved according to their particular contributing negligence or fault.²⁰⁷ However, both approaches also carry considerable disadvantages. They involve a high number of potential defendants and legal proceedings in the first or second tier, ratcheting up legal and other costs related to litigation.²⁰⁸ As experienced in practice with regard to the US-CERCLA,²⁰⁹ joint and several liability also discourages a readiness to negotiate settlements and creates an incentive to hold responsible in the first instance the person with the "deepest pockets", irrespective of this person's actual portion of contributing fault or negligence. This approach, therefore, has been criticised as being ineffective and also as being surrounded by a sense of unfairness.²¹⁰

Since none of these approaches by itself seems to sufficiently take into account the particular features associated with the transboundary movement of hazardous

---

²⁰⁴ This may also involve persons who do not actually possess the wastes, but rather exercise indirect control, such as waste brokers and waste dealers and persons organising and notifying the transport to the State authorities.

²⁰⁵ Murphy, 'Prospective Liability Regimes', 88 *AJIL* (1994), at 51–52.

²⁰⁶ See Murphy, 'Prospective Liability Regimes', 88 *AJIL* (1994), at 52–53; This approach is incorporated, for instance, in the regulatory system of the CLC/Fund Convention for oil pollution damage, which provides for a two-tier system of liability involving the secondary liability of a compensation fund. The same regulatory approach is followed by the 1960 Paris Convention and the 1963 Vienna Convention, which are supplemented by a secondary layer of liability guaranteed by a public fund established by the 1963 Brussels Supplementary Convention and the 1997 Vienna Supplementary Compensation Convention (see *supra*, Sect. "Liability for Nuclear Damage" in Chap. 3).

²⁰⁷ Murphy, 'Prospective Liability Regimes', 88 *AJIL* (1994), at 53.

²⁰⁸ Murphy, 'Prospective Liability Regimes', 88 *AJIL* (1994), at 53–54.

²⁰⁹ US Comprehensive Environmental Response, Compensation, and Liability Act.

²¹⁰ Murphy, 'Prospective Liability Regimes', 88 *AJIL* (1994), at 54.

wastes by sea, the Basel Protocol pursues a different concept.[211] This concept is in principle based on the approach of channelling liabilities to only one person, but this approach is modified with regard to the particular needs and circumstances of the transboundary hazardous waste movement, and to this end also includes elements of the approach of joint and several liability. The Basel Protocol does not focus on one single person throughout the entire movement. It rather always channels liability to another person, depending on which particular stage of the movement is concerned. The Protocol, thus, creates spheres of responsibility which are attributed to the person who in the respective stage of the transboundary movement generally exercises control over the hazardous wastes. As a consequence, the Basel Protocol ensures that each occurrence of damage can be allocated to the sphere of responsibility of one person, depending on which particular stage of the movement the damage occurs.[212] By this means the Basel Protocol allocates strict liability to the notifier (which can be generator or the exporter), the importer or the disposer of the hazardous wastes.[213] It is only in case there are several persons strictly liable at the same time that the Protocol imposes joint and several liability among these persons.

(b) Allocation of Strict Liability Under the Basel Protocol

In general terms, it can be said that the Basel Protocol establishes a temporal break at the moment when the disposer takes possession of the hazardous wastes. Prior to this moment liability is basically imposed on the notifier, whereas incidents occurring after this moment are basically attributed to the sphere of responsibility of the disposer.

(aa) The Regulation in Detail

In the first two sentences of Article 4(1) of the Basel Protocol the general rule allocating strict liability is laid down. According to this, the person who notifies the movement in accordance with Article 6 of the Basel Convention is liable for damage until the disposer has taken possession of the hazardous wastes. Thereafter

---

[211] A concept of allocating strict liability which deviates from the usual pattern is also established by the 2001 Bunker Oil Convention. This Convention channels liability not only to one person, but to a group of persons. Article 1(3) of the Convention defines the "shipowner", who is liable according to Article 3, to be the owner, registered owner, bareboat charterer, manager or operator of the ship.

[212] See also Boyle, 'Globalising Environmental Liability', 17 *J. Envtl. L.* (2005), at 14.

[213] During the negotiations of the Basel Protocol an alternative option of allocating liability, according to which liability should be channelled to the person in operational control of the wastes, was contemplated. This option, however, could not prevail in the end; see Lawrence, 'Negotiation of a Protocol on Liability and Compensation', 7 *RECIEL* (1998), at 252; Soares/Vargas, 'The Basel Liability Protocol', 12 *Yb. Int'l Env. L.* (2001), at 86.

the disposer is liable for damage.[214] It becomes apparent that this provision does not explicitly define the commencement and the end of liability. Therefore, the commencement and the end of liability are determined by the Protocol's scope of application. According to this, liability of the notifier commences at that point where the wastes are loaded on the means of transport in an area under the national jurisdiction of the State of export or, if the State of export opted to shift the starting point of the application to the point where the wastes leave the area of its national jurisdiction, liability commences at that point.[215] Liability ends if damage occurs after the completion of final disposal process.[216] Another issue related to this general rule is that the "person who notifies" is not defined separately by the Basel Protocol or the Basel Convention, so that it is necessary to refer to Article 6 of the Basel Convention governing the duty to notify. According to this, either the generator[217] or the exporter[218] is to be considered the "person who notifies". Although Article 6 of the Basel Convention also allows notifications of the State of export, such constellations are of no significance concerning the allocation of strict liability. Article 4(1) of the Basel Protocol provides that if the State of export is the notifier or if no notification has taken place, the exporter of the wastes will be liable for damage.

In case damage occurs which involves wastes that are considered hazardous only by the State of export not the State of import[219] and provided this damage occurs in an area under the national jurisdiction of the State of export or a State of transit which also considers these wastes as hazardous, Article 4(1) of the Protocol provides that Article 6(5) of the Basel Convention applies *mutatis mutandis* with regard to the allocation of liability. This means that the liability provisions applying to the importer or disposer apply *mutatis mutandis* to the exporter. This, however, does not encompass the provisions applying to the disposer since Article 4(1) furthermore provides that (only) "[t]hereafter the disposer shall be liable for damage". Consequently, the regulatory content of this provision in unclear and it is doubtful whether such constellations will ever become relevant in practice. Liability of an importer is limited, according to Article 4 of the Protocol, to those cases where the wastes are considered hazardous only by the State of import, so that that no liability at all is imposed on the importer regarding the constellations under consideration. A corresponding rule regarding wastes that are considered

---

[214] Basel Protocol, Article 4(1).

[215] See *ibid.*, Article 3(1).

[216] See *ibid.*, Article 3(2)(a) and (b).

[217] The generator is defined by Article 2(18) of the Basel Convention as "any person whose activity produces hazardous wastes [...] or, if that person is not known, the person who is in possession and/or control of those wastes".

[218] Exporter means according to Article 2(15) of the Basel Convention "any person under the jurisdiction of the State of export who arranges for hazardous wastes [...] to be exported".

[219] According to Article 1(1)(b) of the Basel Convention, which also requires that those wastes have been notified according to Article 3 of the Convention.

hazardous only by the State of import[220] is laid down in Article 4(2) of the Basel Protocol. According to this, the importer[221] is liable for damage until the disposer has taken possession of the wastes provided that the State of import is the notifier or that no notification has taken place. Thereafter the disposer will be liable for damage.

With regard to cases of re-importation according to Articles 8 and 9 of the Basel Convention, strict liability is allocated among the persons involved according to Article 4(3) and (4) of the Protocol. When a transboundary movement of hazardous wastes cannot be completed in accordance with the terms of the contract, the State of export is under the obligation to ensure that the wastes are taken back by the exporter within 90 days.[222] Strict liability in this situation is imposed on the original notifier, meaning either the generator or exporter, from the time the hazardous wastes leave the disposal site until the wastes are taken into possession by the original exporter or the alternative disposer.[223] When a transboundary movement of hazardous wastes is deemed illegal as the result of conduct on the part of the exporter or generator, the State of export is under the obligation to ensure that the wastes in question are taken back by the exporter, the generator or, if necessary, by the State of export itself.[224] In this case, strict liability is imposed on the respective person[225] who is under the obligation to re-import until the wastes in question are taken into possession by the original exporter or by the alternate disposer.[226] Liability is also imposed on the person who re-imports hazardous wastes if responsibility for illegal trafficking cannot be assigned to either the exporting or the importing side and the States involved through co-operation reach the result that the wastes in question are to re-imported to the State of export.[227]

Finally, the Basel Protocol establishes joint and several liability among the persons strictly liable under the Protocol. In case two or more persons are strictly liable[228] the claimant is entitled to seek full compensation for the damage from any

---

[220] Article 1(1)(b) of the Basel Convention requires that those wastes have been notified according to Article 3 of the Convention.

[221] Basel Convention, Article 2(16) defines the importer as "any person under the jurisdiction of the State of import who arranges for hazardous wastes [...] to be imported".

[222] Basel Convention, Article 8.

[223] Basel Protocol, Article 4(3).

[224] Basel Convention, Article 9(2)(a).

[225] Since the term "person" also includes legal persons, liability may also be imposed on the original State of export itself if this State acted in private capacity; *supra*, Sect. "No Subsidiary Liability of the State".

[226] Basel Protocol, Article 4(4).

[227] *Ibid.*, Article 4(4), referring to Basel Convention, Article 9(4).

[228] This might be the case if, for instance, several persons notify the transport or if the wastes should be delivered to several disposal companies.

or all of the persons liable.[229] The person from whom compensation is sought may then take recourse action against any other person also liable under the Basel Protocol.[230]

(bb) The Temporal Break When Liability Shifts to the Disposer

The basic idea underlying the allocation of strict liability according to Article 4 of the Basel Protocol is that liability shifts from the notifier or exporter to the disposer at that moment when the disposer takes possession of the hazardous wastes. However, the wording of the provision in this context is ambiguous and requires further clarification.

First, Article 4 does not specify whether it is the moment when the incident occurs or when the damage occurs which is relevant for the allocation of liability. This issue may become relevant in cases where the damage does not occur simultaneously with the incident, but only after some passage of time.[231] Although the wording of Article 4, by repeatedly referring to the term "damage", implies that the moment when damage occurs is crucial for the allocation of liability, such an understanding can hardly be brought in line with the legal context and the purpose of this provision. The allocation of liability to either the notifier or the disposer takes account of the fact that only the respective person is able to execute operational control over the wastes at that time and, thus, is the only person able to prevent as far as possible the occurrence of an incident resulting in damage. If, for instance, the disposer could be held liable for damage that was caused by an incident which took place even before the disposer had taken possession of the wastes, this would mean that the disposer has no chance to prevent the occurrence of damage and, thus, the imposition of liability. The correct understanding of Article 4 must, therefore, be to make the allocation of liability dependent on the moment when the incident occurs.

A further ambiguity of Article 4 is to be seen in the definition of the "possession of the disposer", since according to the wording of this provision it remains unclear under which conditions the disposer has actually taken possession of the hazardous wastes. As a general rule, the parties to a contract concerning the sale of hazardous wastes are free to decide at what time delivery of the wastes is to be deemed to be performed.[232] Depending on the respective terms of delivery the individual carriers involved in the chain of transportation are to be considered the agents of either the seller or the buyer. Thus, if hazardous wastes are sold to a disposer abroad, it is basically possible that delivery within the meaning of the

---

[229] Basel Protocol, Article 4(6).

[230] *Ibid.*, Article 8, see also *infra*, Sect. "Right of Resource, Article 8".

[231] This may be the case, for instance, if the wastes are temporarily stored during the transportation and it is only after they have been delivered to the disposer that residues emanating from the wastes cause damage by accumulation.

[232] Several conditions of delivery are defined, for instance, by the INCOTERMS, the International Commercial Terms, elaborated by the International Chamber of Commerce (ICC).

sales contract is performed already upon delivery to the freight forwarder at the seller's premises or upon loading of the wastes on board the seagoing vessel. This attribution of the conduct of agents to either the seller or buyer leads to appropriate solutions regarding the level of the sales contract, however, with regard to the allocation of liability in the context of transboundary movements of hazardous wastes it cannot be considered suitable. This outcome would rather undermine the sophisticated allocation of liability as provided for by Article 4 and would allow the persons involved in a hazardous waste movement to circumvent the liability provisions of the Basel Protocol by means of a simple contractual clause.

As to the resolution of this ambiguity, the legal definition of the term "disposer" does not help either. Under the Basel Convention and Protocol the disposer is defined as "any person to whom hazardous wastes [...] are shipped and who carries out the disposal of such wastes".[233] This definition does not contain any clear statement about whether or not agents are included by the term "disposer". However, it cannot be assumed that agents are per se excluded from the scope of this definition. This is due to the fact that in many cases the contracting disposer is not the person actually operating the disposal site to which the hazardous wastes are delivered. In case an agreement is in place between the contractual disposer and the operator of a disposal site, delivery to this disposal site must be deemed to constitute delivery to the contractual disposer. Therefore, it is necessary that delivery to the operator of the disposal site, who is deemed to be an agent of the contractual disposer, is considered sufficient within the meaning of the term "disposer". Consequently, it must be summarised that the term "disposer" must, on the one hand, include agents within the meaning of subcontractors, who are for example actually operating the disposal plant; whereas on the other hand, it seems appropriate to exclude agents from this definition who are acting on behalf of the disposer regarding the transportation of the hazardous wastes. As a result, it thus seems appropriate to construe the definition of the term "disposer" and thus "delivery to the disposer" in a geographical fashion, meaning that it includes the delivery to the operator of the disposal site, irrespective of whether this operator is the contractual disposer or an agent of this person, but disregards any previous processes of delivery to agents of the disposer as agreed upon in the sales contract.

(cc) Non-establishment of a General Secondary Liability of the Generator

A major point of criticism voiced against the allocation of strict liability under the Basel Protocol involves the Protocol's failing to establish a general secondary liability of the generator of the hazardous wastes. It is argued that in the present shape of the liability provisions the generator is enabled and encouraged to circumvent strict liability by simply selling or handing over the wastes to a third company, which in turn undertakes to arrange the exportation of the hazardous

---

[233] Basel Convention, Article 2(19).

wastes and to submit the formal notification.[234] In this case the Protocol's liability provisions impose strict liability solely on the notifying exporting company, whereas the generator is discharged from any liability. The same outcome applies if the State of export is the notifier or if no notification has taken place. Also in this case strict liability is exclusively imposed on the exporter by the Protocol. The practical consequence of this regulation is that the generator can effectively evade any liability by simply entering into a contract with a third company. This company or person will most likely be a waste broker, waste dealer, a letter-box trading company or any other company having been set-up only for the purpose of incurring liabilities, which in cases of significant liability might lack sufficient funds for compensation, become insolvent or simply be dissolved. This possibility for the waste generator or exporter to effectively circumvent strict liability by handing over the wastes to a third company that at least formally undertakes to arrange the exportation of the wastes and the related notification must, in fact, be considered a major loophole of the Basel Protocol.[235] Shifting liability from the company which derives the benefit from the process resulting in the generation of hazardous wastes to a third company, which only answers the purpose of absorbing liabilities, undermines the liability regime of the Basel Protocol and conflicts with the polluter-pays principle. It may be argued that the exporter is under a duty to obtain insurance coverage for any liability incurring under Article 4 of the Protocol and that a transboundary movement must not commence unless evidence of coverage has been produced.[236] This, however, does not encompass illegal traffic and any other cases in which the exporter, in fact, failed to obtain insurance coverage and to maintain the same throughout the entire journey. Moreover, insurance coverage may fail to apply when the insurer has the right to deny indemnification such in case the insured event was caused by a wilful or grossly negligent act of the insured.

A more favourable solution, therefore, would have been to establish a secondary liability of the generator of the hazardous wastes or to include the generator

---

[234] 'Compensation and Liability Protocol Adopted', 30 *Envtl. Pol'y & L.* (2000), at 44; Choksi, '1999 Protocol on Liability and Compensation', 28 *Ecology L. Q.* (2001), at 524; Long, 'Protocol on Liability and Compensation', 11 *Colo. J. Int'l Envtl. L. & Pol'y* (2000), at 258; see also 'No Agreement on Draft Protocol', 29 *Envtl. Pol'y & L.* (1999), at 154.

[235] Another criticism that has been voiced with regard to possible discrepancies with existing US laws is that the absence of a general secondary liability of the waste generator would contradict the liability provisions of the US Comprehensive Environmental Response, Compensation, and Liability Act of 1980 (CERCLA), according to which the waste generator is subject to joint and several liability; see 'Compensation and Liability Protocol Adopted', 30 *Envtl. Pol'y & L.* (2000), at 44; Long, 'Protocol on Liability and Compensation', 11 *Colo. J. Int'l Envtl. L. & Pol'y* (2000), at 258. As to the liability of the waste generator under US CERCLA see Obstler, 'Toward a Working Solution to Global Pollution', 16 *Yale J. Int'l L.* (1991), at 98 et seq.; Greenfield, 'CERCLA's Application Abroad', 19 *Emory Int'l L. Rev.* (2005), at 1704 et seq.

[236] See Basel Protocol, Article 14, and *infra*, Sect. "Compulsory Insurance or Similar Guarantees". According to Article 14(3) of the Protocol, a document reflecting the coverage of the liability is to accompany the notification referred to in Article 6 of the Basel Convention.

Incidents Covered by the Basel Protocol 257

in a scheme of joint and several liability covering the period until the disposer takes possession of the wastes. The first solution seems even more appropriate. It increases the incentive of the generator to contract only with the most careful exporters and carriers and, thus, to indirectly invest in damage precaution. On the other hand, since liability is imposed on the generator only at the second tier, the risk of being directly exposed to claims for compensation is assessable and hence does not impair or hinder the economic activity of the generator. Such second-tier strict liability of the generator would also be subject to the limitation of liability provisions.[237]

In this context one could assume that the Contracting States to the Basel Protocol could eliminate this loophole in the Protocol by establishing a rule in domestic law that enlarges the scope of strict liability by including a secondary liability of the generator of the hazardous wastes. Such approach, however, would contradict Article 20(2) of the Basel Protocol, which provides that States have to ensure that claims for compensation based on a strict liability of the notifier are impermissible other than in accordance with the Basel Protocol.

(dd) Liability of the Generator for Aftercare Operations

In the early days of the Protocol some critics went even further and criticised that the Protocol failed to establish liability of the waste generator for aftercare operations. It was argued that the waste generators, who are the only persons benefitting from the activity which generates the hazardous wastes, should also assume financial responsibility for possible long-term damage (such as ground water contamination) occurring after the final disposal of such wastes.[238] This criticism, however, may not really be persuasive. First, such allocation of liabilities would disregard the import State's sovereign decision to either allow or prohibit the importation of hazardous wastes as recognised by the Basel Convention. If a State decides to participate in the global waste trade, it seems appropriate that this State also bears the remaining potential risk that even after the final disposal of the hazardous wastes further contamination cannot be excluded with absolute certainty.[239] Second, the incentive of import States to prevent damage in the first place would be diminished and import States would not be encouraged to implement an effective regime of liability and enforcement into

---

[237] See also the very instructive paper of Murphy, 'Prospective Liability Regimes', 88 *AJIL* (1994), at 71.

[238] 'No Agreement on Draft Protocol', 29 *Envtl. Pol'y & L.* (1999), at 154; Choksi, '1999 Protocol on Liability and Compensation', 28 *Ecology L. Q.* (2001), at 525, 532; Sharma, 'The Basel Protocol', 26 *Delhi L. Rev.* (2004), at 189–190, 194–195.

[239] Murphy, 'Prospective Liability Regimes', 88 *AJIL* (1994), at 66. This applies all the more since the traditional north-south pattern of legal hazardous waste transports seems to be outdated, see *supra*, Sect. "Quantities and Typical Patterns of Hazardous Wastes Movements" in Chap. 2.

their domestic laws.²⁴⁰ And finally, during the aftercare period the generator of the hazardous wastes would lack any opportunity to execute actual control over the wastes and, thus, to avoid the occurrence of damage.²⁴¹ Therefore, as regards legal movements of hazardous wastes the imposition of strict liability on the waste generator for aftercare operations does not seem to be an appropriate approach.²⁴² As regards illegal traffic, by contrast, the Basel Convention primarily focuses on the obligation to re-import, in the course of which strict liability is primarily channelled to the person who re-imports until the wastes are taken into possession by the exporter or alternative disposer.

(c) Exceptions to Strict Liability

Liability under the Basel Protocol is not conceived as absolute liability, but as strict liability that allows for a limited number of exclusions of liability when damage was caused due to circumstances beyond the control of the liable parties. By providing for exclusions of liability the Protocol aims to ensure that liability imposed on the responsible party remains calculable and does not unduly hamper or practically prevent the commercial activity. The establishment of fixed exclusions of liability is an ameliorating counterpart to the imposition of strict liability.

(aa) Article 4(5)

According to Article 4(5) of the Basel Protocol strict liability does not attach to the person otherwise liable in four cases, namely if "the damage was:

(a) The result of an act of armed conflict, hostilities, civil war or insurrection;
(b) The result of a natural phenomenon of exceptional, inevitable, unforeseeable and irresistible character;
(c) Wholly the result of compliance with a compulsory measure of a public authority of the State where the damage occurred; or
(d) Wholly the result of the wrongful intentional conduct of a third party, including the person who suffered the damage".²⁴³

These exclusions of liability, however, only apply if liability is imposed according to Article 4, Paragraphs (1) and (2) of the Basel Protocol. It does not

---

[240] Murphy, 'Prospective Liability Regimes', 88 *AJIL* (1994), at 66; see also the economic analyses of Helm, 'How Liable Should An Exporter Be?', 28 *Int'l Rev. L. & Econ.* (2008), at 263 *et seq.*

[241] Murphy, 'Prospective Liability Regimes', 88 *AJIL* (1994), at 66–67.

[242] A comparable approach which includes the generator of hazardous wastes in a compensation regime applying to the aftercare period after the final disposal of the hazardous wastes has been established by the US CERCLA "Superfund" regulation. This, however, mostly concerns domestic cases and may, therefore, not simply be transferred to the international level.

[243] Basel Protocol, Article 4(5).

apply, by contrast, if, in case of re-importation of hazardous wastes pursuant to Articles 8 and 9 of the Basel Convention, strict liability is imposed on the notifier, re-importer, original exporter or alternate disposer of the hazardous wastes according to Article 4, Paragraphs (3) or (4).[244] The burden of proof regarding the application of these exclusions of liability lies with the person who invokes this defence.[245] In case one of the defences covered by Article 4(5) of the Protocol is successfully established, the person suffering damage may seek compensation from the Technical Cooperation Fund.[246]

This approach of the Basel Protocol to exclude strict liability in case damage was caused due to circumstances that are beyond the control of the otherwise liable person is consistent with most other civil liability regimes. Differences exist, however, regarding the particular scenarios that result in an exclusion from liability as well as in respect of the particular scope of these exclusions.[247]

It was argued that that the wording of the latter two exclusions suggests the understanding that only in those cases must damage be "wholly the result" of the mentioned circumstances, whereas with regard to the circumstances listed under (a) and (b) liability is excluded even if the mentioned circumstances are of cumulative, parallel or complementary nature.[248] Although it is correct that the former two exemptions from liability differ it their applicable test of causation from the latter two exemptions, the conclusion that with regard to the former exemptions even a cumulative, parallel or complementary causation is deemed sufficient to exclude liability is not correct. The applicable test of causation regarding the circumstances listed under (a) and (b) is rather to be seen in the *causa proxima* principle, according to which it is only the dominant or proximate cause of pollution that determines whether this situation comes under the particular exclusion of liability or not.[249] The test of causation applying to the cases listed under (c) and (d) must be even stricter. The mentioned circumstances must have exclusively caused the pollution and as soon as there is any other contributory

---

[244] *Ibid.*, Article 4(5).

[245] *Ibid.*, Article 4(5).

[246] As to Decision V/32 of COP5 see *infra*, Sect. "Background and Basic Legal Features of the Convention" in Chap. 3.

[247] See Article 7(2) of the HNS Convention, Article 5(4) of the CRTD Convention, Article III(2) of the CLC, Article 3(3) of the Bunker Oil Convention, Article 4(2) of the Kiev Protocol to the Helsinki Conventions, Article 3(3) of the Mineral Resources Convention. By contrast, absolute civil liability is imposed by Article II(1) of the Nuclear Ships Convention.

[248] Soares/Vargas, 'The Basel Liability Protocol', 12 *Yb. Int'l Env. L.* (2001), at 89.

[249] The *causa proxima* principle, which originates from the law on marine insurance, is accepted and applied also in the context of other civil liability conventions, such as the CLC and the Bunker Oil Convention, see de la Rue/Anderson, *Shipping and the Environment* (2nd ed., 2009), at 99; Zhu, *Compulsory Insurance and Compensation for Bunker Oil Pollution Damage* (2007), at 101–102.

cause, however small, pollution damage is considered not to be "wholly the result" of the specified exemptions.[250]

The exceptions to liability have also been criticised as being ambiguous regarding their particular scope. According to a literal understanding of this provision one could argue that it is only with regard to natural phenomena that the causation of damage must be unforeseeable, whilst the other exceptions to liability would apply even in case they have been foreseen by the otherwise liable person. This would mean that the liable person would not be able to invoke the defence of an exceptional natural phenomenon if he sends the vessel carrying hazardous wastes through the predicted path of a hurricane, whereas liability would be excluded if he sends the vessel through an area where it is likely that the vessel would become involved in an armed conflict, in hostilities, acts of sabotage, or where it is likely that the vessel would be arrested by regulatory action.[251]

On a closer examination, however, such understanding is not really convincing. First of all, Article 4(5) has to be seen in the light of the Protocol's liability regime as a whole. In this context, Article 5 ensures that unlimited liability is imposed on any person responsible for the occurrence of damage by means of an act of negligence or fault. Such fault-based liability would also apply to the cases discussed earlier: If a person sends a vessel carrying hazardous wastes through an area where it is likely that the vessel would become caught in hostilities, acts of sovereign powers, etc., and provided this must have been taken into account by the responsible person, it is likely that negligence or fault on the part of the acting person can be established so that this person is liable according to Article 5 anyway. From this consideration of the liability regime of the Basel Protocol as a whole it, furthermore, becomes apparent that Article 4(5) of the Protocol does not intend to exclude liability in the described situations per se, but intends to exclude liability only in cases where damage was caused due to circumstances that are beyond the control of the otherwise liable person. With this legal intention in mind, it seems appropriate to understand Article 4(5) in a way that the actual situation leading to an exclusion of liability must be unforeseeable in any case.[252] The fact that the requirement of unforeseeability is expressly mentioned only with regard to natural phenomena is to be explained by the function of this term in defining in more detail the nature of the required phenomenon. Not any meteorological event will be covered by this exemption, but only exceptional and unforeseeable intensive events. Such additional qualification of the phenomenon allowing for an exclusion of liability is not required in respect of the other cases listed in Article 4(5), e.g. armed conflicts, hostilities, compulsory measures of public authorities or acts of sabotage.

---

[250] de la Rue/Anderson, *Shipping and the Environment* (2nd ed., 2009), at 100–101, who refers to the position of the IOPC Fund. Different, however, Zhu, *Compulsory Insurance and Compensation for Bunker Oil Pollution Damage* (2007), at 101–102.

[251] Tsimplis, 'The 1999 Protocol to the Basel Convention', 16 *IJMCL* (2001), at 313.

[252] See also Tsimplis, 'The 1999 Protocol to the Basel Convention', 16 *IJMCL* (2001), at 314.

(bb) Article 6(2)

A further exception to strict liability is provided for by Article 6(2) of the Protocol with regard to damage occurring in the course of preventive measures. Article 6(2) of the Protocol theoretically also applies to fault-based liability. This constellation, however, is of no practical relevance.[253]

Article 6(1) of the Basel Protocol constitutes the obligation of any person in operational control of the hazardous wastes at the time of an incident to take all reasonable measures to mitigate damage arising from this incident.[254] If in response to an incident any person in possession and/or control of the hazardous wastes takes reasonable preventive measures for the sole purpose of preventing, minimising or mitigating loss or damage, or in order to effect environmental cleanup,[255] and provided that in the course of such measures further or different damage is caused, the acting person is excluded from liability under the Basel Protocol.[256]

3. Fault-Based Liability According to Article 5

The liability regime of the Basel Protocol provides for a secondary tier of liability whereby strict liability is supplemented by an additional fault-based liability according to Article 5. According to this, any person who causes or contributes to damage by his lack of compliance with the provisions implementing the Convention or by his wrongful intentional, reckless or negligent acts or omission is liable for such damage.[257] The Basel Protocol, hence, ensures that any person that comes in contact with the hazardous wastes or is involved in the exportation or notification process can be held liable on the grounds of fault or negligence. By establishing fault-based liability as well, the idea of the polluter-pays principle is strengthened and the acting persons are encouraged to act diligently and to comply with the requirements of the Basel Convention.[258]

Fault-based liability thus requires the determination of two conditions. First, it must be established that a legal rule has been neglected.[259] In this context,

---

[253] Fault-based liability according to Article 5 of the Basel Protocol requires an additional subjective element of personal fault or negligence (see *infra*, Sect. "Fault-Based Liability According to Article 5"), so that there are virtually no situations conceivable in which a person can be liable on the grounds of personal fault or negligence and at the same time acted have acted reasonably according to Article 6(2).

[254] The particular content and scope of this obligation needs to be determined in more detail by domestic legislation, see Basel Secretariat, *Implementation Manual* (2005), at 12.

[255] For the definition of "preventive measures" see Basel Protocol, Article 2(2)(e).

[256] Basel Protocol, Article 6(2).

[257] *Ibid.*, Article 5.

[258] A similar approach to establish a supplementary tier of fault-based liability can be found only in 2003 Kiev Protocol to the Helsinki Conventions (Article 5).

[259] Wolfrum/Langenfeld/Minneron, *Environmental Liability in International Law* (2005), at 504.

Article 5 mentions two different obligations that come into consideration. This comprises non-compliance with the provisions of the Basel Convention on the one hand and any wrongful intentional, reckless or negligent act or omission on the other hand. The second condition of fault-based liability is the establishment of fault. It is unclear whether fault in the meaning of this provision is conceived as objective fault comprising any conduct that is actually sub-standard, or whether it is conceived as subjective fault additionally requiring negligence, defined as the omission to act with due care.[260] At first sight, Article 5 seems to suggest a different understanding of fault depending on whether non-compliance with the Basel Convention or an intentional or negligent act is the reason for liability. Whereas with regard to wrongful intentional, reckless or negligent acts or omissions a subjective understanding of fault is already implied by the wording of this rule, fault-based liability on the basis of non-compliance with the Basel Convention suggests at first sight that it is necessary to additionally establish the personal negligence of the liable party. However, given the very general wording of this rule, an objective understanding of fault would mean that any formal mistake during the notification process of a hazardous waste movement would result in the imposition of unlimited fault-based liability, for which no insurance coverage is available. This seems to be a too heavy burden, even taking account of the further requirement that a causal link must be established between negligent conduct and the damage for which compensation is sought. In conclusion it can be said that—in any event—fault-based liability within the meaning of Article 5 of the Protocol additionally requires personal negligence on the part of the liable person.

The wording of Article 5 has been criticised as being ambiguous also with regard to the precise coverage of the term "any person". It is unclear whether this term also covers third persons, or whether "any person" is restricted to any of those persons that are involved in the hazardous waste movement. While the latter group of persons[261] is unequivocally covered, doubts arise with regard to servants and agents, since the Protocol provides that the domestic law of the Contracting States governing liability of servants and agents is not affected.[262] This reference to the domestic laws, however, does not mean that the domestic laws are to be decisive as to whether or not liability is imposed upon servants and agents. It rather concerns only the question in which way liability is allocated in the relationship between servants and agents and their principals. Consequently, nothing in the provisions of the Basel Convention and Protocol suggest that fault-based liability is to be restricted to only those persons actually involved in the transboundary

---

[260] Wolfrum/Langenfeld/Minnerop, *Environmental Liability in International Law* (2005), at 504.
[261] Such as the generator, exporter, carrier, importer or disposer.
[262] Basel Protocol, Article 5.

Incidents Covered by the Basel Protocol 263

movement of hazardous wastes. It must be concluded that Article 5 rather applies to any person, including any third person,[263] which comes in contact with the hazardous wastes.

4. The Requirement of a Causal Link

The obligation to pay compensation for damage caused by hazardous wastes does not only require the determination that an incident occurred and that this incident can be allocated to the sphere of responsibility of one particular person, or the determination of fault or negligence on the part of any person. An additional requirement of liability is that a causal link can be established between the conduct or the incident on the one hand and the actual damage on the other hand. This requirement of a causal link differs depending on whether strict liability or fault-based liability is concerned.

(a) Differences Between Strict and Fault-Based Liability

The most beneficial feature of strict liability is that the claimant is relieved from the requirement of proving fault on the part of the respondent. By this means lengthy disputes and litigation concerned with the determination of proof on the part of the respondent are avoided and prompt and efficient compensation should be ensured. The absence of the requirement of fault thus places special emphasis on the determination of the causal link. With regard to strict liability a person is deemed liable solely on grounds that an incident has occurred which can be linked back to a certain risk for which the person is deemed responsible, but regardless of the particular quality of the conduct of this person in terms of subjective fault or negligence.

In respect of fault-based liability, the grounds of liability are that a particular act or omission of a person violates an applicable legal rule and that, furthermore, from a subjective perspective this person acted intentionally or at least negligently. In order to claim compensation on the basis of fault, it is thus necessary to establish a causal link not between the risk and the damage, but between the particular conduct of the liable person, on the one hand, and the actual damage, on the other hand.

The determination of a causal link may pose considerable difficulties with regard to environmental damage. Pollution damage often occurs as long-term damage whose allocation to only one source of pollution is often impossible. Hazardous wastes that are, for example, released to the marine environment may spread over a wide area and may persist and multiply in flora and fauna. The actual hazards and long-term effects of contamination with hazardous wastes cannot be

---

[263] According to Basel Convention, Article 2(14), the term "person" includes any natural or legal person.

foreseen or estimated. Environmental pollution, thus, always involves a melange of different causes, an aspect which has to be taken into account when ascertaining a causal chain. The Basel Protocol does not prescribe which test of causation is to be applied by the competent courts, so that it is left to the national courts to develop by case law which tests of causation will be applied so as to eliminate those causes which are too indirect or remote.[264] Nevertheless, some general rules, which have been developed with regard to other civil liability regimes, should be mentioned at this point.

First of all, it should be stressed that the damage for which compensation is sought must have been caused by hazardous wastes, whereas damage which occurred only on the occasion of a transboundary movement of hazardous wastes, but which was not caused by hazardous wastes, is not eligible for compensation.[265] The first test of causation to be applied is the *conditio sine qua non* requirement. According to this, the crucial test for establishing a causal link is whether the particular damage would have failed to materialise if one assumes away either the risk (in case of strict liability) or the negligent conduct (in case of fault-based liability).[266] This test may eliminate factors which are not at all causal for damage in a factual sense. It fails, however, to provide a solution for cases in which different factors contributed to the occurrence of damage.[267] Therefore, normative or legal criteria have been developed to determine under which conditions damage is allocated to a particular factor. In this regard, common law systems mainly operate with the proximate cause test, according to which only the primary, proximate or most direct cause is deemed to be legally relevant. In contrast, civil law systems tend to apply the adequate causation test, which asks from the perspective of the particular risk (in case of strict liability) or the conduct (in case of fault-based liability) whether the consequent damage was foreseeable and had to be taken into account by the acting person.[268] Any further details regarding the determination of a causal link will depend on the case law established by the respective competent courts.

(b) Combined Causes of Damage, Article 7

A special case of the determination of a causal link is provided by Article 7 of the Protocol. It has been mentioned in the previous section that in principle the

---

[264] See also Churchill, 'Facilitating Civil Liability Litigation', 12 *Yb. Int'l Env. L.* (2001), at 33.

[265] This can be inferred from Article 7 of the Protocol, providing for special rules in case damage is caused by wastes covered by the Protocol as well as wastes not covered by the Protocol.

[266] See Bergkamp, *Liability and Environment* (2001), at 281–282; Gunasekera, *Civil Liability for Bunker Oil Pollution Damage* (2010), at 153.

[267] Gunasekera, *Civil Liability for Bunker Oil Pollution Damage* (2010), at 154.

[268] Bergkamp, *Liability and Environment* (2001), at 285–297; Gunasekera, *Civil Liability for Bunker Oil Pollution Damage* (2010), at 154–157; de la Rue/Anderson, *Shipping and the Environment* (2nd ed., 2009), at 236–239.

damage for which compensation is sought must have been caused by hazardous wastes, whereas damage which occurred only on the occasion of a transboundary movement of hazardous wastes, but which was not caused by hazardous wastes, is not eligible for compensation. Article 7 modifies and specifies this principle with regard to damage which is caused jointly by hazardous and non-hazardous wastes.

In case damage is caused by wastes covered by the Protocol as well as by wastes not covered by the Protocol, a person liable according to the provisions of the Protocol will be liable for compensation only to the proportion of the contribution attributable to the hazardous wastes.[269] The distinction between the proportion of the contribution made by hazardous wastes and non-hazardous wastes is to be made with regard to the volume and properties of the wastes involved and the type of damage that occurred.[270] If it is not possible to determine what contributing percentages should be attributed, respectively, to the hazardous and non-hazardous wastes vis-à-vis the damage, the Basel Protocol provides that all damage is to be considered as being covered by the Protocol.[271]

The Basel Protocol, by contrast, fails to determine the consequences if damage is caused by a combination of hazardous wastes covered by the Protocol and other hazardous or noxious substances that cannot be considered "wastes". With regard to such cases it seems appropriate to apply *mutatis mutandis* the rules laid down in Article 7.[272]

5. Contributory Fault, Article 9

A further defence that is given to the person liable under the Basel Protocol is that he may invoke the contributory fault of the person who suffered damage. The Protocol provides that compensation may be reduced or disallowed if the person who suffered damage, or a person for whom he is responsible under domestic law, has by his own fault caused or contributed to the damage having regard to all circumstances.[273] The burden of proof regarding contributory fault thereby lies with the liable person invoking this defence.

6. Right of Recourse, Article 8

The Basel Protocol provides, on the one hand, for limited strict liability of the persons involved in a transboundary hazardous wastes movement and, on the other hand, for unlimited fault-based liability of any person on the grounds of fault or

---

[269] Basel Protocol, Article 7(1).
[270] *Ibid.*, Article 7(2).
[271] *Ibid.*, Article 7(3).
[272] See also Tsimplis, 'The 1999 Protocol to the Basel Convention', 16 *IJMCL* (2001), at 312.
[273] Basel Protocol, Article 9.

negligence. The Protocol combines these two approaches to ensure that prompt and efficient compensation can be sought from the strictly liable person without it being necessary for the claimant to prove both the violation of a legal rule and the personal fault or negligence on the parts of the respondent. This concept and the inclusion of several persons in the strict liability scheme, however, require that the preliminary outcome can be adjusted at a second tier.

According to this, any person liable under the Protocol is entitled to a right of recourse in three cases: (a) if another person is also liable under the Protocol[274]; (b) if another person is liable according to a contractual arrangement[275]; and (c) if the law of the competent court provides for a right of recourse.[276] The most relevant scenarios concern recourse claims asserted by the person strictly liable according to Article 4 of the Protocol. This person either intends to take recourse against another person strictly liable according to Article 4 of the Protocol,[277] or he pursues recovery from the person who is liable on the basis of fault or negligence according to Article 5 of the Protocol.[278] However, the wording of Article 8 does not exclude recourse claims to be asserted by persons under fault-based liability according to Article 5. Thus it is basically possible that a person liable on the grounds of fault takes recourse from another person liable according to Article 5 or even from a person strictly liable according to Article 4 of the Protocol.

The Basel Protocol, however, fails to establish just how to determine the respective portion of liability that can be claimed by the person claiming recourse. As far as recourse is taken among persons strictly liable according to Article 4 of the Protocol, there are strong arguments for distributing liability per capita. Such distribution would take into account that strict liability is imposed solely due to the involvement of the liable person in the hazardous waste movement, regardless of his particular conduct or any subjective factors like fault or negligence. In contrast, as far as recourse is taken against a person who is liable on the basis of fault or negligence or among persons who are liable on the basis of fault, the legal notion contained in Article 9 of the Protocol can be brought to bear. According to this, the degree in which the negligent conduct of the respondent has contributed to the

---

[274] *Ibid.*, Article 8(1)(a).

[275] *Ibid.*, Article 8(1)(b). The Basel Secretariat, *Implementation Manual* (2005), at 14, gives the following example: The notifying exporter has entered into a contractual agreement with the carrier, according to which the carrier is responsible for any damage caused during the movement. After the notifier has been held strictly liable according to Article 4 of the Protocol and has compensated the claimant, he is entitled to recourse against the carrier pursuant to Article 8(1)(b).

[276] Basel Protocol, Article 8(2).

[277] In case two or more persons are strictly liable according to Article 4, these persons are subject to joint and several liability pursuant to Article 4(6), thus giving the claimant the right to seek full compensation from any or all of the liable persons.

[278] These scenarios are explicitly mentioned by the Basel Secretariat, *Implementation Manual* (2005), at 13.

causation of damage should be determined or assessed, and this figure should be used to determine the proportion of the total claim which can be asserted vis-à-vis the respondent.

7. Summary

The liability regime of the Basel Protocol is to a great extent modelled on the standards of the existing civil liability regimes and contains most of the typical legal features of those regimes. However, in some respects the Protocol pursues rather modern approaches by which it attempts to take into due consideration the specific circumstances and particular features of transboundary movement of hazardous wastes.

The general concept of liability under the Basel Protocol is characterised by the combination of limited[279] strict liability and unlimited fault-based liability. In case of damage caused by hazardous wastes, it is thus possible that one or more persons are strictly liable, while at the same time other or the same persons are liable on the basis of fault or negligence. In such cases the claimant may choose against whom to claim either partial or full compensation. Liability based on fault has the advantage that it is unlimited. Hence, the claimant may choose to proceed on the basis of fault if the respondent is financially strong and if the circumstances of the incident are known so that there are good chances to establish fault on the part of the respondent. In most cases, however, it will be difficult to clear up the particular circumstances of an incident, so that litigation will take a substantial period of time and involve considerable efforts. Therefore, in most cases it seems preferable for the claimant to seek compensation, in the first instance, from the person strictly liable. Strict liability has the advantage that it relieves the claimant from the burden of proving fault on the part of the respondent and, furthermore, that it is covered by compulsory insurance. Only as a complementary measure may the claimant at the same time file a claim against the person liable on the basis of fault in respect of the damage amount exceeding the amount covered by the limited strict liability. By this means, it is be ensured that, on the one hand, at least a partial if limited compensation is available for the victim of pollution without undue delay occasioned by long-lasting litigation. On the other hand, it is ensured that the "real" perpetrator of pollution may be turned to for full compensation, albeit involving the requirement that full evidence of fault on his part is produced within the framework of a lawsuit. A further element of the general concept of liability of the Basel Protocol is that both types of liability are interconnected by the right of recourse, according to which any person liable under the Protocol may take recourse against any other person also liable in proportion to their respective actual contributions to the causation of damage. By this legal instrument it is ensured that subsequent to the claim issued by the victim of pollution a further

---

[279] The issue of limitation of liability is outlined in the following section.

allocation of liability among the persons liable under the Protocol can be achieved, which is rather focused on the actual contributions to the causation of damage. This instrument, therefore, represents a legal compensation for the establishment of joint and several liability under the Basel Protocol.

Strict liability under the Basel Protocol basically follows the approach of channelling liability to only one person. However, in the transboundary movement of hazardous wastes several persons are involved, including the exporting and the importing side as well as the carrier and the disposer, all of which exercise control over the wastes at various stages of the movement. The Basel Protocol, therefore, establishes spheres of responsibility contingent on these stages of the movement and channels strict liability to the respective person who is deemed responsible for the stage of the movement during which the incident occurred. In this way, strict liability under the Basel Protocol is channelled to only one person, this particular person differing depending on which stage of the movement is concerned. As a general rule, the Basel Protocol establishes a temporal break at that moment when the disposer takes possession of the hazardous wastes. Prior to this moment liability is imposed on the notifier (which can be the generator or exporter of the hazardous wastes), whereas incidents occurring after this moment are attributed to the disposer. Relevant for the allocation of liability is the moment when the incident occurs, whereas the moment when the damage actually materialises is of no significance for the allocation of liability. Liability basically covers incidents that occur during the period when the wastes are loaded on the means of transport until the final disposal of the wastes. It seems appropriate to include a geographical element in the determination of the moment when the disposer takes possession of the wastes. The conduct of agents and servants of the disposer has to be attributed to the disposer. This, however, may not be construed to allow the disposer to enter into a contractual agreement with the exporter according to which delivery to the disposer is to be assumed at any other place than at the premises of the disposal plant. The Protocol, furthermore, provides that if the State of export is the notifier or if no notification has taken place, it is solely the exporter and not the generator who can be held liable. With regard to wastes considered hazardous only by the State of export or only by the State of import, the provisions of the Basel Protocol are inconsistent and far from clear. If the wastes are considered hazardous only by the State of export the liability provisions applying to the importer apply *mutatis mutandis* to the exporter until the disposer takes possession of the wastes. The importer, however, is not liable for damage, unless the wastes are considered hazardous only by the State of import and the State of import is the notifier or no notification has taken place. Although the practical relevance of such constellations is obviously limited, it is nevertheless an example of unfortunate law-making. In cases of re-importation of hazardous wastes according to Articles 8 of the Basel Convention, strict liability is imposed on the original notifier. In cases of illegal trafficking, strict liability is imposed on either the original exporter or the generator, depending on which person is under the obligation to re-import the hazardous wastes pursuant to Article 9(2) and (4) of the Basel Convention.

A main defect of the Basel Protocol is to be seen in the absence of a secondary liability of the generator of the hazardous wastes, covering the period until the disposer takes possession of the wastes. In its present shape the liability regime creates a strong incentive for waste generators to delegate the exportation and notification tasks to third companies in order to circumvent any strict liability under the Basel Protocol. The notifying third company, which might solely be set up for the purpose of incurring liabilities, may prove elusive, become insolvent or simply lack sufficient insurance coverage or funds for compensation. Therefore, it would have been better to also include a secondary strict liability of the generator of hazardous wastes, covering the period until the disposer takes possession of the wastes. This secondary liability would have to apply in case compensation cannot be sought from the person otherwise strictly liable and would also have to be limited according to the general limits established by the Protocol.

Strict liability under the Basel Protocol is furthermore subject to several exemptions. Those exclusions of liability cover causes of damage which are usually excluded by civil liability conventions, such as acts of war, exceptional natural phenomena, administrative orders and acts of sabotage. All of these circumstances must have been unforeseeable for the liable party in order to exclude liability. Liability is furthermore excluded if damage was caused in the course of preventive measures.

Fault-based liability is imposed by Article 5 of the Basel Protocol on any person who causes or contributes to damage by his lack of compliance with the provisions implementing the Basel Convention or by his wrongful intentional, reckless or negligent acts or omissions. The scope of this provision is not limited to any person involved in the hazardous waste movement, but covers any person whose conduct has an actual influence on the occurrence of damage. The imposition of fault-based liability underlies the concept of subjective fault, meaning that in any case of an objectively sub-standard conduct an additional element of subjective fault or negligence on the part of the acting person must be determined.

A further requirement of liability is the determination of a causal link between the incident (in case of strict liability) or the negligent conduct (in case of fault-based liability), on the one hand, and the particular damage on the other hand. The applicable test of causation includes first the test of the *conditio sine qua non*. The test which is applied in the second step so as to consider normative elements of causation will depend on the law of the competent court. It is likely that courts in common law jurisdictions will apply the proximate cause test, while in civil law jurisdictions the adequate cause test is prevalent. As regards the specific case where damage is caused by wastes covered by the Protocol as well as by wastes not covered by the Protocol, Article 7 provides that liability is imposed on the liable person in proportion to the contribution of the hazardous wastes to the occurrence of damage. If the person suffering damage has contributed to the damage by his own fault, his claim for compensation is reduced to the extent of his contributory fault.

Finally, the Basel Protocol entitles any person liable under the Protocol to a right of recourse against any other person also liable under the Protocol, liable according to a contractual agreement or liable pursuant to the domestic law of the competent court. This right of recourse is not only assigned to persons strictly liable, but also to persons liable on the basis of fault, albeit this will be of less significance. The degree to which recourse can be taken depends on the respective proportion to which the conduct of the other liable person has contributed to the occurrence of damage. As far as the allocation of damage among two or more persons strictly liable is concerned, the proportion of liability of the respective persons should be determined per capita.

IV. The Regime of Limitation of Liability

1. Financial Limitation of Liability

Limitation of liability regarding the recoverable amount is a common feature in civil liability conventions.[280] As regards maritime claims there is even an international convention in force dealing exclusively with the limitation of liability for such claims.[281] By means of establishing financial limits of liability it is acknowledged that there is a general need for carrying out activities which are potentially hazardous or even ultra-hazardous, but which are also beneficial and necessary for societies as a whole. If the persons conducting such activities were to face unlimited strict liability, it is feared that this commercial activity would be hindered or rendered practically impossible, which would eventually lead to the conduct of such activities being left to less responsible and unscrupulous operators who are based in less stringent jurisdictions.[282] A further argument sometimes produced is that there is a need for limitation of liability in order to make potential claims assessable and, thus, to facilitate the acquisition of insurance coverage for carrying out this activity.[283] Finally, by establishing financial limits of liability a counterbalance to the imposition of strict liability is created, thereby providing a certain legal compensation for having set aside any requirement to prove fault on the part of the liable person.[284] In the end, however, it must be admitted that the

---

[280] The only convention that provides for unlimited strict liability is the 1993 Lugano Convention which, however, never entered into force. A plea for strict and unlimited liability is presented by Louka, 'Bringing Polluters Before Transnational Courts', 22 *Denv. J. Int'l L. & Pol'y* (1993/1994), at 63 *et seq.*

[281] As to the 1976 LLMC Convention, see *supra*, Sect. "The LLMC Convention" in Chap. 3.

[282] Churchill, 'Facilitating Civil Liability Litigation', 12 *Yb. Int'l Env. L.* (2001), at 35–36; Kummer, *The Basel Convention* (1995), at 241–242.

[283] This argument, however, is not entirely convincing since unlimited liability would not prevent the liable persons from obtaining limited insurance coverage, which would moreover create an additional incentive for the liable person to act with particular care; see Churchill, 'Facilitating Civil Liability Litigation', 12 *Yb. Int'l Env. L.* (2001), at 36; Bocken, *et al.*, *Limitations of Liability and Compulsory Insurance, Report at the Request of the Basel Secretariat*, at 12.

[284] Churchill, 'Facilitating Civil Liability Litigation', 12 *Yb. Int'l Env. L.* (2001), at 36.

establishment of a financial limitation of liability cannot be seen as being indispensible for meeting the essential needs of societies as a whole, but rather is to be viewed as a politically driven decision to facilitate such activities.[285]

The basic conditions underlying the transboundary movement of hazardous wastes differ from the usual pattern underlying civil liability conventions in respect of the fact that hazardous waste is, in principle, an unwanted good. There is no intrinsic economic interest of the person involved in the movement to ensure that the wastes actually arrive at the place of destination. Furthermore, the Contracting States have the sovereign right to decide to either allow hazardous waste movements taking place in their national territory or to ban any such movements from their territory. This political decision is acknowledged in the Basel Convention, whereby a State's prohibiting the importation of hazardous wastes imposes a corresponding obligation upon the other Contracting States not to export hazardous wastes to that State. These diverging political intentions of the involved States is taken into account also by the limitation of liability regime of the Basel Protocol.

Under the Basel Protocol strict liability is limited according to the financial limits determined by each Contracting State in accordance with the provisions contained in Annex B to the Protocol.[286] Fault-based liability, by contrast, is exempted from the limitation provisions and, moreover, cannot be limited in advance by means of a contractual agreement.[287]

(a) Potential Conflicts with the LLMC Convention

It has been outlined above[288] that claims for damage resulting from the transboundary movement of hazardous wastes by sea are basically eligible for limitation under the 1976 LLMC Convention. It is, therefore, possible that a claim raised under the Basel Protocol is subject to two regimes of limitation of liability at the same time, i.e. the limitation of liability regimes of the Basel Protocol and the LLMC Convention.[289] It has been argued that this overlap may lead to conflicts, particularly in case a shipowner is liable on the basis of fault according to Article 5 of the Basel Protocol. In this case Article 12(2) of the Basel Protocol provides that liability of the shipowner is unlimited, whereas under the LLMC Convention the shipowner has a right to limit his liability.[290] This criticism, however, is not entirely correct. The potential conflicts between both regimes of limitation of liability are rather limited.

---

[285] Churchill, 'Facilitating Civil Liability Litigation', 12 *Yb. Int'l Env. L.* (2001), at 36–37.
[286] Basel Protocol, Article 12(1).
[287] *Ibid.*, Article 12(2).
[288] See *supra*, Sect. "The LLMC Convention" in Chap. 3.
[289] *Supra*, Sect. "Potential Conflicts with the LLMC Convention".
[290] Tsimplis, 'The 1999 Protocol to the Basel Convention', 16 *IJMCL* (2001), at 330.

In order to determine in which constellations a potential conflict between these two regimes may arise, it is necessary to first determine in which constellation a simultaneous application of both regimes may become relevant. Since the LLMC Convention exclusively applies to the liability of shipowners,[291] and since shipowners usually do not fall under the strict liability scheme of the Basel Protocol, relevant constellations may only involve fault-based liability of the shipowner according to Article 5 of the Basel Protocol. Fault-based liability is basically not excluded from the scope of the LLMC Convention.[292] However, Article 4 of the Convention provides that a person is not entitled to limit his liability if the loss resulted from his personal act or omission, committed with the intent to cause such loss, or recklessly and with knowledge that such loss would probably result.[293] Consequently, the limitation regime of the LLMC Convention does not apply to claims raised on the basis of fault, provided a certain increased degree of fault has been reached by the liable person. It can furthermore be argued that this degree of fault is higher than the degree of fault giving rise to fault-based liability as required by Article 5 of the Basel Protocol,[294] so that there remains a certain zone, in which a shipowner who is engaged in wilful misconduct is liable on the basis of fault according to Article 5 of the Protocol but has not yet reached the degree of fault under Article 4 of the LLMC Convention at which point the right to limit liability is lost.[295] Hence, it becomes clear that in general neither the Basel Protocol nor the LLMC Convention allows for a limitation of liability if damage or loss was caused by a deliberate or grossly negligent act of the liable person. Conflicts may only arise in the rare cases when the liable person intentionally committed a wrongful act or acted negligently, however, without the awareness that this conduct would probably cause damage or loss. It is obvious that such constellations are of a rather theoretical nature.

A further aspect to be considered is that even in such a rare case not every type of damage eligible for compensation under the Basel Protocol may be subject to limitation according to the LLMC Convention.[296] If, after all, it is nevertheless established that both regimes will apply simultaneously regarding this one

---

[291] This term includes the owner, charterer, manager and operator of a seagoing ship; see LLMC Convention, Article 1(2). The Convention also covers the liability of salvors, which, however, is of no relevance at this point.

[292] LLMC Convention, Article 2(1).

[293] *Ibid.*, Article 4.

[294] It can be argued that the degree of fault required by Article 4 of the LLMC Convention is higher than the degree of fault required by Article 5 of the Basel Protocol since Article 4 requires the intent to cause loss or at least the knowledge that loss would probably result.

[295] Since the shipowner is usually not involved in the exportation and notification procedures, the provision in Article 5 of the Protocol, according to which also a lack of compliance with the provisions implementing the Basel Convention gives rise to fault-based liability, is of no relevance.

[296] For example, although eligible for compensation under the Basel Protocol, not all costs incurred due to measures of reinstatement are subject to limitation according to Article 2(1)(d) of the LLMC Convention.

particular claim, it must be taken into account that both Conventions operate independently and that there is no mutual dependence apart from the fact that the LLMC Convention does not establish its own basis of liability. Therefore, there is no reason not to apply both regimes. Hence, the shipowner has the right to limit his liability, even if this is based on fault, pursuant to the provisions of the LLMC Convention and contrary to Article 12(2) of the Basel Protocol.[297]

(b) The Legal Arrangement of Limitation of Liability

Fault-based liability under the Basel Protocol is necessarily unlimited.[298] This also means that it is not possible for a potentially liable person to limit fault-based liability in advance by means of a contractual agreement. By contrast, strict liability under the Basel Protocol is subject to limitation according to the financial limits determined by each Contracting State in accordance with the provisions laid down in Annex B to the Protocol. This limitation, however, does not include any interests or costs awarded by the competent court.[299]

The financial limits of liability are not determined by the Protocol itself. In this respect, the Protocol differs from the limitation of liability pattern usually applied by civil liability conventions. Whereas the most common structure sees fixed limits of liability incorporated in the convention text itself,[300] the Basel Protocol assigns the Contracting States the task of specifying the financial limits in their respective domestic law.[301] However, the Protocol restricts the scope of possible domestic arrangements in a manner so as to stipulate minimum limits of liability[302] and provide that the domestic limits determined by the Contracting States must not fall below these minimum limits.[303] Unlike most of the civil liability conventions, the Basel Protocol does not provide for a rule according to which the

---

[297] The same result is provided by Tsimplis, 'The 1999 Protocol to the Basel Convention', 16 *IJMCL* (2001), at 330.

[298] Basel Protocol, Article 12(2).

[299] *Ibid.*, Article 12(1).

[300] See in this respect Article 9(1) of the HNS Convention, Article 9 of the CRTD Convention, Article V(1) of the CLC, Article 4(4) of the FUND Convention, Article V of the 1963 Vienna Convention, Article III(1) of the Nuclear Ships Convention, Article 6 of the Mineral Resources Convention, Article 9 in connection with Annex II of the 2003 Kiev Protocol to the Helsinki Conventions, Article 6 of the Bunker Oil Convention refers to the LLMC Convention, Articles 6 and 7 of the LLMC Convention. See also Article 4(5) of the Hague/Visby Rules, Article 6 of the Hamburg Rules, Article 59 of the Rotterdam Rules, and Article 8 of the 1974 Athens Convention.

[301] Basel Protocol, Annex B, Paragraph 1.

[302] *Ibid.*, Annex B, Paragraph 2.

[303] A similar approach is provided for by Article 7 of the 1997 Protocol to amend Article V of the 1963 Vienna Convention as well as by Article 7 of the 1974 Athens Convention as regards liability for personal injury. These provisions determine certain minimum limits of liability accompanied by the right of the Contracting States to domestically determine higher limits of liability. A further different approach is adopted by Article 7 of the 1960 Paris Convention,

liable person loses his right to limit liability if he acted with the intention to cause damage or with the certain knowledge that damage would result.[304] Such rule, however, would be redundant under the Basel Protocol since fault-based liability is excluded from the right to limit liability in any case.[305]

The minimum limits of liability determined by the Basel Protocol are always with reference to one single incident. Different limits are established for the notifier, exporter and importer, on the one hand, and for the disposer on the other hand.[306] For the disposer the minimum limit of liability amounts to a lump sum of 2 million SDR[307] per incident.[308] In contrast, the minimum limits of liability for a notifier, exporter or importer are determined with reference to the respective weight of the shipment.[309] However, the limits of liability per tonne of the shipment do not increase linearly with the increasing weight of the shipment. The Protocol rather establishes certain incremental steps in the course of which the limit of liability per tonne decreases as the total weight of the shipment increases.[310] This decrease in the limit of liability per tonne is to be explained by the fact that the most dangerous wastes are shipped only in small volumes and that also in respect of smaller shipments sufficient funds must be available for expenses such as preventive measures.[311] In this respect, it has been criticised that the minimum limits of liability are based on the volume of the shipment rather than on the degree

---

(Footnote 303 continued)
according to which the Contracting States are free to domestically determine the limits of liability within a certain range (between 5 and 15 million SDR).

[304] See Article 9(2) of the HNS Convention, Article 10(1) of the CRTD Convention, Article V(2) of the CLC, Article 4(3) of the FUND Convention, Article 6(4) of the Mineral Resources Convention and Article 4 of the LLMC Convention.

[305] The same consideration applies to the 2003 Kiev Protocol to the Helsinki Conventions which, in Article 9(3), also excludes fault-based liability from the limitation of liability.

[306] Basel Protocol, Annex B, Paragraph 2(a) and (b).

[307] The "unit of account" is defined by Article 2(2)(j) of the Protocol as the Special Drawing Right (SDR) as defined by the International Monetary Fund.

[308] Basel Protocol, Annex B, Paragraph 2(b).

[309] *Ibid.*, Annex B, Paragraph 2(a).

[310] *Ibid.*, Annex B, Paragraph 2(a) sets out the following increments:

(i) 1 million SDR for shipments up to and including 5 tonnes;
(ii) 2 million SDR for shipments exceeding 5 tonnes, up to and including 25 tonnes;
(iii) 4 million SDR for shipments exceeding 25 tonnes, up to and including 50 tonnes;
(iv) 6 million SDR for shipments exceeding 50 tonnes, up to and including 1,000 tonnes;
(v) 10 million SDR for shipments exceeding 1,000 tonnes, up to and including 10,000 tonnes;
(vi) Plus an additional 1,000 SDR for each additional tonne up to a maximum of 30 million SDR.

According to these figures, the limit of liability per tonne decreases from 200,000 SDR (in respect of smaller shipments of up to 5 tonnes) to 1,000 SDR per tonne (in respect of larger shipments of more than 10,000 tonnes).

[311] Tsimplis, 'The 1999 Protocol to the Basel Convention', 16 *IJMCL* (2001), at 319.

of hazardousness of the wastes.[312] However, it needs to be considered that the determination of the actual hazardous quality of each particular shipment would involve considerable costs related to the required technical capacities and correspondingly impose an additional administrative burden on the States. Thus, from a practical perspective an alternative approach focused on the degree of hazardousness hardly seems feasible.

(c) Practical Implications

According to the regulations of the Basel Protocol there is no restriction in what way and to what particular amount the limits of liability can be determined by the Contracting States, provided the minimum limits established by the Protocol are heeded. Consequently, it is possible that a Contracting State could decide, for example, to impose a uniform limit of liability per tonne, irrespective of the actual weight of the shipment. It would in principle also be possible that a Contracting State stipulate a fixed lump sum as the ceiling amount for strict liability.[313] It is more likely, however, that for political reasons individual States will be inclined to impose higher limits, or even limits so high that they amount to unlimited strict liability as a practical matter.[314]

It could be expected, first of all, that such "very high" domestic limits of strict liability would lead to difficulties regarding the availability of sufficient coverage on the insurance market. Since a legal obligation to obtain and maintain insurance coverage exists only to the extent of the minimum limits of liability as laid down in Annex B to the Protocol,[315] the practical consequence of "very high" domestic limits would be either that the persons strictly liable would have to bear higher costs for insurance premiums or that the gap between the insured minimum limits as laid down by the Protocol and the domestic limits of liability would remain uninsured. In the end, one could anticipate a situation in which the legal purpose related to the limitation of liability would be practically eliminated since the persons involved in hazardous waste movements would be faced with considerable additional costs and risks and thus find themselves encouraged to delegate the notification and disposal procedures to cheaper and less responsible companies.

However, it should be kept in mind that the particular significance of limits of liability being determined domestically and, thus, being subject to significant

---

[312] See in this respect Basel Secretariat, *Implementation Manual* (2005), at 15; Soares/Vargas, 'The Basel Liability Protocol', 12 *Yb. Int'l Env. L.* (2001), at 102.

[313] This would, of course, require that this lump sum include all minimum limits of liability as laid down in Annex B, Paragraph 2 of the Basel Protocol.

[314] In contrast, it might not be possible for a State to completely refrain from determining any limits of liability since such a regulation would undermine the general decision of the Basel Protocol to establish a limitation of strict liability. If a State fails to determine limits of liability by domestic law, the limits contained in Annex B to the Protocol will rather directly apply.

[315] Basel Protocol, Article 14(1).

variations between the Contracting States results only from the interconnection of this provision with the right of the claimant to bring his claim to any of those jurisdictions where either the damage was suffered, the incident occurred or where the defendant has his place of business ("forum shopping").[316] Since transboundary movements of hazardous waste usually take place across several territories—and taking account also of the fact that pollution may spread over large areas—it is hardly possible to determine in advance what the respective domestic limits of liability are that will apply in case of damage. This situation where the domestic limits of liability in case of damage are uncertain and may vary between the minimum limits as stipulated by Annex B to the Protocol and "very high" limits which practically amount to unlimited strict liability will have considerable negative effects on the availability of insurance coverage. Insurance coverage comprising the entire range of strict liability will either become unaffordable, or on the potential basis of strict liability a person will be faced with the prospect of being personally liable for any damage exceeding the minimum limits of liability established in Annex B to the Protocol. In order to narrow this uncertainty over the respective domestic limit of liability in case of damage and thus to make insurance premiums calculable so as to enable potentially liable persons to obtain full insurance coverage, it would have been a more favourable solution to also stipulate maximum limits of liability in the Basel Protocol.[317] Hence, the Basel Protocol would predetermine a certain range within which the Contracting States would be able to determine the respective domestic limits of liability.

Notwithstanding this conclusion, however, there might also be a positive effect associated with the imposition of "very high" domestic limits of liability by some of the Contracting States. It has initially been outlined that the sovereign political decision of any State to either participate in the transboundary movement of hazardous wastes or to ban all such movements from its territory is acknowledged by the Basel Convention and must be considered by all the other Contracting States. These different political interests of the respective States are duly taken into account by the Basel Protocol in its allowing different domestic limits of liability. The Contracting States may, according to their respective political intention, decide to impose low limits of liability in order to facilitate hazardous waste movements and to support their domestic waste trading industry. They may, by contrast, also decide to impose "very high" limits of liability in order to interpose an obstacle for private companies considering undertaking hazardous waste movements to or from this State or through areas under the national jurisdiction of this State. These different domestic limits of liability in conjunction with the right of a claimant to bring his claim to courts of several jurisdictions involved in the movement will have the practical effect that any person potentially liable under the Basel Protocol must anticipate that in the event of damage the claimant as a general rule will bring his claim to the courts of the jurisdiction which has imposed

---

[316] See *ibid.*, Article 17(1).
[317] See also Tsimplis, 'The 1999 Protocol to the Basel Convention', 16 *IJMCL* (2001), at 319.

the highest limits of liability. In order to avoid the imposition of "very high" limits of liability the persons engaged in hazardous waste movements will, therefore, have a strong economic interest to export hazardous wastes only to those countries, and to only cross the territories of those countries, which have stipulated low or moderate limits of liability. Hence, the determination of the domestic limits of limitation can be seen as an important and effective political parameter which States can deploy to influence the economic decision of private actors to forego or pursue hazardous waste movements conducted in their national territory.

(d) Procedural Issues

The Basel Protocol does not contain any rules regarding the legal procedures according to which the limitation of liability can be invoked by the liable person. The Protocol does, in particular, not contain a provision according to which the limitation of liability must be invoked by the establishment of a limitation fund with the court in which the claim for compensation has been brought.[318] Thus, the respective national laws of the court in which the claim for compensation under the Basel Protocol has been brought remain applicable as to the procedural requirements for the establishment of the limitation of liability.[319]

In general, the limitation of liability may be invoked by the liable party in two different manners. First, if a claim has been filed, the liable person may invoke the limitation of liability as a defence against this claim. The liable party may thus avoid having to pay costs and interests on a limitation fund. However, if a second claim is subsequently raised against the liable party, the payments made in respect of the first claim are not included when the limitation of liability is invoked in respect of this second claim; the liable party thus risks eventually having to pay beyond the limit of liability.[320] Therefore, it seems preferable for the liable person to set up a limitation fund—the second means of invoking the limitation of liability—in any case where it cannot be envisaged that only one claim will be brought against the liable person.

(e) Revision of the Financial Limits Established by Annex B

The minimum financial limits established by Annex B of the Basel Protocol were subject to constant criticism even during the negotiation process of the Protocol.[321]

---

[318] Such provision can be found in Article 9(3) of the HNS Convention, Articles 10(3) and 11 of the CRTD Convention, Article V(3) of the CLC, Article 6(5) of the Mineral Resources Convention, and Articles 10 and 11 *et seq.* of the LLMC Convention.

[319] Basel Protocol, Article 19.

[320] See Meeson/Kimbell, *Admiralty Jurisdiction and Practice* (4th ed., 2011), at 314.

[321] See e.g. Long, 'Protocol on Liability and Compensation', 11 *Colo. J. Int'l Envtl. L. & Pol'y* (2000), at 260; Sharma, 'The Basel Protocol', 26 *Delhi L. Rev.* (2004), at 186.

In order to nevertheless garner the support of the States to adopt the Protocol at the 5th Conference of the Parties to the Basel Convention (COP5) and to gain time for a more in-depth consideration of the financial limits, it was decided to include a provision according to which at COP6 the limits set out in Paragraph 2 of Annex B could be amended, even prior to the entry into force of the Basel Protocol.[322] At COP5 the Parties to the Basel Convention took note of Article 23 of the Protocol and requested the Legal and Technical Working Groups to consider the limits set out in Annex B to the Protocol with a view to presenting their recommendations at COP6.[323] However, the subsequent discussions of the Legal and Technical Working Groups as well as several statements of the States Parties involved in the elaboration of the Protocol revealed that there was broad disappointment with the results achieved during the elaboration of the Basel Protocol.[324] Consequently, no amendments of the financial limits set out in Paragraph 2 of Annex B to the Basel Protocol were made during COP6.

The Basel Protocol, however, provides that the minimum limits of liability stipulated in Annex B are to be reviewed by the Contracting States on a regular basis. In so doing, the States are required to take into account, *inter alia*, the potential risks posed to the environment by the movement of hazardous wastes, waste recycling, and the nature, quantity and hazardous properties of the wastes.[325] Whether this provision will prove to be an efficient instrument for increasing the acceptance of the Basel Protocol may be doubted. It must be concluded that the States were given a useful instrument for reviewing the financial limits set out in Annex B to the Protocol but failed to use this instrument to at least undertake a more in-depth consideration of these limits and evaluate the related advantages and disadvantages. Such consideration could have led to a broader acceptance of this provision among States. However, this opportunity has been missed.

(f) Summary

Under the Basel Protocol any person strictly liable pursuant to Article 4 is entitled to limit his liability, whereas fault-based liability under to Article 5 of the Protocol is exempted from the limitation of liability. The regime of limitation of liability under the Protocol does not follow the traditional convention approach of explicitly establishing fixed limits of liability, but rather mandates the Contracting States to determine financial limits in their respective domestic laws. The Protocol, however,

---

[322] Basel Protocol, Article 23. See also Basel Secretariat, *Implementation Manual* (2005), at 15–16; Soares/Vargas, 'The Basel Liability Protocol', 12 *Yb. Int'l Env. L.* (2001), at 102–103.

[323] Decision V/31 of COP5, 10 December 1999 (Doc. UNEP/CHW.5/29) at 57. See also Daniel, 'Civil Liability Regimes as a Complement to MEAs', 12 *RECIEL* (2003), at 231; Soares/Vargas, 'The Basel Liability Protocol', 12 *Yb. Int'l Env. L.* (2001), at 102.

[324] Daniel, 'Civil Liability Regimes as a Complement to MEAs', 12 *RECIEL* (2003), at 231; Soares/Vargas, 'The Basel Liability Protocol', 12 *Yb. Int'l Env. L.* (2001), at 103.

[325] Basel Protocol, Annex B, Paragraph 3.

establishes certain minimum limits of liability which domestic law may not go below. These minimum limits differentiate between the liability of the disposer, which is determined by a lump sum, and the liability of the notifier, exporter or importer, which is determined in certain increments depending on the weight of the shipment. In this respect, the minimum limits of liability per tonne of the shipment decrease with the increasing weight of the shipment, which is supposed to take account of the fact that the most hazardous wastes are shipped only in small volumes. The alternate approach of calculating the limits of liability according to the hazardous quality of the wastes would be impossible to implement in practice.

The Contracting States are basically free to decide in what way and to what particular amount the domestic limits of liability are determined. Therefore, it is possible that some States will opt to impose "very high" domestic limits of liability. Taken in conjunction with the right of any victim of pollution to bring his claim to the courts of different jurisdictions involved in the particular movement ("forum shopping"), the imposition of "very high" limits in certain jurisdiction would mean that the actual limits of strict liability in case of damage cannot be assessed in advance. The potentially liable person would, hence, be faced with considerable difficulties in obtaining and maintaining full insurance coverage. In order to facilitate the availability of reasonably priced insurance coverage, it would have been a more favourable approach for the Basel Protocol to have also established maximum limits of liability. In so doing, the States would be able to determine domestic limits of liability within a certain range, so that full insurance coverage would be available and premiums would be calculable. On the other hand, affording discretion in the setting of domestic limits of liability is a potential means of ensuring that the decisions of a respective sovereign State to either participate in the transboundary movement of hazardous wastes or to ban such movements from its territory are sufficiently taken into account. The imposition of "very high" domestic limits, thus, could act as an economic lever on the private actors so as to enforce this political decision. If a person engaged in a hazardous waste movement must anticipate that he will become subject to a "very high" limit of liability in case he arranges for a movement to or from a State, or crossing the territory of a State establishing such "very high" domestic limits, he is given a strong economic incentive to stay well clear of the territory of that particular State. In the end, however, these considerations are pending their test in practice.

2. Temporal Limitation of Liability

The Basel Protocol also provides for a statute of limitation. According to this, claims for compensation based on strict liability pursuant to Article 4 and claims based on fault pursuant to Article 5 of the Protocol[326] become time-barred if no

---

[326] Soares/Vargas, 'The Basel Liability Protocol', 12 *Yb. Int'l Env. L.* (2001), at 96, argue that Article 13 of the Basel Protocol only refers to strict liability according to Article 4, while fault-

action is brought within a period of 5 years from the date the claimant knew or ought to have reasonably known of the damage.[327] In addition, the Protocol provides that claims become time-barred in all cases after a period of 10 years from the date of the incident.[328] The Protocol also provides for some interpretive rules, according to these rules when the incident consists of a series of occurrences having the same origin, the date of the last of such occurrences is deemed to initiate the period of limitation. In respect of incidents consisting of a continuous occurrence, time limits run from the end of the continuous occurrence.[329] In conclusion it can be said that the limit limits imposed by the Basel Protocol seem to be appropriate and sufficient in consideration of the underlying conditions of hazardous waste movements.[330]

The Basel Protocol specifies the statute of limitation in a comprehensive manner. It provides that claims are not admissible after a period of 10 or 5 years, respectively, which means that these provisions establish a preclusive time period.[331] Consequently, there is no room for an additional application of the national laws of the court in which the claim for compensation is brought. Hence, even if the respective domestic law provides for rules according to which the period of limitation is suspended or disrupted,[332] such rules will not apply to claims raised under the Basel Protocol. Consequently, an extension of time by private agreement is not possible under the Basel Protocol. There is also no special rule concerning recourse claims, which would allow such claim being filed even after the expiration of the limitation period, provided the claim is filed within a certain period after the claimant either has himself settled the claim or has been served with process in the action against him.

(Footnote 326 continued)
based liability according to Article 5 remains unlimited. However, this opinion contradicts the express wording of this provision and is furthermore not in line with the systematic context of the provision.

[327] Basel Protocol, Article 13(2).

[328] *Ibid.*, Article 13(1).

[329] *Ibid.*, Article 13(3).

[330] The limitation period established by the Basel Protocol (5/10 years) corresponds with the time limits established by other civil liability conventions. Those limits are: 3/6 years in Article VIII of the CLC and in Article 8 of the Bunker Oil Convention; 5 years in Article 3 of the Mineral Recourses Convention; 3/10 years in Article 37 of the HNS Convention and in Article 18 of the CRTD Convention; 3/10 years or 20 years in Article V of the Nuclear Ships Convention; 3/15 years in Article 10 of the Kiev Protocol to the Helsinki Conventions.

[331] In respect of the 1992 CLC, see Kappet, *Tankerunfälle und der Ersatz ökologischer Schäden* (2006), at 65.

[332] German law, for example, provides that if negotiations take place between the parties to the claim, the limitation period will be suspended until they are considered to have failed. In addition, the claim will become time-barred at the earliest three months after the end of negotiations. See Section 203 of the German Civil Code (Bürgerliches Gesetzbuch—BGB).

V. Further Financial Instruments Implemented by the Basel Protocol

1. Compulsory Insurance or Similar Guarantees

Any person potentially liable according to Article 4 of the Basel Protocol is required to establish and maintain insurance, bonds or other financial guarantees during the entire timeframe of potential liability. Such insurance or financial guarantees must cover at least the minimum limits of liability as established by Annex B, Paragraph 2 of the Basel Protocol.[333] This obligation is restricted, however, to claims raised under the Basel Protocol.[334] The requirement of compulsory insurance or similar financial guarantees is part of the legal concept introducing strict but limited liability on any person engaged in the transboundary movement of hazardous wastes. It intends to strengthen the protection of both the victims of pollution and the persons potentially liable under the Protocol. By means of compulsory insurance, victims of pollution can be assured that there will be sufficient financial means available on the part of the liable respondent party so that the victim of pollution does not have to bear the risk of insolvency of the liable person. The liable person, in turn, can spread the risk of liability through insurance, and thus he can be sure that he is not burdened with claims for compensation exceeding his financial capacities.[335] Compulsory insurance, furthermore, increases the total number of insurance contracts (insurance pool) and, as a consequence, reduces the amount of insurance premiums.[336]

In order to safeguard the claim of the victim of pollution, the Protocol provides that contractual provisions between the insurer and the insured regarding deductibles or co-payments may not be interposed as a defence against a claimant seeking compensation.[337] A further instrument to safeguard the claim of the victim of pollution is the establishment of a right of direct action against the person providing insurance, bonds or other financial guarantees.[338] The Contracting

---

[333] Basel Protocol, Article 14(1). The Protocol provides further that if a State is considered to be the liable person pursuant to the rules in Article 4, it may fulfil its obligation to provide insurance or other financial guarantee by a declaration of self-insurance.

[334] *Ibid.*, Article 14(2).

[335] See also Røsæg, 'Compulsory Maritime Insurance', 258 *SIMPLY* (2000), at 181 *et seq*.

[336] The availability of insurance and other financial guarantees for liabilities in the context of the transboundary movement of hazardous wastes is outlined by Bocken et al., *Limitations of Liability and Compulsory Insurance, Report at the Request of the Basel Secretariat*, at 30 *et seq*.

[337] Basel Protocol, Article 14(1).

[338] *Ibid.*, Article 14(4). This provision also states that the insurer has the right to require the liable person to be joined in the proceedings and that the insurer may invoke any defence the liable person would be entitled to invoke.

States, however, have the right to opt against the domestic implementation of a right of a direct action against the insurer.[339]

The Basel Protocol, furthermore, imposes on the person submitting the notification of the proposed hazardous waste movement[340] an obligation to include with the notification proof of his coverage against liability by insurance or similar financial guarantees. The disposer, by contrast, is obliged to submit to the competent authorities of the State of import proof of his coverage against liability by insurance or similar financial guarantees.[341] Compliance with this provision is of importance for persons strictly liable since it is arguable that any failure to comply with this obligation constitutes fault within the meaning of Article 5 of the Basel Protocol, which will lead to an exemption from the limitation of liability, provided this failure contributed to the occurrence of damage.[342]

2. The Non-establishment of a Compensation Fund

(a) The Need for an Additional Financial Instrument

Although the liability regime of the Basel Protocol attempts to be comprehensive in covering the entire period of the hazardous waste movement, beginning with the moment when the wastes are loaded on the means of transport and ending with the completion of the final disposal process, there are nevertheless situations in which damage caused by hazardous wastes remains uncompensated. Strict liability fails to attach, for example, if the damage was caused by circumstances excluding liability pursuant to Article 4(5) of the Protocol, and even where it does attach it may not result in full compensation if the applicable domestic limits of liability are exceeded by the actual amount of damage, or if the liable person fails to obtain and maintain insurance coverage for whatever reason. In cases where there are, furthermore, persons liable on the basis of fault it is likely that those persons will not possess sufficient financial means for compensation. This may be explained by the fact that unscrupulous persons or companies are involved, or it may be due to the lack of insurance covering fault-based liability.

---

[339] *Ibid.*, Article 14(5). In this case, however, the Contracting State is required to notify the Depositary of the non-implementation at the time of signature, ratification, or approval of, or accession to the Protocol. The Secretariat is to maintain a record of the Contracting States which have opted against the implementation of a direct action against the insurer.

[340] According to Article 6 of the Basel Convention this can be the notifier, the exporter or the importer.

[341] Basel Protocol, Article 14(3).

[342] See Tsimplis, 'The 1999 Protocol to the Basel Convention', 16 *IJMCL* (2001), at 321, who gives the following example: If the importation of hazardous wastes is refused because of insufficient insurance coverage and if during the re-importation of these wastes an incident occurs causing considerable damage, the exporting party is liable on the basis of fault according to Article 5 of the Basel Protocol and, thus, is not entitled to invoke the limitation of liability.

Therefore, it was highly disputed during the negotiation process of the Basel Protocol whether or not an additional financial instrument should be implemented into the legal framework of the Basel Protocol, in order to provide financial aid particularly to developing countries in case of emergency. It was argued that a compensation fund was needed to enable the developing countries to take immediate pollution response action in the event of damage and to provide compensation in case liability fails to attach for any reason whatsoever. Others contested the need for an additional compensation fund and rather referred to the existing financial mechanisms.[343] While in the first Draft Articles of the Basel Protocol a provision establishing a compensation fund was included,[344] this provision was deleted in the later course of the negotiations over the Basel Protocol.[345] Instead of this, Article 15 in its present form was included in the Convention text, according to which existing financial mechanisms are to be used to ensure adequate and prompt compensation in the event compensation under the

---

[343] See 'Compensation and Liability Protocol Adopted', 30 *Envtl. Pol'y & L.* (2000), at 44; Choksi, '1999 Protocol on Liability and Compensation', 28 *Ecology L. Q.* (2001), at 518–519; Lawrence, 'Negotiation of a Protocol on Liability and Compensation', 7 *RECIEL* (1998), at 252; Sharma, 'The Basel Protocol', 26 *Delhi L. Rev.* (2004), at 194; Soares/Vargas, 'The Basel Liability Protocol', 12 *Yb. Int'l Env. L.* (2001), at 94. See also the references in Footnote 345.

[344] Between the 1st and 6th Session of the Ad Hoc Working Group to Consider and Develop a Draft Protocol, Draft Article 8 was concerned with the establishment of a compensation fund. Draft Article 8 read as follows:

"1. The Parties to this Protocol shall establish an international fund, hereinafter 'the Fund', for immediate response measures in an emergency situation and for compensation to the extent that compensation for damage under the civil liability regime is inadequate or not available.

2. The Parties to this Protocol shall adopt as soon as possible, the legal instrument establishing the Fund."

After the 6th Session, the establishment of a compensation fund was made subject of Draft Article 16, which read:

"1. The Parties to this Protocol commit themselves to the establishment of an international Fund, hereinafter 'the Fund', as a means to ensure that compensation will be available at all events and entrust the Fund to be created with the following functions:

(a) to minimize damage from accidents arising from transboundary movement of hazardous wastes [...] or during the disposal of the wastes;
(b) to provide for compensation when the liable person is or remains unknown, or is or may become financially incapable of meeting his or her obligations;
(c) to provide for compensation when the liable person is exempted from liability in conformity with Article 4 paragraph 3.

2. The Parties to this Protocol shall endeavour to adopt, as soon as possible, the legal instrument required for the establishment of the Fund. [...]"

[345] The Draft Article implementing the compensation fund was finally deleted during the 9th Session of the Ad Hoc Working Group. As to the discussing on this issue during the negotiation process, see Reports of the Ad Hoc Working Group, AHWG Docs. UNEP/CHW.1/WG.1/1/5, at 5–6; UNEP/CHW.1/WG.1/4/2, at 4–5; UNEP/CHW.1/WG.1/5/5, at 6–7; UNEP/CHW.1/WG.1/6/2, at 7; UNEP/CHW.1/WG.1/7/2, at 8; UNEP/CHW.1/WG.1/8/5, at 7; UNEP/CHW.1/WG.1/9/2, at 5–8; UNEP/CHW.1/WG.1/10/2, at 7.

Protocol is insufficient to cover the costs of damage.[346] In addition to this, the Contracting States are to regularly review the need for and possibility of either improving existing mechanisms or establishing new mechanisms.[347]

(b) The Utilisation of the Technical Cooperation Trust Fund

In conjunction with the adoption of the Basel Protocol, the 5th Conference of the Parties to the Basel Convention (COP5) in 1999 decided to enlarge the scope of the existing Technical Cooperation Trust Fund[348] on an interim basis.[349] By means of Decision V/32 of COP5, the Parties to the Basel Convention intended to provide an interim solution in response to the non-establishment of a compensation fund by the Basel Protocol and also to comply with their obligation to set up an emergency fund for the purpose of minimising damage in the case of an accident occurring during a transboundary movement of hazardous wastes.[350]

Decision V/32 of COP5 also stipulates the conditions for payment under the enlarged Technical Cooperation Trust Fund. According to this, the Trust Fund may exclusively be used to assist developing countries and countries with economies in transition.[351] The Decision further provides that the Secretariat of the Basel Convention is entitled to use the Trust Fund for the achievement of three purposes: These include, first, that the Trust Fund can be used to estimate the magnitude of damage and the measures needed to prevent further damage, to take appropriate emergency measures to prevent or mitigate damage and to help to find those persons who are in a position to give the assistance needed.[352] Second, the Trust Fund may be used to provide compensation for damage to and reinstatement of the environment up to the limits provided for in the Protocol, provided such compensation and reinstatement is not adequate under the Protocol. This second purpose, however, is made conditional on the entry into force of the Basel Protocol.[353] Additionally, it has been criticised as ambiguous with regards to the precise meaning of the restriction "up to the limits provided for in the Protocol". It

---

[346] Basel Protocol, Article 15(1).

[347] *Ibid.*, Article 15(2).

[348] The Technical Cooperation Trust Fund had been set up in the wake of Decision I/7 of COP1 (Doc. UNEP/CHW.1/24) in order "to support developing countries and other countries in need of technical assistance in the implementation of the Basel Convention".

[349] Decision V/32 of COP5 (Doc. UNEP/CHW.5/29).

[350] The obligation to set up an emergency fund is laid down in Article 14(2) of the Basel Convention. The Ad Hoc Working Group, which was established by Decision I/5 of COP1 (Doc. UNEP/CHW.1/24) to develop a Protocol on Liability and Compensation, had been requested by Decision I/14 of COP1 to also consider the elements necessary for the establishment of an emergency fund.

[351] Decision V/32 of COP5, Paragraph 1 (Doc. UNEP/CHW.5/29).

[352] Decision V/32 of COP5, Paragraph 2 (Doc. UNEP/CHW.5/29).

[353] Decision V/32 of COP5, Paragraph 3 (Doc. UNEP/CHW.5/29).

is argued that it is unclear whether this wording refers to the minimum limits of liability as stipulated in Annex B to the Protocol, or whether it refers to the domestic limits of liability established by the Contracting States pursuant to Annex B to the Protocol. Moreover, it was considered unclear whether the referenced "limits" also comprise unlimited fault-based liability.[354] Since, however, the domestic limits are not "provided for in the Protocol" and since in respect of fault-based liability no "limits" are established at all, the only appropriate interpretation of this provision can be that it refers to the minimum limits of liability as established in Annex B, Paragraph 2 of the Protocol.[355] The third purpose for which the Technical Cooperation Fund may be used is to develop capacity building and the transfer of technology as well as to put in place measures to prevent accidents and damage to the environment.[356]

After the adoption of Decision V/32 of COP5, the Parties to the Basel Convention reached agreement on further measures to implement this Decision. With Decision VI/14 of COP6 the Parties adopted Interim Guidelines for the Implementation of Decision V/32 on Enlargement of the Scope of the Trust Fund.[357] These guidelines are divided into three Parts in accordance with the possible purposes of payments made by the Trust Fund as outlined above. They contain detailed rules on the functioning and administration of the Fund, the application for and concession of payments and the use of contributions to the Fund. At COP9 the Parties to the Basel Convention, furthermore, adopted a standard form for request for emergency assistance from the Technical Cooperation Trust Fund.[358] In October 2012, the Secretariat of the Basel Convention published a Draft Report on the Implementation of Decision V/32.[359] This Draft Report was prepared under the lingering impression of the *M/V "Probo Koala"* incident in the Ivory Coast in 2006[360] and revealed that, after the government of the Ivory Coast had issued a formal request for assistance, a period of 3 months elapsed until financial aid could be disbursed. The reason for this delay was seen in the detailed requirements of a formal request, the necessary consultations and the earmarking of available funds and contribution. [361]

---

[354] See Tsimplis, 'The 1999 Protocol to the Basel Convention', 16 *IJMCL* (2001), at 326.

[355] See in particular the "Interim Guidelines for the Implementation of Decision V/32", COP6 Doc. UNEP/CHW.6/40, at 62.

[356] Decision V/32 of COP5, Paragraph 4 (Doc. UNEP/CHW.5/29).

[357] The Interim Guidelines for the Implementation of Decision V/32 were approved by Decision VI/14 of COP6 (Doc. UNEP/CHW.6/40) and are annexed to this Decision at 51–72. They are also printed in COP6 Doc. UNEP/CHW.6/10.

[358] Decision IX/22 of COP9 (Doc. UNEP/CHW.9/39). The standard form is annexed to this Decision and is also available at www.basel.int.

[359] Draft Report on the Implementation of Decision V/32 in Responding to Emergency Situations. The Report is available at www.basel.int.

[360] See *supra*, Sect. "The Factual Perspective: Transboundary Movements of Hazardous Wastes by Sea" in Chap. 1.

[361] See Draft Report on the Implementation of Decision V/32, at 4–5.

The greatest weakness of the present interim solution extending the scope of the Technical Cooperation Trust Fund, however, is that this Fund is based on voluntary contributions of the Parties to the Basel Convention as well as other donors. These contributions are, moreover, mainly earmarked for particular purposes, so that the amount of the Trust Fund is not fully available for every purpose.[362] Due to the voluntary nature of the contributions to the Technical Cooperation Trust Fund the available resources for emergency assistance are actually considerably limited. For example, as of June 2012, the financial means available under the Trust Fund amounted to approximately USD 330,000.00.[363]

(c) Interim Solution Insufficient Only upon Entry into Force of the Basel Protocol

The interim solution provided for by Decision V/32 to enlarge the scope of the Technical Cooperation Trust Fund has been criticised by several authors as being inadequate and incapable of meeting the needs of developing countries in case of an emergency. These authors consider the failure of the Basel Protocol to establish its own compensation fund as being one of its main weaknesses.[364] Formulated in such a very general sense, however, this criticism is not entirely convincing. In fact, there are two issues that must be clearly separated from each other.

Decision V/32 of COP5 explicitly states that the enlargement of the Technical Cooperation Trust Fund is deemed to be an interim solution. From the context of this rule it follows that this interim solution is to have effect until either an existing financial mechanism has been satisfactorily improved or a new financial mechanism has been established in accordance with Article 15(2) of the Basel Protocol. In this regard it should be kept in mind that the enlargement of the Technical Cooperation Trust Fund by virtue of Decision V/32 of COP5 is intended to fulfil two tasks. The first one is to comply with the obligation arising from Article 14(2) of the Basel Convention to set up an emergency fund for the purpose of minimising damage in the case of an accident occurring during the transboundary movement of hazardous wastes. This particular task is addressed in Paragraphs 2 and 4 of Decision V/32 of COP5. The second purpose to be achieved by the enlargement of the Technical Cooperation Trust Fund has its origin in Article 15(1) of the Basel Protocol and is addressed in Paragraph 3 of Decision V/32 of COP5. According to this, a financial instrument supplementary to the Basel

---

[362] According to Decision V/32 of COP5, Paragraph 8 (Doc. UNEP/CHW.5/29), contributions can be earmarked for any of the purposes mentioned in Paragraphs 2–4. They can, furthermore, be earmarked for such purpose in general or for specific activities.

[363] See Draft Report on the Implementation of Decision V/32, at 5.

[364] Long, 'Protocol on Liability and Compensation', 11 *Colo. J. Int'l Envtl. L. & Pol'y* (2000), at 258–259; Sharma, 'The Basel Protocol', 26 *Delhi L. Rev.* (2004), at 130; Soares/Vargas, 'The Basel Liability Protocol', 12 *Yb. Int'l Env. L.* (2001), at 94; Tsimplis, 'The 1999 Protocol to the Basel Convention', 16 *IJMCL* (2001), at 329; Widawsky, 'In My Backyard', 38 *Envtl. L.* (2008), at 603–604, 619–623.

Protocol is to be established which intends to ensure compensation of the victims of pollution in case liability and compensation under the Basel Protocol fails to attach or proves to be inadequate to cover the entire damage. The supplementary character of this second task and its relation to the Basel Protocol becomes obvious by means of the restriction made in Paragraph 3 of Decision V/32, according to which this Paragraph becomes operational only on the date the Basel Protocol enters into force. This, however, means that as long as the Basel Protocol has not entered into force, this second aim connected with the enlargement of the Technical Cooperation Trust Fund is suspended. One must clearly distinguish, therefore, whether the suitability and appropriateness of the interim solution of enlarging the scope of the Technical Cooperation Trust Fund is assessed regarding the period prior to the entry into force of the Basel Protocol or whether the period after its entry into force is being considered.

Concerning the period prior to the entry into force of the Basel Protocol, it must be concluded that the criticism voiced against the interim solution is mostly inapplicable. Since in this period the Technical Cooperation Trust Fund may be used only for the purposes of providing emergency funds in case of an accident and for facilitating capacity building and technology transfer, it can be argued that the limited financial means available and the earmarking of contributions to particular purposes or activities will not constitute a major obstacle in practice. Since the measures usually associated with these purposes do not require financial means to an extent comparable to the amount of funds required for compensating victims of pollution, the financial means available can be considered sufficient and it is arguably unnecessary to require compulsory contributions to the Trust Fund. Finally, it may be argued that emergency assistance, capacity building and technology transfer must be subsidised only where assistance is sought by developing countries and countries with economies in transition, so that the exclusion of developed countries from the scope of Paragraphs 2 and 4 of Decision V/32 of COP5 seems justifiable. In conclusion it can therefore be said that concerning the period prior to the entry into force of the Basel Protocol the interim solution of Decision V/32 of COP5 to enlarge the scope of the Technical Cooperation Trust Fund seems appropriate in view of the purposes to be achieved during this period.

In respect of the period after the entry into force of the Basel Protocol, however, a different conclusion must be drawn. At the moment the Basel Protocol enters into force, the second purpose intended by the enlargement of the scope of the Technical Cooperation Trust Fund becomes operational. Paragraph 3 of Decision V/32 provides that, from that moment, the Trust Fund may also be used to provide compensation for damage if liability and compensation under the Basel Protocol fails to attach or proves to be inadequate. This means that the financial resources required by the Trust Fund to fulfil its tasks will be much higher than before. The Fund in its present shape will then fail to ensure that sufficient financial means are available for compensation. From that moment, the voluntary nature of the contributions made to the Trust Fund and the earmarking of contributions to particular purposes and activities will prove to be a major obstacle for a sufficient functioning of the Technical Cooperation Trust Fund. In addition, as soon as the Trust Fund is used as a means of providing supplementary compensation in instances where liability and compensation under the Basel Protocol fail to attach or prove to be inadequate, the restriction of the Trust Fund, according to which developed countries are excluded from payments under the Fund, can no longer be justified. It must be concluded, therefore, that it is only at the point when the Basel Protocol enters into force that the interim solution provided for by Decision V/32 of COP5 to enlarge the scope of the Technical Cooperation Trust Fund will prove to be a weak instrument that fails to ensure that sufficient compensation will be available for victims of pollution in case compensation under the Basel Protocol is not adequate. This interim solution, thus, cannot be considered a suitable financial mechanism within the meaning of Article 15 of the Basel Protocol.

Consequently, as soon as the Basel Protocol enters into force, this interim solution must be replaced by a newly developed permanent solution. Such solution may in principle be found by elaborating a completely new financial instrument, such as the elaboration of a Compensation Fund under the Basel Protocol in accordance with Article 15(2) of the Protocol, or it may be found by redesigning the existing Technical Cooperation Trust Fund. Some parameters, however, should be taken into account in any event while elaborating such instrument. The prospective financial mechanism should establish a second tier of compensation supplementing liability and compensation provided for by the liability regime of the Basel Protocol which applies to any event of damage, irrespective of whether the victim is based in a developed or in a developing country. A distinction between damage suffered in an area under the jurisdiction of a developed country and damage suffered in an area under the jurisdiction of a developing country or a country with an economy in transition would not be consistent with the general concept of liability under the Basel Protocol. This particularly applies since the victims of pollution are mainly individuals and private companies whose financial capabilities are largely independent of whether the State is a developing or a developed country. Moreover, only the equal treatment of all States will ensure a wide-ranging acceptance of such a revised Trust Fund among potential Contracting States. A second prerequisite for the prospective Trust Fund is that it must be ensured that contributions are made by both States and the main private parties which are involved in and benefitting from the transboundary movement of hazardous wastes. Such contributions must be compulsory and regularly made.[365] And finally, the prospected Trust Fund must be furnished with legal standing so that it can be sued and is also entitled to take recourse action against any other person also liable under the Basel Protocol.[366]

The establishment of a Trust Fund supplementing the liability regime of the Basel Protocol in case compensation under the Protocol proves to be inadequate

---

[365] Daniel, 'Civil Liability Regimes as a Complement to MEAs', 12 *RECIEL* (2003), at 240; Kummer, *The Basel Convention* (1995), at 255; Tsimplis, 'The 1999 Protocol to the Basel Convention', 16 *IJMCL* (2001), at 329.

[366] Murphy, 'Prospective Liability Regimes', 88 *AJIL* (1994), at 58.

seems to be a suitable and appropriate tool to ensure full and prompt compensation of victims of pollution. In this respect, however, the purpose connected with the establishment of a Compensation Fund differs from the purpose of the existing Technical Cooperation Trust Fund. It is, therefore, doubtful whether the enlargement of the scope of the Technical Cooperation Trust Fund, even if this instrument is further amended, is the appropriate solution. In view of these two purposes it would be the better approach to establish a Compensation Fund as a new and autonomous instrument.[367]

As a final remark, it should also be added that the approach of the Basel Protocol to not initially include the mechanism of a Compensation Fund into the Protocol itself seems to be the correct strategy. By making the entry into force of the Basel Protocol in any way conditional on the actual establishment of a Compensation Fund, a further major obstacle for the States' ratification of the Protocol would be created. As can be seen in the context of the 1996/2010 HNS Convention, the issue of who must contribute to the Trust Fund is of major concern to States, particularly in the absence of any reliable data on the amount of expected contributions. The Basel Protocol may be compared with the HNS Convention with regard to the underlying conditions in the sense that a great number of private players are involved in both of the involved activities. Thus, it can be expected that it will be hard to find agreement on the question of who will have to pay levies to the Fund. In order to facilitate the entry into force of the Basel Protocol, these issues related to the establishment of the Compensation Fund should be postponed to the period after the entry into force of the Basel Protocol. In the meantime, States should be urged to ratify the Basel Protocol, and these States should emphasise by means of an express declaration that the entry into force of the Basel Protocol carries with it a fixed schedule for the elaboration of a new supplementary mechanism of a Compensation Fund.

It can therefore be concluded that regarding the period prior to the entry into force of the Basel Protocol, the present interim solution provided for by Decision V/32 of COP5 represents a suitable and appropriate solution in view of the intended purposes. Regarding the period after the entry into force of the Basel Protocol, this interim solution will, however, prove to be insufficient and inadequate. The Parties to the Basel Convention should, therefore, emphasise their determination to elaborate within a certain period of time a new financial mechanism of a Compensation Fund that is specifically designed to provide compensation in case liability and compensation under the Basel Protocol fails to attach or proves inadequate.

---

[367] See also Kummer, *The Basel Convention* (1995), at 253.

VI. Rules of Procedures

The Basel Protocol contains rules of procedures regarding competent courts, regarding the applicable law and regarding the mutual recognition and enforcement of judgments.

1. Rules on Competent Courts

Legal proceedings regarding claims for compensation under the Basel Protocol may be brought before the courts of a Contracting State either where the damage was suffered, where the incident occurred, or where the defendant has his habitual residence or principal place of business.[368] These jurisdictions are basically alternative and exclusive.[369] Since hazardous waste movements are usually conducted in the territories of different States and since in the case of an incident usually several victims and sometimes more than one liable person are involved, it is likely that the courts of several jurisdictions will be competent to hear the case. The possibility for the claimant to bring the claim before the courts of the most favourable jurisdiction (so-called "forum shopping")[370] is, however, limited to the extent that in case of related actions[371] any court other than the court first seized may, while the actions are pending at first instance, stay its proceedings,[372] or that on the application of one of the parties a court may decline jurisdiction if the law of that court permits the consolidation of related actions and another court has jurisdiction over both actions.[373]

2. Rules on the Applicable Law

The Basel Protocol provides that all matters of substance or procedure regarding claims before the competent court which are specifically regulated in the Protocol will be governed by the law of that court including any rules of law relating to

---

[368] Basel Protocol, Article 17(1)(a) to (c).

[369] It follows from the expression "only" in Article 17(1) that Contracting States may not establish jurisdiction over claims raised under the Basel Protocol in constellations not mentioned in Article 17(1) of the Protocol.

[370] The specific consequences resulting from the reciprocity of this rule and the different domestic limits of liability have been outlined *supra*, Sect. "Practical Implications".

[371] According to Article 18(3) of the Protocol, actions are deemed to be related where they are so closely connected that it is expedient to hear and determine them together to avoid the risk of irreconcilable judgments resulting from separate proceedings.

[372] Basel Protocol, Article 18(1).

[373] *Ibid.*, Article 18(2).

conflict of laws.[374] It is, furthermore, prescribed that claims for compensation which are raised against the notifier, the exporter or the importer on grounds of strict liability, must exclusively be in accordance with the Protocol.[375] This means by implication that claims for compensation which are raised against the persons strictly liable in case of re-importation or illegal traffic may be based both on the Basel Protocol and on the autonomous domestic law.[376] Claims for compensation on grounds of fault may also be based on the respective domestic law of torts.[377]

3. Mutual Recognition and Enforcement of Judgments

The Basel Protocol also ensures the mutual recognition and enforcement of judgments based on the provisions of the Protocol. Where two Contracting States are Parties to an agreement or arrangement in force which contains rules on mutual recognition and enforcement of judgments, then this agreement will exclusively apply to a judgment falling under its provisions and override the respective provisions of the Basel Protocol.[378] It is only where such agreement or arrangement is not in force that the Protocol provides that any final judgment of a competent court, which is based on the Basel Protocol and which is enforceable in the State of origin, is to be recognised and is to be enforceable in any other Contracting State as soon as the respective formalities required by the other Contracting State have been completed.[379] However, a judgment will not be recognised and enforceable if it is obtained by fraud, if the defendant was not given reasonable notice and a fair opportunity to present his case, if the judgment is irreconcilable with an earlier judgment validly pronounced in another Contracting State with regard to the same cause of action and the same parties, or if the judgment is contrary to the *ordre public* of the State in which its recognition is sought.[380]

---

[374] *Ibid.*, Article 19. As regards the relationship of the provisions of the Basel Protocol to the conflict of laws rules under the Rome II Regulation of the European Union (Regulation (EC) No 864/2007 on the Law Applicable to Non-contractual Obligations) see Basedow, 'Rome II at Sea', 74 *RabelsZ* (2010), at 127–128.

[375] Basel Protocol, Article 20(2). This applies to strict liability imposed by Article 4(1) and (2) of the Basel Protocol.

[376] Tsimplis, 'The 1999 Protocol to the Basel Convention', 16 *IJMCL* (2001), at 323. This applies to strict liability imposed by Article 4(3) and (4) of the Basel Protocol.

[377] Basel Protocol, Article 20(1).

[378] *Ibid.*, Article 21(3). As regards the law of the European Union, relevant rules can be found in Regulation (EC) No 44/2001 on Jurisdiction and the Recognition and Enforcement of Judgments in Civil and Commercial Matters.

[379] Basel Protocol, Article 21(1) and (2).

[380] *Ibid.*, Article 21(1)(a) to (d).

VII. Overview of the Protocol's Final Clauses

The Basel Protocol establishes upon its entry into force a Meeting of the Parties to the Basel Protocol (MOP), which is distinct from the Conference of the Parties to the Basel Convention (COP) and limited to the Parties of the Protocol.[381] The MOP must hold ordinary sessions in conjunction with the Conference of the Parties to the Basel Convention. Extraordinary MOPs are to be held if deemed necessary by the MOP or upon the written request of a Contracting State.[382] The functions of the MOP are to review the implementation of and compliance with the Protocol, to provide for reporting, to consider and adopt, where necessary, proposals for amendment of the Protocol, and to consider and undertake any additional action that may be required for the purposes of the Protocol.[383]

Under the Basel Protocol several functions which are related to administration and preparatory tasks are assigned to the Secretariat of the Basel Convention.[384]

The basic obligation of the Contracting States to adopt the legislative, regulatory and administrative measures necessary to implement the Protocol is laid down earlier in the body of the Protocol in Article 10.[385] This Article also provides that for the purpose of transparency the Contracting States are required to inform the Secretariat of any measures to implement the Protocol, including the domestic limits of liability established pursuant to Annex B, Paragraph 1 of the Protocol.[386]

The Basel Protocol enters into force on the 90th day after the date of deposit of the 20th instrument of ratification, acceptance, formal confirmation, approval or accession.[387] The Depositary of the Protocol is the Secretary-General of the United Nations.[388]

---

[381] Basel Secretariat, *Implementation Manual* (2005), at 22.
[382] Basel Protocol, Article 24(1) and (2).
[383] *Ibid.*, Article 24(3).
[384] *Ibid.*, Article 25. The particular tasks assigned to the Secretariat are listed in Paragraph 1. Paragraph 2 clarifies that these functions are to be carried out by the Secretariat to the Basel Convention.
[385] *Ibid.*, Article 10(1).
[386] *Ibid.*, Article 10(2).
[387] *Ibid.*, Article 29(1). As regards the present state of ratifications, see *supra*, Sect. "Evolution of the Protocol" in Chap. 3.
[388] *Ibid.*, Article 32.

B. Excursus: The Cases of the M/V "Khian Sea" and the M/V "Probo Koala"

At the beginning of this work[389] the cases of the *M/V "Khian Sea"* and the *M/V "Probo Koala"* were outlined as examples of transboundary movements of hazardous wastes that ultimately led to considerable damage to human health and the environment. At this point, it seems appropriate to apply the prospective regime of the Basel Protocol to these cases in order to estimate the practical effect of the Basel Protocol.[390]

The particular issue in the *M/V "Khian Sea"* case was that the true nature of the toxic ash had been concealed so that the hazardous wastes were discharged without the prior informed consent of the Haitian government. This violation of the PIC, ESM and other procedural requirements of the (later adopted) Basel Convention would have rendered the movement illegal within the meaning of Article 9 of the Basel Convention, triggering an obligation of the State of export to ensure the re-importation of the hazardous wastes by the exporter, the generator or, if necessary, by the State of export itself. In the *M/V "Khian Sea"* case, the person initiating and conducting the transboundary movement who was responsible for the misdeclaration of the wastes would be liable on the basis of fault pursuant to Article 5 of the Basel Protocol. This would include any clean-up costs and costs for measures of reinstatement undertaken at the beach in Haiti. By contrast, costs for measures of reinstatement that might have been taken subsequent to the dumping of the remaining cargo into the Indian Ocean would not be recoverable under the Basel Protocol, unless they would have been taken within the EEZ of another State. Limitations of liability would not apply to liability based on fault.

With regard to the *M/V "Probo Koala"* case, it can be stated that the transport represents a transboundary movement, since it is to be assumed that the wastes were generated in an area under the national jurisdiction of a coastal State in the Mediterranean, so that the movement took place involving areas under the national jurisdiction of at least two different States. However, the application of the Basel Convention and Protocol presupposes that the wastes do not derive from the "normal operation of a ship", which would make the MARPOL 73/78 Convention applicable. Although the Basel Secretariat found that in general "specific industrial processes or activities on board a ship (such as refining oil products [...]) might be considered distinct from the 'normal operation of ships'",[391] it nevertheless found itself unable to establish whether or not in the case of the *M/V "Probo Koala"* the particular wastes fell under the MARPOL 73/78 Convention. This is mainly to be explained by the lack of information concerning the

---

[389] See *supra*, Sect. "The Factual Perspective: Transboundary Movements of Hazardous Wastes by Sea" in Chap. 1.

[390] It should be mentioned that the exact facts of these cases are not entirely known, so that this excursus can only give a rough overview.

[391] See *supra*, Sect. "Wastes Excluded from the Scope of the Protocol".

generation and the precise source of the hazardous wastes.[392] If the Basel Convention and Protocol applied, the movement would constitute illegal traffic only where the PIC requirements of the Basel Convention had not been complied with. In that case, the coastal State in the Mediterranean, in whose area under the national jurisdiction of the coastal State the vessel anchored while conducting the caustic washing, would be considered the State of export. Since an exporter under the jurisdiction of the State of export would not exist, the obligation to re-import the hazardous wastes according to Article 9(2) of the Basel Convention would be with the oil trading company as the generator of the wastes or, if necessary, with the Mediterranean State of export. In case of illegal traffic, the oil trading company would be liable on the basis of fault according to Article 5 of the Basel Protocol for any damage that occurred in consequence of the dumping of the wastes in Abidjan. Fault-based liability would also attach to the importing company or, alternatively, to the disposing company in Abidjan, at least one of which failed to properly handle and dispose of the hazardous wastes. Liability in this case would be joint and several and unlimited. State responsibility of the Mediterranean State of export, of the flag State and of Ivory Coast as the importing State based on a breach of the ESM requirements of the Basel Convention would prove unlikely to be established in court, due to the fact that this obligation is conceived as an obligation of due diligence requiring the knowledge of the State about the circumstances of the case and the ability to prevent the particular conduct of the private persons. If the PIC requirements had been complied with and provided a wrongful intentional, reckless or negligent act or omission of the exporter or generator could not be established, the disposer of the wastes in Abidjan would be strictly liable according to Article 4(1) of the Basel Protocol (in addition to fault-based liability). Strict liability would be limited to 6 million SDR pursuant to the minimum limits of liability as stipulated by the Basel Protocol.

Considering this summary examination of the two cases of the M/V "*Khian Sea*" and the M/V "*Probo Koala*", it becomes apparent that with regard to these cases the Basel Protocol basically provides for reasonable and sufficient solutions regarding the protection of victims of pollution. The most relevant constellations regarding the causation of damage are covered by the Basel Protocol. It is only in rather atypical constellations that the provisions of the Basel Protocol fail to provide a suitable solution. This is the case, for example, if hazardous wastes are generated on board a ship in the EEZ of a Contracting State without being covered by the MARPOL 73/78 Convention. In this case, the coastal State is to be considered the State of export, which is responsible for ensuring the re-importation of the hazardous wastes, even when the generator of the wastes cannot be identified or held responsible by the State of export.

---

[392] A Basel Convention Secretariat's technical assistant mission to Abidjan in Ivory Coast was undertaken in 2006 in accordance with Decision V/32 of COP5 to enlarge the Technical Cooperation Trust Fund. The report of the mission can be found at Doc. UNEP/SBC/BUREAU/8/1/INF/2 and is annexed to Doc. UNEP/CHW/OEWG/6/2.

C. An Assessment of the Basel Protocol

Due to the fact that assessment of the Protocol in practice is still pending, a final and concluding analysis of the Basel Protocol is hardly possible. Many of the new approaches and regulations of the Basel Protocol can only be considered and assessed in theory, so that it is likely that such inquiries will not reveal each problem that might become crucial when the legal instrument is eventually applied in practice. Therefore, the assessment of the Basel Protocol which shall be given in this section can only be schematic and preliminary.

I. Summary of the Major Achievements and the Major Defects of the Basel Protocol

The economic and political circumstances related to the transboundary movement of hazardous wastes by sea are in several respects unique, so that it is not possible to simply transfer practical experiences and legal conclusions made with regard to other civil liability regimes to the legal regime of the Basel Protocol. The transboundary movement of hazardous wastes is characterised by the fact that hazardous wastes are in general an unwanted and dangerous substance, posing substantial risk for human health and the environment. In most cases hazardous wastes are required by law to be treated or disposed of in an environmentally sound manner involving huge costs. Hence, States and private parties usually tend to get rid of hazardous wastes in the cheapest way possible by finding a buyer for these wastes via a waste dealer or waste broker. As a consequence, however, the waste seller has no intrinsic economic interest in seeing that the wastes actually arrive at the place of destination in safe and sound condition. Notwithstanding these adverse characteristics of hazardous wastes, States and private companies may nevertheless have a legitimate interest in engaging in the cross-border trade of hazardous wastes. Those interests may derive from the desire for raw materials found in the hazardous wastes or may result from the fact that not every State is able to provide specific disposal capacities for every kind of hazardous waste.[393] In the end, it must be concluded, therefore, that a rigid ban of all transboundary movements would fail to sufficiently take into account the actual needs of States. The correct approach must rather be to reduce the volume of hazardous waste movements and to provide for a legal regime allocating risks and liabilities associated with the transboundary movement of hazardous wastes to the persons engaged in the movement.

---

[393] As regards the different interests involved in hazardous waste movements, see Chap. 2. The economic background is outlined in Chap. 4.

Such a regime requires in general that liability is attached to the parties on the outgoing side, thus ensuring that potential pollution damage is internalised and, in this fashion, that the exporting side is incentivised to invest in damage precaution. A comparable situation exists with regard to the parties on the incoming side, particularly with regard to the disposal company. The disposal company usually undertakes a commercial activity and, thus, reaps the financial benefits from receiving hazardous wastes. In order to facilitate an environmentally sound disposal of the hazardous wastes, the imposition of strict liability on the incoming side is a further necessary tool of a liability regime covering the entire period of transboundary hazardous waste movements. Finally, such a legal regime must be complemented by a second-tier instrument providing compensation in case liability fails to attach or proves to be inadequate in a given case.

The Basel Protocol attempts to address the specific circumstances of hazardous waste movements and to provide tailor-made solutions on the basis of approved legal methods. In many respects, the solutions found by the Basel Protocol can be considered suitable and appropriate in order to reconcile the involved commercial interests and the legal protection of victims of pollution. In other respects, the provisions and arrangements of the Basel Protocol appear to be rather unclear and confusing. Only a few issues must in fact be criticised as true shortcomings or weaknesses of the Basel Protocol. These findings can be summarised as follows:

Scope of application: The rules of the Basel Protocol governing its scope of application are conceived in an extremely detailed and sophisticated manner. The amount of text that was used to describe in which particular cases the Protocol applies are finally the very reason why this regulation turns out to be unduly complicated and in many respects confusing. Or, to use *Kant*'s words: "many a book would have been much clearer, if it had not been intended to be so very clear.[394]" The reason why the Protocol intends to be so very clear in this respect is that it attempts to regulate each particular case of application. However, due to the fact that hazardous waste movements are usually carried out between different States, including different States of transit, only some of which may be Contracting States to the Basel Protocol, and since many shipments also cross the high seas, there are several constellations conceivable in which the application of the Basel Protocol may be unclear. Moreover, the Protocol also acknowledges the right of each Contracting State to opt for a later commencement of the application of the Protocol. Consequently, it is not possible to establish a rule which is short and crisp. Nevertheless, the inclusion of diverse parameters determining the Protocol's scope of application, such as the definition of the covered wastes, the occurrence of an incident and the occurrence of damage—this being determined in both a temporal and a geographical regard—must be considered as unnecessarily complicated. Yet apart from this criticism, the rules governing the scope of application of the Basel Protocol are in general well-balanced and appropriate. This applies

---

[394] Translation of: "[M]anches Buch wäre viel deutlicher geworden, wenn es nicht so gar deutlich hätte werden sollen", Kant, *Critik der reinen Vernunft* (1781), at 19.

most of all to the partial application of the Protocol to damage suffered on the high seas as well as to the question of who may seek compensation for what kind of measures undertaken.

Relationship to other civil liability regimes: The relationship of the Basel Protocol to other civil liability conventions is dealt with by Article 3(7), which concerns the relationship to other liability agreements regarding particularly the *transboundary movement of hazardous wastes*, as well as by Article 11, which governs the relationship to other *general* agreements on liability and compensation. It has been outlined that an objective justification for this distinction does not exist. What is of even more importance, however, is that Article 11 establishes only a formal, temporal criterion for the overriding application of another convention, instead of relying on a qualitative requirement as provided for by Article 3(7). Although it must be conceded that the practical relevance of this weakness is limited, the absence of any qualitative criterion cannot be considered an appropriate solution. It would have been the better approach to not distinguish between Article 3(7) and Article 11 and instead establish only one rule which is based on the model of Article 3(7). In so doing, potential loopholes created by Article 11 could have been eliminated and a shorter and far clearer regulation would have been established.

Particularly with regard to the relationship of the Basel Protocol to the 1996/2010 HNS Convention one also needs to concede that States are reluctant to undertake the efforts to implement the sophisticated rules of the Basel Protocol when its application to the entire "tackle-to-tackle" period is superseded by the HNS Convention. In order to increase the importance of the Basel Protocol, it would thus be an ideal (but rather illusory) solution to exclude waste substances from the scope of application of the HNS Convention.

The liability regime: The liability regime of the Basel Protocol is characterised by a combination of limited strict liability and unlimited fault-based liability. By means of the right of recourse given to any person liable under the Basel Protocol against any other person liable under the Protocol, a two-tier system of liability is established, according to which on the first tier limited strict liability is promptly available for the claimant, whereas on the second tier the claimant or the person strictly liable may claim partial or full recovery from any other person either strictly liable or liable on grounds of fault.[395] This approach must be considered advantageous and a major achievement of the Basel Protocol.[396] It ensures that prompt, and therefore efficient, compensation is available, and it opens the possibility of achieving a subsequent allocation of liability on the basis of the actual contributions to damage as well as on elements of fault.

The Basel Protocol, furthermore, follows the approach of channelling strict liability to only one person; however, the actual person strictly liable will always

---

[395] The advantages related to this approach are outlined in detail *supra*, Sect. "The Concept of Combining Strict and Fault-Based Liability".

[396] A similar approach was later pursued by the 2003 Kiev Protocol to the Helsinki Conventions.

depend on the respective sage of movement at which the incident occurred. The basic caesura established by the Protocol, at which strict liability shifts from the outgoing to the incoming side, is the moment when the disposer takes possession of the hazardous wastes. This basic approach must, again, be considered a reasonable and well-balanced solution. It transfers the concept of channelling strict liability to the circumstances prevailing in the transboundary movement of hazardous wastes. This conclusion, however, presumes that a geographical element is included in the determination of the moment when the disposer takes possession of the wastes. It must be ensured that liability shifts to the disposer only at that moment when the wastes actually arrive at the place of disposal and that this moment may not be accelerated by means of a simple contractual clause.

One of the main defects of the Basel Protocol must be seen in the absence of a secondary (strict and limited) liability of the waste generator covering the period up until the disposer takes possession of the hazardous wastes. The present shape of the Protocol's liability provisions creates a strong incentive for waste generators to hand over the hazardous wastes to third companies, such as waste brokers or waste dealers, in order to undertake the exporting and notification procedures. By this means, waste generators may effectively evade strict liability while only a shell company may be held liable under the Protocol, the latter perhaps not possessing sufficient funds for compensation and failing to obtain sufficient insurance coverage.

The regime of limitation of liability: Liability under the Basel Protocol is limited with regard to a certain timeframe; strict liability is moreover limited in amount. The concept of limitation under the Basel Protocol does not follow the traditional approach of stipulating global limits of liability, but rather mandates the Contracting States to establish domestic limits which must not go below the minimum limits as laid down by the Basel Protocol. This approach, again, must be considered an innovative adaption of the traditional limitation pattern to the specific circumstances of hazardous waste movements. Since it is likely that States will impose different domestic limits of liability, the persons engaged in hazardous waste movements will presumably avoid sending the wastes to or through areas under the jurisdiction of a State that has opted to impose high domestic limits of liability. By means of determining high or low domestic limits of liability, States are thus, at least in theory, given the possibility to influence the economic decision of private actors in accordance with their respective national policy. It may be assumed, therefore, that this innovative approach might turn out to be a suitable and effective solution that sufficiently takes account of the diverging political interests of the Contracting States as regards the question whether or not to take part in the transboundary movement of hazardous wastes. However, in order to facilitate the availability of reasonably priced insurance coverage, it would have been better to also establish maximum limits of liability within the framework of the Basel Protocol.

The non-establishment of a compensation fund: There seems to be agreement that, due to the fact that liability under the Basel Protocol proves to be inadequate in certain situations, there is a need for a supplementary financial instrument to protect the victims of pollution. However, it is highly contested whether or not the non-establishment of a compensation fund within the legal framework of the Basel Protocol must be considered a major weakness of the Protocol. In the end, one needs to draw a distinction. At COP5, when the Basel Protocol was adopted, the Parties to the Basel Convention also decided to enlarge the scope of the Technical Cooperation Trust Fund on an interim basis to compensate for the non-establishment of an autonomous compensation fund under the Protocol. This interim solution seems in general to be suitable and sufficient until the entry into force of the Basel Protocol. After the Protocol's entry into force, however, the Technical Cooperation Trust Fund will also be used to provide compensation in case liability under the Protocol proves to be inadequate. For that purpose the Technical Cooperation Trust Fund in its present shape is not sufficiently equipped and the interim solution will turn out to be inadequate. This means that as soon as the Basel Protocol enters into force the interim solution must be replaced by a newly developed permanent solution, which should be set up in accordance with Article 15(2) of the Protocol as an autonomous compensation fund under the Basel Protocol. Consequently, it is only in the event that the Contracting States fail to elaborate and set up a permanent compensation fund by the time the Basel Protocol comes into force that the present non-establishment of such a fund must be considered a major weakness of the Protocol.

II. Reasons for the Protocol's not Entering into Force

More than 14 years after the adoption of the Basel Protocol at the 5th Conference of the Parties to the Basel Convention in 1999, the Basel Protocol has not received a sufficient number of ratifications in order to enter into force. The reasons for the reluctance of States to ratify the Protocol are manifold. In this section, a brief overview shall be given of the variety of reasons explaining why States appear to lack an incentive to ratify the Protocol.

1. Lack of Political Incentive to Ratify the Basel Protocol

The Basel Protocol is the product of a political compromise that was eventually reached by the Parties to the Basel Convention, their having been faced with high expectations and a strong pressure to succeed as exerted by the international community. However, as a political compromise it is inevitably the case that the Basel Protocol—already at its inception—has failed to meet the expectations of the respective State groups, NGOs and industrial sectors. This applies all the more to the Basel Protocol as the transboundary movement of hazardous wastes involves a large spectrum of commercial and environmental interests and touches upon different political camps formed by the States that were represented in the negotiation process.[397] Since the positions of no political camp were entirely met by the final compromise regulation of the Basel Protocol, States lack a genuine interest in expediting the ratification of the Protocol. In such a situation it is rather likely that States first tend to be reserved and to await the further progress of ratification until they are in a position to assess what actual advantages can be expected from the entry into force of the Protocol. In general terms, it may be concluded, therefore, that a civil liability convention is unlikely to enter into force if there is a strong indication already during the negotiation phase that the most significant States do not support the basic provisions of the legal instrument.[398] It can also be stated that a civil liability convention is more likely to enter into force if it was adopted by consensus among the States concerned, rather than by vote.[399] When applying these findings to the Basel Protocol, it must be concluded that the compromise regulation of the Protocol failed to the sufficiently reflect the main positions of at least one of the political camps, with the result that an original interest in supporting the Protocol and promoting ratification is not to be found in any particular group of States.

In addition to this reason, the States' reluctance to promote ratification and expedite the entry into force of the Protocol can be explained by a related and similarly absent political incentive. Namely, the process of political decision making is strongly influenced by the public awareness regarding a certain issue as well as by convictions regarding the perceived urgency of the matter. It is only once the genuine need for a new international convention enjoys consensus at the political level that individual States will become likely to make the effort to ratify the convention.[400] As regards the Basel Protocol, the lack of public and political awareness concerning the necessity of this legal instrument has been identified as one of the major problems faced by States.[401] Furthermore, it has been noted that already during the negotiation process the Basel Secretariat faced problems in determining the factual basis of the prospective Protocol, meaning giving a description of the actual incidents to which the Protocol is to apply.[402] A rather pointed conclusion, hence, could be that a major environmental disaster is required

---

[397] Those interests involve most of all the commercial interests of the waste exporting and waste disposing industries, the interests of the insurance sector, the interests related to the protection of human health and the environment and interests related to the conservation of natural resources. Political camps are formed depending on the respective national waste treatment and management policy of States, which is for the most part independent of the States' classification into developed or developing countries.

[398] Churchill, 'Facilitating Civil Liability Litigation', 12 *Yb. Int'l Env. L.* (2001), at 32.

[399] Churchill, 'Facilitating Civil Liability Litigation', 12 *Yb. Int'l Env. L.* (2001), at 32.

[400] Churchill, 'Facilitating Civil Liability Litigation', 12 *Yb. Int'l Env. L.* (2001), at 32.

[401] COP7 Doc. UNEP/CHW.7/INF/11, at 5; COP7 Doc. UNEP/CHW.7/INF/11/Add.1, at 3; OEWG Doc. UNEP/CHW/OEWG/4/INF/4, at 2; COP8 Doc. UNEP/CHW.8/INF/16/Add.1, at 5.

[402] Daniel, 'Civil Liability Regimes as a Complement to MEAs', 12 *RECIEL* (2003), at 231.

to promote the entry into force of the Basel Protocol.[403] However, as even the *M/V "Probo Koala"* incident did not produce a large number of ratifications, one can ask if something more is at issue than simply the need for a disaster of sufficient magnitude. What undoubtedly remains, however, is that States still lack the awareness and the conviction that the risks associated with the transboundary movement of hazardous wastes can materialise at any time, such that they would actually benefit from the early entry into force of the Basel Protocol.

2. Shortcomings of the Protocol and Obstacles for Implementation

In addition to the political disincentives outlined in the previous section there are also a number of perceived or actual legal shortcomings in the Basel Protocol as well as anticipated obstacles for implementation which act as an impediment for ratification of the Basel Protocol.

The most relevant defects and legal shortcomings of the Basel Protocol have been outlined above,[404] so that at this point it can be summarised that there is only one substantial defect in the legal regime of the Basel Protocol which could from a legal perspective justify the reluctance of States to ratify the Protocol. This defect must be seen in the failure to establish a secondary (strict and limited) liability of the waste generator covering the period up until the disposer takes possession of the hazardous wastes. The present shape of the liability provisions and the absence of a secondary liability of waste generators create an incentive for waste generators to delegate the exportation and notification procedures to undercapitalised third- or front companies in order to effectively evade strict liability under the Protocol.[405] A further weakness of the Basel Protocol is to be seen in the non-establishment of an autonomous compensation fund.[406] However, this issue may be overcome by elaborating a new financial mechanism in accordance with Article 15 of the Protocol prior or subsequent to the entry into force of the Basel Protocol. Further objections raised against the legal regime of the Basel Protocol are of

---

[403] The Regional Workshop Aimed at Promoting Ratification of the Protocol held in Addis Ababa, Ethiopia in August/September 2004 under the aegis of the Basel Secretariat revealed that no participating country had experienced any incident which would have been dealt with under the Protocol, see COP7 Doc. UNEP/CHW.7/INF/11, at 3.

[404] See *supra*, Sect. "Summary of the Major Achievements and the Major Defects of the Basel Protocol".

[405] This issue was raised already during the negotiation phase of the Basel Protocol, see 'No Agreement on Draft Protocol', 29 *Envtl. Pol'y & L.* (1999), at 154; 'Compensation and Liability Protocol Adopted', 30 *Envtl. Pol'y & L.* (2000), at 44; Choksi, '1999 Protocol on Liability and Compensation', 28 *Ecology L. Q.* (2001), at 532.

[406] See also 'No Agreement on Draft Protocol', 29 *Envtl. Pol'y & L.* (1999), at 154; 'Compensation and Liability Protocol Adopted', 30 *Envtl. Pol'y & L.* (2000), at 44; Choksi, '1999 Protocol on Liability and Compensation', 28 *Ecology L. Q.* (2001), at 532; Tsimplis, 'The 1999 Protocol to the Basel Convention', 16 *IJMCL* (2001), at 329.

minor significance or are ultimately not accurate.[407] Regarding the legal content of the Basel Protocol in general, it can be summarised that the rules and provisions of the Protocol are drafted in a very complex and often complicated way. The Protocol attempts to prescribe a regulation for each individual case instead of relying on rather basic principles. As a result, this large volume of rules may act as an obstacle for the proper implementation into the domestic laws by the potential Contracting States and, moreover, has laid to the Protocol open to attack by critics who tend to find imbalances in the detailed provisions of the Protocol.

Apart from the perceived and actual legal shortcomings of the Basel Protocol, there are also anticipated obstacles for implementation which prevent States from ratifying the Protocol. This involves, first, the finding that private sector insurance companies tend to be reluctant to provide insurance coverage required under the Basel Protocol.[408] Further major obstacles for implementation are seen in the lack of financial resources, the lack of technical and legal expertise, and the lack of existing administrative infrastructures in order to sufficiently implement the rules and provisions of the Basel Protocol into domestic law.[409] This applies all the more to the Basel Protocol, considering the fact that the provisions of the Protocol are drafted in a very complex manner, including the mandate placed on Contracting States to establish domestic limits of liability in accordance with Annex B, Paragraph 1, of the Basel Protocol.[410]

3. Summary

There are several reasons that have been identified which discourage States from ratifying the Basel Protocol. However, most of these reasons may be characterised as a lack of any incentive to ratify, rather than an affirmative disincentive to do so.

---

[407] This concerns, for instance, the rather theoretical possibility of circumventing either the application of the Protocol or its level of protection by means of a bilateral or multilateral agreement that comes within the ambit of Article 3(7) or Article 11 of the Protocol, as well as the alleged failure to address the aftercare of hazardous wastes subsequent to their final disposal. See 'No Agreement on Draft Protocol', 29 *Envtl. Pol'y & L.* (1999), at 154; 'Compensation and Liability Protocol Adopted', 30 *Envtl. Pol'y & L.* (2000), at 44; Sharma, 'The Basel Protocol', 26 *Delhi L. Rev.* (2004), at 194–195.

[408] COP7 Doc. UNEP/CHW.7/INF/11, at 5; OEWG Doc. UNEP/CHW/OEWG/4/INF/4, at 2; COP7 Doc. UNEP/CHW/OEWG/5/INF/6, at 4; COP8 Doc. UNEP/CHW.8/INF/16, at 4; COP8 Doc. UNEP/CHW.8/INF/16/Add.1, at 5.

[409] COP7 Doc. UNEP/CHW.7/INF/11, at 4–5; OEWG Doc. UNEP/CHW/OEWG/4/INF/4, at 2; OEWG Doc. UNEP/CHW/OEWG/5/INF/6, at 4; COP8 Doc. UNEP/CHW.8/INF/16, at 4; COP8 Doc. UNEP/CHW.8/INF/16/Add.1, at 4; Daniel, 'Civil Liability Regimes as a Complement to MEAs', 12 *RECIEL* (2003), at 236.

[410] COP8 Doc. UNEP/CHW.8/INF/16/Add.1, at 5; Daniel, 'Civil Liability Regimes as a Complement to MEAs', 12 *RECIEL* (2003), at 231; Soares/Vargas, 'The Basel Liability Protocol', 12 *Yb. Int'l Env. L.* (2001), at 85.

This applies first of all to the lack of political incentives for ratification of the Basel Protocol. States must become aware and convinced of the fact that the compromise regulation represented by the Basel Protocol, even if it does not entirely match their respective expectations, constitutes a substantial and current benefit for them. It is only upon the States' realisation that the liability provisions of the Basel Protocol do not hamper the commercial activities of their waste management industry, it becomes likely that these States can begin to appreciate that the benefits of the Protocol are greater than any perceived drawbacks for their national economies.

From a legal perspective, the Basel Protocol is in fact better than the criticisms levelled by many authors.[411] It is indeed to be conceded that the legal provisions of the Protocol are unduly complex and unclear in many regard, which has laid the Protocol open to attack by criticism voiced in respect of minor imbalances in the details. However, the only major defect and actual loophole of the Basel Protocol is found in the absence of a secondary (strict and limited) liability of the waste generator covering the period until the disposer takes possession of the hazardous wastes and the resulting incentive to evade strict liability by delegating the exportation and notification procedures to third companies. The further legal shortcomings of the Protocol are of minor significance and may be overcome if there is a corresponding intention in the potential Contracting States to resolve them.

The same basically applies to the perceived obstacles of implementation. Within the framework of the Basel Convention there are some efficient mechanisms in place, such as the Technical Cooperation Trust Fund and the Regional Workshops organised by the Basel Secretariat, which could bring potential Contracting States into a position to ratify the Basel Protocol in addition to the Basel Convention. In the end, it turns out to be only a matter of political will and determination to take the efforts to ratify and implement the Basel Protocol as well as to engage in the negotiation of a new and appropriate permanent compensation fund.

III. Consequences of the Protocol's not Entering into Force

If it turns out that States fail to overcome the perceived obstacles for ratification and if they fail to develop a genuine political interest in ratifying the Basel Protocol, it is likely that this Protocol will not enter into force. In that case it is to be expected that in the foreseeable future the issue of liability and compensation for

---

[411] For example, Sharma, 'The Basel Protocol', 26 *Delhi L. Rev.* (2004), at 196, asserts that "the treaty offers very little that is positive and much that is highly negative." And furthermore: "What was adopted at Basel in 1999 [...] represents a successful attack on the Basel Convention's own fundamental principles and a dangerous international precedent". However, this statement represents a polemic exaggeration which cannot be traced back to a factual basis.

damage resulting from the transboundary movement of hazardous wastes will remain unregulated. A further consequence could be that the approach of establishing a global convention is abandoned and that focus is laid on the development and strengthening of regional approaches, such as in North America, in the EU, in East Asia, or in Africa. Such solution could provide sufficient result as regards regional movements of hazardous wastes, but it will prove inadequate with regard to transboundary movements that are conducted on a global scale. A failure of the Basel Protocol to enter into force would also mean that other treaty compliance mechanisms applicable in the framework of the Basel Convention, such as capacity building or environmental monitoring and effectiveness reviews, must be enforced through some other means and developed further.[412] However, the most significant consequence of the Protocol's failure to enter into force must be seen in the symbolic effect emanating from this failure of the community of nations. The Basel Protocol could emerge as an international precedent impeding and deferring the negotiation processes of other multilateral environmental agreements.[413]

---

[412] See Daniel, 'Civil Liability Regimes as a Complement to MEAs', 12 *RECIEL* (2003), at 237.

[413] See also Kummer, 'The Basel Convention: Ten Years On', 7 *RECIEL* (1998), at 232.

Chapter 6
Concluding Summary

The results of the legal research conducted in respect of the different aspects of the issue of liability and compensation for damage resulting from the transboundary movement of hazardous wastes by sea have been outlined in detail at the end of the respective chapters. At this point, therefore, only a compilation of the major results shall be given and a recommendation shall be made as to the further steps which should be taken.

1. After having outlined the economic background and the commercial and political interests involved in the transboundary movement of hazardous wastes in the Chap. 2, the following Chap. 3 was concerned with the examination of the rules and provisions of current international law applying *de lege lata* to certain aspects of liability related to hazardous waste movements. This analysis resulted in the conclusion that the rules and provisions of international law, which are currently valid and in force, cannot serve as a basis for a comprehensive and sufficient legal framework governing liability and compensation with regard to damage resulting in the context of hazardous waste movements. In general, there are two concepts of international responsibility and liability that need to be distinguished. These include, first, the international responsibility and liability of States and, second, the approach of imposing civil liabilities on private persons adopted by international uniform law. Regarding the international responsibility and liability of States, it has been shown that a customary principle of State liability for lawful but injurious acts is currently not acknowledged by international law, whereas the principle of State responsibility for internationally wrongful acts represents a valid principle of international law. The essential requirement for the imposition of State responsibility is the determination of a breach of an international obligation by the State. A relevant primary obligation may basically arise from any international or regional convention or agreement. Therefore, a comprehensive investigation into the existing international legal regimes has been conducted, which revealed that

only in a very limited number of constellations may a State be held internationally responsible for the causation of damage in the context of transboundary movements of hazardous wastes by sea. This finding is basically due to the fact that most waste shipments are initiated and conducted by private persons, for which a State cannot be held "directly" responsible. A State may be "indirectly" responsible for the conduct of private persons only if it infringes an international obligation of its own with regard to the conduct of private persons. This may be the case, for example, if the State fails to comply with the general obligation to implement the respective PIC, ESM or other procedural requirements of international conventions into its national laws, or if it fails to subsequently regulate, control and enforce compliance with these rules. In this context, relevant rules arise from the Basel Convention and its regional counterparts, as well as from the MARPOL 73/78 Convention, the London Dumping Convention and the Stockholm POPs Convention. Since, however, most of these general obligations are conceived as rules of due diligence, the imposition of State responsibility, furthermore, requires that the State knew or ought to know about the actual circumstances of the case and the illegal conduct of the private person resulting in damage. Moreover, the particular damage for which compensation is sought must be considered as having been caused by the internationally wrongful act of the State. A further major disadvantage of resorting to the principle of State responsibility is that compensation based on this principle may only be sought by States, which are, however, reluctant to bring environmental claims before the international courts due to diplomatic and political reasons. Consequently, the principle of State responsibility cannot be seen as a suitable approach to ensure the compensation of victims of pollution resulting from the incidents involving hazardous wastes. A similar outcome applies to the analysis of the existing and nascent international conventions establishing a uniform regime of civil liability. In this context, above all the 1996/2012 HNS Convention must be mentioned, which is, however, not yet in force and only covers certain aspects of hazardous waste movements, failing to apply to any land leg of transport and the final stage of disposal. As a general result, it must be summarised, therefore, that the present state of international law as regards liability and compensation in the context of hazardous waste movements is fragmentary and unsatisfactory.

2. Furthermore, it has been shown that it is necessary, in light of the current inadequacy of international law, to establish an international civil liability convention dealing in particular with liability and compensation for damage resulting from the transboundary movement of hazardous wastes. In the Chap. 4, it was first established that non-financially oriented treaty compliance mechanisms are inadequate to ensure sufficient protection of pollution victims. This finding applies particularly to the mechanisms already existing with regard to the Basel Convention, which, however, fail to provide for financial compensation and, furthermore, are addressed at States Parties only. The Chap. 4, furthermore, outlines the economic relevance and implications associated with a legal regime of liability. The particular advantage of an international uniform

law imposing civil liabilities is that it ensures direct compensation of victims of pollution and, therefore, creates a direct economic incentive for the acting persons to invest in damage precaution in order to avoid incurring liability. Following this consideration, this work also determines some basic parameters and general requirements that must be taken into account by the prospective regime of civil liability. This includes, for example, that strong reasons exist to establish a global regime instead of pursuing regional or national approaches.
3. The Chap. 5 then sets the focus of the legal research on the prospective regime of the Basel Protocol on Liability and Compensation. Although this Protocol was adopted already in 1999, it has yet not entered into force and, furthermore, has not received ratification from any of those States which must be considered the most important players in the international trade in hazardous wastes. Since the provisions of the Basel Protocol have not been applied in practice, their legal assessment can only be provisional. However, in conclusion it can be stated that the Basel Protocol generally provides for a suitable and appropriate solution reconciling the involved commercial and political interests and achieving a high standard of protection for the victims of pollution. In some respects, however, the provisions and arrangements of the Protocol appear to be rather unclear and confusing. However, notwithstanding the Basel Protocol's occasionally less than ideally formulated terms, only a handful of areas warrant criticism as genuine weaknesses or substantive shortcomings. In detail: The scope of application of the Basel Protocol is conceived in an extremely detailed and unnecessarily complex manner. The Protocol intends to regulate each particular instance of application, an endeavour which, due to the enormous number of constellations involving Contractual and non-Contractual States and the high seas, is almost impossible and eventually turns out to be a source of uncertainties. It must be concluded that legal clarity would have been furthered and less reasons for dispute would have arisen if the scope of application had been conceived more in a sense of a general clause. A further unduly complicated regulation of the Basel Protocol concerns its relationship to other civil liability regimes. The Protocol distinguishes between its relationship to other liability agreements regarding, particularly, the transboundary movement of hazardous wastes as opposed to its relationship to general agreements on liability and compensation. In this respect it would have been reasonable to combine both of the pertinent rules into a single provision which stipulates that another civil liability agreement may prevail over the Basel Protocol only if it contains a regime of liability that fully meets or exceeds the protection level of the Basel Protocol.

A major achievement of the Protocol, by contrast, is that it combines limited strict liability and unlimited fault-based liability and provides for a right of recourse, so that a two-tier system of liabilities is established. Thus, on a first tier of liability, victims of pollution are in a position to obtain prompt but limited compensation from the person strictly liable, whereas they are required to prove fault on the part of the liable person to move beyond these limits.

On the second tier the claimant or the person strictly liable is given the right to claim partial or full recovery from any other person liable under the Basel Protocol either on the basis of strict liability or on grounds of fault. By means of this second tier of liability a final allocation of liabilities according to aspects like the respective magnitude of fault or the actual contribution to the causation of damage is ensured. Another well-balanced solution of the Protocol is the approach of channelling strict liability to only one person, but to deviate from the common structure of channelling strict liability by always designating another person depending on the respective stage of the movement at which the incident occurred. However, one of the main defects of the Basel Protocol is also related to the allocation of strict liability. This defect is to be seen in the absence of a secondary (strict and limited) liability of the waste generator covering the period until the disposer takes possession of the hazardous wastes. The present shape of the Protocol's liability provisions create a strong incentive in waste generators to delegate the exportation and notification procedures to third or front companies in order to effectively evade any strict liability under the Protocol. Since it may be suspected that accordingly deployed third or front companies lack sufficient funds for compensation or fail to maintain sufficient insurance coverage, the absence of a secondary liability of waste generators creates a major loophole in the Basel Protocol.

The concept of limitation of strict liability under the Protocol follows a rather innovative approach. The Basel Protocol itself only establishes minimum limits of liability and mandates the Contracting States to impose domestic limits of liability which may not be below the limits of the Protocol. By means of determining either very high or rather low domestic limits of liability, the Contracting States are given an economic lever to influence the commercial decisions of the persons engaged in hazardous waste movements. Depending on the respective domestic limits of liability established by the Contracting states, private parties initiating and conducting waste movements have an incentive to either avoid or to intensify the shipment of hazardous wastes to or through areas under the national jurisdiction of the respective Contracting States.

Finally, a further weakness of the Basel Protocol is to be seen in the current non-establishment of an autonomous compensation fund in the legal framework of the Basel Protocol. However, it has been outlined that the establishment of a compensation fund, which entails further potential for conflict, directly within the Basel Protocol seems not to be the most favourable approach. States should rather use the possibility offered by Article 15 of the Basel Protocol to elaborate and set up a permanent compensation fund as a legally free-standing instrument—which is coupled with the liability provisions of the Basel Protocol—prior to or subsequent to the entry into force of the Protocol. Hence, the failure of States to reach agreement upon the establishment of a permanent and autonomous compensation fund during the negotiations of the Basel Protocol will turn out to be a major weakness of the Protocol only if the States are not be able to agree upon the establishment of a compensation fund or at least upon a fixed schedule for the negotiation and adoption of a compensation fund upon or shortly after the entry into force of the Basel Protocol. Prior to the entry into force of the Basel Protocol, the interim solution found by the Parties to the Basel Convention at COP5 to enlarge the scope of the existing Technical Cooperation Trust Fund may be considered sufficient to achieve its current purposes.

4. This work, finally, attempted to identify those reasons which prevent States from ratifying the Basel Protocol. It has been concluded that most of these reasons can be understood as an absence of incentive to ratify the Protocol, rather than specific disincentives to ratification. The Protocol represents a political compromise between the opposed commercial and political interests involved in hazardous waste movements. In consequence of this compromise regulation no political camp has a genuine interest in promoting ratification and in expediting the entry into force of the Basel Protocol. Moreover, it must be stated that there is still a lack of public and political awareness concerning the necessity of liability provisions governing the transboundary movement of hazardous wastes. It is a sad fact that even the major environmental disaster of the *M/V "Probo Koala"* case in Ivory Coast in 2006 did not cause a change in public and political awareness and, thus, failed to prompt States to ratify the Basel Protocol. From a legal perspective, the Protocol offers in general a well-balanced and reasonable system of allocating civil liability among the persons engaged in hazardous waste movements. A substantial obstacle for ratification must only be seen in the absence of a secondary liability of the waste generator. In addition, the very complex and often complicated provisions of the Protocol in combination with the lack of financial resources and technical and legal expertise required for the proper implementation of the Protocol into the respective domestic laws act as a further obstacle for ratification, which, however, could be overcome by means of utilising and strengthening the existing Technical Cooperation Trust Fund. In conclusion, it must be summarised that most of the reasons preventing States from ratifying the Basel Protocol are politically based. Only the absence of a secondary liability of the generator of hazardous wastes constitutes a legal issue that needs to be resolved by a revision of the Basel Protocol. Apart from that, the perceived obstacles for ratification may be overcome by a corresponding political will to do so in the States concerned.

The case of the *M/V "Probo Koala"* has shown that States and private persons are in a position to find solutions for obtaining compensation for victims of pollution by means of bringing claims before the national courts based on the domestic law. Such *ad hoc* procedures might lead to reasonable results with regard to major environmental disasters which are reported in the international press. However, as regards minor cases of pollution lacking the weight of public pressure, such procedures will prove to be inadequate. In those cases, which represent by far the largest number of incidents, the claimants are usually not in a position to institute legal proceedings in a foreign jurisdiction. The need for sufficient protection of the victims of pollution, the protection of the environment and the conservation of natural resources, protection from corruption and the protection against powerful commercial interests thus require the establishment of an international uniform regime of civil liability governing the transboundary movement of hazardous wastes. This work has shown that in consequence of the Basel Protocol not entering into force, the approach of establishing a global regime will most likely be abandoned and that regional approaches may be pursued in the future instead. It has also been shown that the Basel Protocol represents a basically appropriate, suitable and effective global regime of liability, notwithstanding its need to be revised in one respect or another. In conclusion, it must be stated that the advantages of the Basel Protocol still considerably outweigh its disadvantages. A defect in the Basel Protocol that would justify abandoning the approach of establishing a global regime of liability and compensation in the context of hazardous waste movements does not exist. Thus, it is up to the States and it is up to the public to develop an awareness of the actual need for a liability regime in the context of hazardous waste movements. It should not be the case that only a further major environmental disaster in an even more visible area prompts the States to finally support the Basel Protocol. States simply need to overcome their political convenience. Otherwise, it must actually be concluded that *Klaus Töpfer* was wrong and the Basel Protocol does not represent "a major breakthrough", but only "dead letters in the sea".[1]

---

[1] This term was used with regard to the liability provisions of regional sea conventions by Lefeber, *The Liability Provisions of Regional Sea Conventions*, in: Vidas/Østreng (ed.) (1999), at 507 *et seq.*

Appendix I
Text of the Basel Convention (Excerpts)

Basel Convention on the Control of Transboundary Movements of Hazardous Wastes and Their Disposal

ARTICLE 1
Scope of the Convention

1. The following wastes that are subject to transboundary movement shall be "hazardous wastes" for the purposes of this Convention:

    (a) Wastes that belong to any category contained in Annex I, unless they do not possess any of the characteristics contained in Annex III; and

    (b) Wastes that are not covered under paragraph (a) but are defined as, or are considered to be, hazardous wastes by the domestic legislation of the Party of export, import or transit.

2. Wastes that belong to any category contained in Annex II that are subject to transboundary movement shall be "other wastes" for the purposes of this Convention.
3. Wastes which, as a result of being radioactive, are subject to other international control systems, including international instruments, applying specifically to radioactive materials, are excluded from the scope of this Convention. [...]

ARTICLE 2
Definitions
For the purposes of this Convention:

1. "Wastes" are substances or objects which are disposed of or are intended to be disposed of or are required to be disposed of by the provisions of national law; [...]
3. "Transboundary movement" means any movement of hazardous wastes or other wastes from an area under the national jurisdiction of one State to or through an area under the national jurisdiction of another State or to or through an area not under the national jurisdiction of any State, provided at least two States are involved in the movement;

4. "Disposal" means any operation specified in Annex IV to this Convention; [...]
8. "Environmentally sound management of hazardous wastes or other wastes" means taking all practicable steps to ensure that hazardous wastes or other wastes are managed in a manner which will protect human health and the environment against the adverse effects which may result from such wastes;
9. "Area under the national jurisdiction of a State" means any land, marine area or airspace within which a State exercises administrative and regulatory responsibility in accordance with international law in regard to the protection of human health or the environment;
10. "State of export" means a Party from which a transboundary movement of hazardous wastes or other wastes is planned to be initiated or is initiated;
11. "State of import" means a Party to which a transboundary movement of hazardous wastes or other wastes is planned or takes place for the purpose of disposal therein or for the purpose of loading prior to disposal in an area not under the national jurisdiction of any State;
12. "State of transit" means any State, other than the State of export or import, through which a movement of hazardous wastes or other wastes is planned or takes place;
13. "States concerned" means Parties which are States of export or import, or transit States, whether or not Parties;
14. "Person" means any natural or legal person;
15. "Exporter" means any person under the jurisdiction of the State of export who arranges for hazardous wastes or other wastes to be exported;
16. "Importer" means any person under the jurisdiction of the State of import who arranges for hazardous wastes or other wastes to be imported;
17. "Carrier" means any person who carries out the transport of hazardous wastes or other wastes;
18. "Generator" means any person whose activity produces hazardous wastes or other wastes or, if that person is not known, the person who is in possession and/or control of those wastes;
19. "Disposer" means any person to whom hazardous wastes or other wastes are shipped and who carries out the disposal of such wastes; [...]
21. "Illegal traffic" means any transboundary movement of hazardous wastes or other wastes as specified in Article 9. [...]

ARTICLE 4
General Obligations

1.
   (a) Parties exercising their right to prohibit the import of hazardous wastes or other wastes for disposal shall inform the other Parties of their decision pursuant to Article 13.

(b) Parties shall prohibit or shall not permit the export of hazardous wastes and other wastes to the Parties which have prohibited the import of such wastes, when notified pursuant to subparagraph (a) above.

(c) Parties shall prohibit or shall not permit the export of hazardous wastes and other wastes if the State of import does not consent in writing to the specific import, in the case where that State of import has not prohibited the import of such wastes.

2. Each Party shall take the appropriate measures to:

(a) Ensure that the generation of hazardous wastes and other wastes within it is reduced to a minimum, taking into account social, technological and economic aspects;

(b) Ensure the availability of adequate disposal facilities, for the environmentally sound management of hazardous wastes and other wastes, that shall be located, to the extent possible, within it, whatever the place of their disposal;

(c) Ensure that persons involved in the management of hazardous wastes or other wastes within it take such steps as are necessary to prevent pollution due to hazardous wastes and other wastes arising from such management and, if such pollution occurs, to minimize the consequences thereof for human health and the environment;

(d) Ensure that the transboundary movement of hazardous wastes and other wastes is reduced to the minimum consistent with the environmentally sound and efficient management of such wastes, and is conducted in a manner which will protect human health and the environment against the adverse effects which may result from such movement;

(e) Not allow the export of hazardous wastes or other wastes to a State or group of States belonging to an economic and/or political integration organization that are Parties, particularly developing countries, which have prohibited by their legislation all imports, or if it has reason to believe that the wastes in question will not be managed in an environmentally sound manner, according to criteria to be decided on by the Parties at their first meeting;

(f) Require that information about a proposed transboundary movement of hazardous wastes and other wastes be provided to the States concerned, according to Annex V A, to state clearly the effects of the proposed movement on human health and the environment;

(g) Prevent the import of hazardous wastes and other wastes if it has reason to believe that the wastes in question will not be managed in an environmentally sound manner;

(h) Co-operate in activities with other Parties and interested organizations, directly and through the Secretariat, including the dissemination of information on the transboundary movement of hazardous wastes and other wastes, in order to improve the environmentally sound management of such wastes and to achieve the prevention of illegal traffic.

3. The Parties consider that illegal traffic in hazardous wastes or other wastes is criminal.
4. Each Party shall take appropriate legal, administrative and other measures to implement and enforce the provisions of this Convention, including measures to prevent and punish conduct in contravention of the Convention.
5. A Party shall not permit hazardous wastes or other wastes to be exported to a non Party or to be imported from a non-Party.
6. The Parties agree not to allow the export of hazardous wastes or other wastes for disposal within the area south of 60° South latitude, whether or not such wastes are subject to transboundary movement.
7. Furthermore, each Party shall:
    (a) Prohibit all persons under its national jurisdiction from transporting or disposing of hazardous wastes or other wastes unless such persons are authorized or allowed to perform such types of operations;
    (b) Require that hazardous wastes and other wastes that are to be the subject of a transboundary movement be packaged, labelled, and transported in conformity with generally accepted and recognized international rules and standards in the field of packaging, labelling, and transport, and that due account is taken of relevant internationally recognized practices;
    (c) Require that hazardous wastes and other wastes be accompanied by a movement document from the point at which a transboundary movement commences to the point of disposal.
8. Each Party shall require that hazardous wastes or other wastes, to be exported, are managed in an environmentally sound manner in the State of import or elsewhere. Technical guidelines for the environmentally sound management of wastes subject to this Convention shall be decided by the Parties at their first meeting.
9. Parties shall take the appropriate measures to ensure that the transboundary movement of hazardous wastes and other wastes only be allowed if:
    (a) The State of export does not have the technical capacity and the necessary facilities, capacity or suitable disposal sites in order to dispose of the wastes in question in an environmentally sound and efficient manner; or
    (b) The wastes in question are required as a raw material for recycling or recovery industries in the State of import; or
    (c) The transboundary movement in question is in accordance with other criteria to be decided by the Parties, provided those criteria do not differ from the objectives of this Convention.
10. The obligation under this Convention of States in which hazardous wastes and other wastes are generated to require that those wastes are managed in an environmentally sound manner may not under any circumstances be transferred to the States of import or transit.

Appendix I: Text of the Basel Convention... 315

11. Nothing in this Convention shall prevent a Party from imposing additional requirements that are consistent with the provisions of this Convention, and are in accordance with the rules of international law, in order better to protect human health and the environment.
12. Nothing in this Convention shall affect in any way the sovereignty of States over their territorial sea established in accordance with international law, and the sovereign rights and the jurisdiction which States have in their exclusive economic zones and their continental shelves in accordance with international law, and the exercise by ships and aircraft of all States of navigational rights and freedoms as provided for in international law and as reflected in relevant international instruments.
13. Parties shall undertake to review periodically the possibilities for the reduction of the amount and/or the pollution potential of hazardous wastes and other wastes which are exported to other States, in particular to developing countries. [...]

ARTICLE 6
Transboundary Movement between Parties

1. The State of export shall notify, or shall require the generator or exporter to notify, in writing, through the channel of the competent authority of the State of export, the competent authority of the States concerned of any proposed transboundary movement of hazardous wastes or other wastes. Such notification shall contain the declarations and information specified in Annex V A, written in a language acceptable to the State of import. Only one notification needs to be sent to each State concerned.
2. The State of import shall respond to the notifier in writing, consenting to the movement with or without conditions, denying permission for the movement, or requesting additional information. A copy of the final response of the State of import shall be sent to the competent authorities of the States concerned which are Parties.
3. The State of export shall not allow the generator or exporter to commence the transboundary movement until it has received written confirmation that:

    (a) The notifier has received the written consent of the State of import; and
    (b) The notifier has received from the State of import confirmation of the existence of a contract between the exporter and the disposer specifying environmentally sound management of the wastes in question.

4. Each State of transit which is a Party shall promptly acknowledge to the notifier receipt of the notification. It may subsequently respond to the notifier in writing, within 60 days, consenting to the movement with or without conditions, denying permission for the movement, or requesting additional information. The State of export shall not allow the transboundary movement to commence until it has received the written consent of the State of transit. [...]

5. In the case of a transboundary movement of wastes where the wastes are legally defined as or considered to be hazardous wastes only:

   (a) By the State of export, the requirements of paragraph 9 of this Article that apply to the importer or disposer and the State of import shall apply mutatis mutandis to the exporter and State of export, respectively;

   (b) By the State of import, or by the States of import and transit which are Parties, the requirements of paragraphs 1, 3, 4 and 6 of this Article that apply to the exporter and State of export shall apply mutatis mutandis to the importer or disposer and State of import, respectively; or

   (c) By any State of transit which is a Party, the provisions of paragraph 4 shall apply to such State. [...]

9. The Parties shall require that each person who takes charge of a transboundary movement of hazardous wastes or other wastes sign the movement document either upon delivery or receipt of the wastes in question. They shall also require that the disposer inform both the exporter and the competent authority of the State of export of receipt by the disposer of the wastes in question and, in due course, of the completion of disposal as specified in the notification. If no such information is received within the State of export, the competent authority of the State of export or the exporter shall so notify the State of import. [...]

11. Any transboundary movement of hazardous wastes or other wastes shall be covered by insurance, bond or other guarantee as may be required by the State of import or any State of transit which is a Party.

ARTICLE 7
Transboundary Movement from a Party through States which are not Parties

Paragraph 1 of Article 6 of the Convention shall apply mutatis mutandis to transboundary movement of hazardous wastes or other wastes from a Party through a State or States which are not Parties.

ARTICLE 8
Duty to Re-import

When a transboundary movement of hazardous wastes or other wastes to which the consent of the States concerned has been given, subject to the provisions of this Convention, cannot be completed in accordance with the terms of the contract, the State of export shall ensure that the wastes in question are taken back into the State of export, by the exporter, if alternative arrangements cannot be made for their disposal in an environmentally sound manner, within 90 days from the time that the importing State informed the State of export and the Secretariat, or such other period of time as the States concerned agree. To this end, the State of export and any Party of transit shall not oppose, hinder or prevent the return of those wastes to the State of export.

ARTICLE 9
Illegal Traffic

1. For the purpose of this Convention, any transboundary movement of hazardous wastes or other wastes:

    (a) without notification pursuant to the provisions of this Convention to all States concerned; or
    (b) without the consent pursuant to the provisions of this Convention of a State concerned; or
    (c) with consent obtained from States concerned through falsification, misrepresentation or fraud; or
    (d) that does not conform in a material way with the documents; or
    (e) that results in deliberate disposal (e.g. dumping) of hazardous wastes or other wastes in contravention of this Convention and of general principles of international law, shall be deemed to be illegal traffic.

2. In case of a transboundary movement of hazardous wastes or other wastes deemed to be illegal traffic as the result of conduct on the part of the exporter or generator, the State of export shall ensure that the wastes in question are:

    (a) taken back by the exporter or the generator or, if necessary, by itself into the State of export, or, if impracticable,
    (b) are otherwise disposed of in accordance with the provisions of this Convention,

    within 30 days from the time the State of export has been informed about the illegal traffic or such other period of time as States concerned may agree. To this end the Parties concerned shall not oppose, hinder or prevent the return of those wastes to the State of export.

3. In the case of a transboundary movement of hazardous wastes or other wastes deemed to be illegal traffic as the result of conduct on the part of the importer or disposer, the State of import shall ensure that the wastes in question are disposed of in an environmentally sound manner by the importer or disposer or, if necessary, by itself within 30 days from the time the illegal traffic has come to the attention of the State of import or such other period of time as the States concerned may agree. To this end, the Parties concerned shall co-operate, as necessary, in the disposal of the wastes in an environmentally sound manner.

4. In cases where the responsibility for the illegal traffic cannot be assigned either to the exporter or generator or to the importer or disposer, the Parties concerned or other Parties, as appropriate, shall ensure, through co-operation, that the wastes in question are disposed of as soon as possible in an environmentally sound manner either in the State of export or the State of import or elsewhere as appropriate.

5. Each Party shall introduce appropriate national/domestic legislation to prevent and punish illegal traffic. The Parties shall cooperate with a view to achieving the objects of this Article. [...]

ARTICLE 11
Bilateral, Multilateral and Regional Agreements

1. Notwithstanding the provisions of Article 4 paragraph 5, Parties may enter into bilateral, multilateral, or regional agreements or arrangements regarding transboundary movement of hazardous wastes or other wastes with Parties or non-Parties provided that such agreements or arrangements do not derogate from the environmentally sound management of hazardous wastes and other wastes as required by this Convention. These agreements or arrangements shall stipulate provisions which are not less environmentally sound than those provided for by this Convention in particular taking into account the interests of developing countries.
2. Parties shall notify the Secretariat of any bilateral, multilateral or regional agreements or arrangements referred to in paragraph 1 and those which they have entered into prior to the entry into force of this Convention for them, for the purpose of controlling transboundary movements of hazardous wastes and other wastes which take place entirely among the Parties to such agreements. The provisions of this Convention shall not affect transboundary movements which take place pursuant to such agreements provided that such agreements are compatible with the environmentally sound management of hazardous wastes and other wastes as required by this Convention.

ARTICLE 12
Consultations on Liability
The Parties shall co-operate with a view to adopting, as soon as practicable, a protocol setting out appropriate rules and procedures in the field of liability and compensation for damage resulting from the transboundary movement and disposal of hazardous wastes and other wastes. [...]

Appendix II
Text of the Basel Protocol

Protocol on Liability and Compensation for Damage Resulting from Transboundary Movements of Hazardous Wastes and Their Disposal

The Parties to the Protocol,

Having taken into account the relevant provisions of Principle 13 of the 1992 Rio Declaration on Environment and Development, according to which States shall develop international and national legal instruments regarding liability and compensation for the victims of pollution and other environmental damage,

Being Parties to the Basel Convention on the Control of Transboundary Movements of Hazardous Wastes and their Disposal,

Mindful of their obligations under the Convention,

Aware of the risk of damage to human health, property and the environment caused by hazardous wastes and other wastes and the transboundary movement and disposal thereof,

Concerned about the problem of illegal transboundary traffic in hazardous wastes and other wastes,

Committed to Article 12 of the Convention, and emphasizing the need to set out appropriate rules and procedures in the field of liability and compensation for damage resulting from the transboundary movement and disposal of hazardous wastes and other wastes,

Convinced of the need to provide for third party liability and environmental liability in order to ensure that adequate and prompt compensation is available for damage resulting from the transboundary movement and disposal of hazardous wastes and other wastes,

Have agreed as follows:

ARTICLE 1
Objective
The objective of the Protocol is to provide for a comprehensive regime for liability and for adequate and prompt compensation for damage resulting from the transboundary movement of hazardous wastes and other wastes and their disposal including illegal traffic in those wastes.

ARTICLE 2
Definitions

1. The definitions of terms contained in the Convention apply to the Protocol, unless expressly provided otherwise in the Protocol.
2. For the purposes of the Protocol:
    (a) "The Convention" means the Basel Convention on the Control of Transboundary Movements of Hazardous Wastes and their Disposal;
    (b) "Hazardous wastes and other wastes" means hazardous wastes and other wastes within the meaning of Article 1 of the Convention;
    (c) "Damage" means:
        (i) Loss of life or personal injury;
        (ii) Loss of or damage to property other than property held by the person liable in accordance with the present Protocol;
        (iii) Loss of income directly deriving from an economic interest in any use of the environment, incurred as a result of impairment of the environment, taking into account savings and costs;
        (iv) The costs of measures of reinstatement of the impaired environment, limited to the costs of measures actually taken or to be undertaken; and
        (v) The costs of preventive measures, including any loss or damage caused by such measures, to the extent that the damage arises out of or results from hazardous properties of the wastes involved in the transboundary movement and disposal of hazardous wastes and other wastes subject to the Convention;
    (d) "Measures of reinstatement" means any reasonable measures aiming to assess, reinstate or restore damaged or destroyed components of the environment. Domestic law may indicate who will be entitled to take such measures;
    (e) "Preventive measures" means any reasonable measures taken by any person in response to an incident, to prevent, minimize, or mitigate loss or damage, or to effect environmental cleanup;
    (f) "Contracting Party" means a Party to the Protocol;
    (g) "Protocol" means the present Protocol;
    (h) "Incident" means any occurrence, or series of occurrences having the same origin that causes damage or creates a grave and imminent threat of causing damage;
    (i) "Regional economic integration organization" means an organization constituted by sovereign States to which its member States have transferred competence in respect of matters governed by the Protocol and which has been duly authorized, in accordance with its internal procedures, to sign, ratify, accept, approve, formally confirm or accede to it;
    (j) "Unit of account" means the Special Drawing Right as defined by the International Monetary Fund.

ARTICLE 3
Scope of application

1. The Protocol shall apply to damage due to an incident occurring during a transboundary movement of hazardous wastes and other wastes and their disposal, including illegal traffic, from the point where the wastes are loaded on the means of transport in an area under the national jurisdiction of a State of export. Any Contracting Party may by way of notification to the Depositary exclude the application of the Protocol, in respect of all transboundary movements for which it is the State of export, for such incidents which occur in an area under its national jurisdiction, as regards damage in its area of national jurisdiction. The Secretariat shall inform all Contracting Parties of notifications received in accordance with this Article.
2. The Protocol shall apply:
   (a) In relation to movements destined for one of the operations specified in Annex IV to the Convention other than D13, D14, D15, R12 or R13, until the time at which the notification of completion of disposal pursuant to Article 6, paragraph 9, of the Convention has occurred, or, where such notification has not been made, completion of disposal has occurred; and
   (b) In relation to movements destined for the operations specified in D13, D14, D15, R12 or R13 of Annex IV to the Convention, until completion of the subsequent disposal operation specified in D1 to D12 and R1 to R11 of Annex IV to the Convention.
3.
   (a) The Protocol shall apply only to damage suffered in an area under the national jurisdiction of a Contracting Party arising from an incident as referred to in paragraph 1;
   (b) When the State of import, but not the State of export, is a Contracting Party, the Protocol shall apply only with respect to damage arising from an incident as referred to in paragraph 1 which takes place after the moment at which the disposer has taken possession of the hazardous wastes and other wastes. When the State of export, but not the State of import, is a Contracting Party, the Protocol shall apply only with respect to damage arising from an incident as referred to in paragraph 1 which takes place prior to the moment at which the disposer takes possession of the hazardous wastes and other wastes. When neither the State of export nor the State of import is a Contracting Party, the Protocol shall not apply;
   (c) Notwithstanding subparagraph (a), the Protocol shall also apply to the damages specified in Article 2, subparagraphs 2 (c) (i), (ii) and (v), of the Protocol occurring in areas beyond any national jurisdiction;
   (d) Notwithstanding subparagraph (a), the Protocol shall, in relation to rights under the Protocol, also apply to damages suffered in an area under the national jurisdiction of a State of transit which is not a Contracting Party provided that such State appears in Annex A and has acceded to a

multilateral or regional agreement concerning transboundary movements of hazardous waste which is in force. Subparagraph (b) will apply mutatis mutandis.

4. Notwithstanding paragraph 1, in case of re-importation under Article 8 or Article 9, subparagraph 2 (a), and Article 9, paragraph 4, of the Convention, the provisions of the Protocol shall apply until the hazardous wastes and other wastes reach the original State of export.
5. Nothing in the Protocol shall affect in any way the sovereignty of States over their territorial seas and their jurisdiction and the right in their respective exclusive economic zones and continental shelves in accordance with international law.
6. Notwithstanding paragraph 1 and subject to paragraph 2 of this Article:

   (a) The Protocol shall not apply to damage that has arisen from a transboundary movement of hazardous wastes and other wastes that has commenced before the entry into force of the Protocol for the Contracting Party concerned;

   (b) The Protocol shall apply to damage resulting from an incident occurring during a transboundary movement of wastes falling under Article 1, subparagraph 1 (b), of the Convention only if those wastes have been notified in accordance with Article 3 of the Convention by the State of export or import, or both, and the damage arises in an area under the national jurisdiction of a State, including a State of transit, that has defined or considers those wastes as hazardous provided that the requirements of Article 3 of the Convention have been met. In this case strict liability shall be channelled in accordance with Article 4 of the Protocol.

7.

   (a) The Protocol shall not apply to damage due to an incident occurring during a transboundary movement of hazardous wastes and other wastes and their disposal pursuant to a bilateral, multilateral or regional agreement or arrangement concluded and notified in accordance with Article 11 of the Convention if:

   (i) The damage occurred in an area under the national jurisdiction of any of the Parties to the agreement or arrangement;

   (ii) There exists a liability and compensation regime, which is in force and is applicable to the damage resulting from such a transboundary movement or disposal provided it fully meets, or exceeds the objective of the Protocol by providing a high level of protection to persons who have suffered damage;

   (iii) The Party to the Article 11 agreement or arrangement in which the damage has occurred has previously notified the Depositary of the nonapplication of the Protocol to any damage occurring in an area under its national jurisdiction due to an incident resulting from movements or disposals referred to in this subparagraph; and

(iv) The Parties to the Article 11 agreement or arrangement have not declared that the Protocol shall be applicable;

(b) In order to promote transparency, a Contracting Party that has notified the Depositary of the non-application of the Protocol shall notify the Secretariat of the applicable liability and compensation regime referred to in subparagraph (a) (ii) and include a description of the regime. The Secretariat shall submit to the Meeting of the Parties, on a regular basis, summary reports on the notifications received;

(c) After a notification pursuant to subparagraph (a) (iii) is made, actions for compensation for damage to which subparagraph (a) (i) applies may not be made under the Protocol.

8. The exclusion set out in paragraph 7 of this Article shall neither affect any of the rights or obligations under the Protocol of a Contracting Party which is not party to the agreement or arrangement mentioned above, nor shall it affect rights of States of transit which are not Contracting Parties.

9. Article 3, paragraph 2, shall not affect the application of Article 16 to all Contracting Parties.

ARTICLE 4
Strict liability

1. The person who notifies in accordance with Article 6 of the Convention, shall be liable for damage until the disposer has taken possession of the hazardous wastes and other wastes. Thereafter the disposer shall be liable for damage. If the State of export is the notifier or if no notification has taken place, the exporter shall be liable for damage until the disposer has taken possession of the hazardous wastes and other wastes. With respect to Article 3, subparagraph 6 (b), of the Protocol, Article 6, paragraph 5, of the Convention shall apply mutatis mutandis. Thereafter the disposer shall be liable for damage.

2. Without prejudice to paragraph 1, with respect to wastes under Article 1, subparagraph 1 (b), of the Convention that have been notified as hazardous by the State of import in accordance with Article 3 of the Convention but not by the State of export, the importer shall be liable until the disposer has taken possession of the wastes, if the State of import is the notifier or if no notification has taken place. Thereafter the disposer shall be liable for damage.

3. Should the hazardous wastes and other wastes be re-imported in accordance with Article 8 of the Convention, the person who notified shall be liable for damage from the time the hazardous wastes leave the disposal site, until the wastes are taken into possession by the exporter, if applicable, or by the alternate disposer.

4. Should the hazardous wastes and other wastes be re-imported under Article 9, subparagraph 2 (a), or Article 9, paragraph 4, of the Convention, subject to Article 3 of the Protocol, the person who reimports shall be held liable for

damage until the wastes are taken into possession by the exporter if applicable, or by the alternate disposer.

5. No liability in accordance with this Article shall attach to the person referred to in paragraphs 1 and 2 of this Article, if that person proves that the damage was:

    (a) The result of an act of armed conflict, hostilities, civil war or insurrection;
    (b) The result of a natural phenomenon of exceptional, inevitable, unforeseeable and irresistible character;
    (c) Wholly the result of compliance with a compulsory measure of a public authority of the State where the damage occurred; or
    (d) Wholly the result of the wrongful intentional conduct of a third party, including the person who suffered the damage.

6. If two or more persons are liable according to this Article, the claimant shall have the right to seek full compensation for the damage from any or all of the persons liable.

ARTICLE 5
Fault-based liability

Without prejudice to Article 4, any person shall be liable for damage caused or contributed to by his lack of compliance with the provisions implementing the Convention or by his wrongful intentional, reckless or negligent acts or omissions. This Article shall not affect the domestic law of the Contracting Parties governing liability of servants and agents.

ARTICLE 6
Preventive measures

1. Subject to any requirement of domestic law any person in operational control of hazardous wastes and other wastes at the time of an incident shall take all reasonable measures to mitigate damage arising therefrom.
2. Notwithstanding any other provision in the Protocol, any person in possession and/or control of hazardous wastes and other wastes for the sole purpose of taking preventive measures, provided that this person acted reasonably and in accordance with any domestic law regarding preventive measures, is not thereby subject to liability under the Protocol.

ARTICLE 7
Combined cause of the damage

1. Where damage is caused by wastes covered by the Protocol and wastes not covered by the Protocol, a person otherwise liable shall only be liable according to the Protocol in proportion to the contribution made by the wastes covered by the Protocol to the damage.
2. The proportion of the contribution to the damage of the wastes referred to in paragraph 1 shall be determined with regard to the volume and properties of the wastes involved, and the type of damage occurring.

3. In respect of damage where it is not possible to distinguish between the contribution made by wastes covered by the Protocol and wastes not covered by the Protocol, all damage shall be considered to be covered by the Protocol.

ARTICLE 8
Right of recourse

1. Any person liable under the Protocol shall be entitled to a right of recourse in accordance with the rules of procedure of the competent court:

    (a) Against any other person also liable under the Protocol; and
    (b) As expressly provided for in contractual arrangements.

2. Nothing in the Protocol shall prejudice any rights of recourse to which the person liable might be entitled pursuant to the law of the competent court.

ARTICLE 9
Contributory fault

Compensation may be reduced or disallowed if the person who suffered the damage, or a person for whom he is responsible under the domestic law, by his own fault, has caused or contributed to the damage having regard to all circumstances.

ARTICLE 10
Implementation

1. The Contracting Parties shall adopt the legislative, regulatory and administrative measures necessary to implement the Protocol.
2. In order to promote transparency, Contracting Parties shall inform the Secretariat of measures to implement the Protocol, including any limits of liability established pursuant to paragraph 1 of Annex B.
3. The provisions of the Protocol shall be applied without discrimination based on nationality, domicile or residence.

ARTICLE 11
Conflicts with other liability and compensation agreements

Whenever the provisions of the Protocol and the provisions of a bilateral, multilateral or regional agreement apply to liability and compensation for damage caused by an incident arising during the same portion of a transboundary movement, the Protocol shall not apply provided the other agreement is in force for the Party or Parties concerned and had been opened for signature when the Protocol was opened for signature, even if the agreement was amended afterwards.

ARTICLE 12
Financial limits

1. Financial limits for the liability under Article 4 of the Protocol are specified in Annex B to the Protocol. Such limits shall not include any interest or costs awarded by the competent court.
2. There shall be no financial limit on liability under Article 5.

ARTICLE 13
Time limit of liability

1. Claims for compensation under the Protocol shall not be admissible unless they are brought within ten years from the date of the incident.
2. Claims for compensation under the Protocol shall not be admissible unless they are brought within five years from the date the claimant knew or ought reasonably to have known of the damage provided that the time limits established pursuant to paragraph 1 of this Article are not exceeded.
3. Where the incident consists of a series of occurrences having the same origin, time limits established pursuant to this Article shall run from the date of the last of such occurrences. Where the incident consists of a continuous occurrence, such time limits shall run from the end of that continuous occurrence.

ARTICLE 14
Insurance and other financial guarantees

1. The persons liable under Article 4 shall establish and maintain during the period of the time limit of liability, insurance, bonds or other financial guarantees covering their liability under Article 4 of the Protocol for amounts not less than the minimum limits specified in paragraph 2 of Annex B. States may fulfil their obligation under this paragraph by a declaration of self-insurance. Nothing in this paragraph shall prevent the use of deductibles or co-payments as between the insurer and the insured, but the failure of the insured to pay any deductible or co-payment shall not be a defence against the person who has suffered the damage.
2. With regard to the liability of the notifier, or exporter under Article 4, paragraph 1, or of the importer under Article 4, paragraph 2, insurance, bonds or other financial guarantees referred to in paragraph 1 of this Article shall only be drawn upon in order to provide compensation for damage covered by Article 2 of the Protocol.
3. A document reflecting the coverage of the liability of the notifier or exporter under Article 4, paragraph 1, or of the importer under Article 4, paragraph 2, of the Protocol shall accompany the notification referred to in Article 6 of the Convention. Proof of coverage of the liability of the disposer shall be delivered to the competent authorities of the State of import.
4. Any claim under the Protocol may be asserted directly against any person providing insurance, bonds or other financial guarantees. The insurer or the person providing the financial guarantee shall have the right to require the person liable under Article 4 to be joined in the proceedings. Insurers and persons providing financial guarantees may invoke the defences which the person liable under Article 4 would be entitled to invoke.
5. Notwithstanding paragraph 4, a Contracting Party shall, by notification to the Depositary at the time of signature, ratification, or approval of, or accession to the Protocol, indicate if it does not provide for a right to bring a direct action

Appendix II: Text of the Basel Protocol 327

pursuant to paragraph 4. The Secretariat shall maintain a record of the Contracting Parties who have given notification pursuant to this paragraph.

ARTICLE 15
Financial mechanism

1. Where compensation under the Protocol does not cover the costs of damage, additional and supplementary measures aimed at ensuring adequate and prompt compensation may be taken using existing mechanisms.
2. The Meeting of the Parties shall keep under review the need for and possibility of improving existing mechanisms or establishing a new mechanism.

ARTICLE 16
State responsibility

The Protocol shall not affect the rights and obligations of the Contracting Parties under the rules of general international law with respect to State responsibility.

PROCEDURES

ARTICLE 17
Competent courts

1. Claims for compensation under the Protocol may be brought in the courts of a Contracting Party only where either:
   (a) The damage was suffered; or
   (b) The incident occurred; or
   (c) The defendant has his habitual residence, or has his principal place of business.
2. Each Contracting Party shall ensure that its courts possess the necessary competence to entertain such claims for compensation.

ARTICLE 18
Related actions

1. Where related actions are brought in the courts of different Parties, any court other than the court first seized may, while the actions are pending at first instance, stay its proceedings.
2. A court may, on the application of one of the Parties, decline jurisdiction if the law of that court permits the consolidation of related actions and another court has jurisdiction over both actions.
3. For the purpose of this Article, actions are deemed to be related where they are so closely connected that it is expedient to hear and determine them together to avoid the risk of irreconcilable judgements resulting from separate proceedings.

ARTICLE 19
Applicable law
All matters of substance or procedure regarding claims before the competent court which are not specifically regulated in the Protocol shall be governed by the law of that court including any rules of such law relating to conflict of laws.

ARTICLE 20
Relation between the Protocol and the law of the competent court
1. Subject to paragraph 2, nothing in the Protocol shall be construed as limiting or derogating from any rights of persons who have suffered damage, or as limiting the protection or reinstatement of the environment which may be provided under domestic law.
2. No claims for compensation for damage based on the strict liability of the notifier or the exporter liable under Article 4, paragraph 1, or the importer liable under Article 4, paragraph 2, of the Protocol, shall be made otherwise than in accordance with the Protocol.

ARTICLE 21
Mutual recognition and enforcement of judgements
1. Any judgement of a court having jurisdiction in accordance with Article 17 of the Protocol, which is enforceable in the State of origin and is no longer subject to ordinary forms of review, shall be recognized in any Contracting Party as soon as the formalities required in that Party have been completed, except:
    (a) Where the judgement was obtained by fraud;
    (b) Where the defendant was not given reasonable notice and a fair opportunity to present his case;
    (c) Where the judgement is irreconcilable with an earlier judgement validly pronounced in another Contracting Party with regard to the same cause of action and the same parties; or
    (d) Where the judgement is contrary to the public policy of the Contracting Party in which its recognition is sought.
2. A judgement recognized under paragraph 1 of this Article shall be enforceable in each Contracting Party as soon as the formalities required in that Party have been completed. The formalities shall not permit the merits of the case to be re-opened.
3. The provisions of paragraphs 1 and 2 of this Article shall not apply between Contracting Parties that are Parties to an agreement or arrangement in force on mutual recognition and enforcement of judgements under which the judgement would be recognizable and enforceable.

ARTICLE 22
Relationship of the Protocol with the Basel Convention
Except as otherwise provided in the Protocol, the provisions of the Convention relating to its Protocols shall apply to the Protocol.

ARTICLE 23
Amendment of Annex B

1. At its sixth meeting, the Conference of the Parties to the Basel Convention may amend paragraph 2 of Annex B following the procedure set out in Article 18 of the Basel Convention.
2. Such an amendment may be made before the Protocol enters into force.

FINAL CLAUSES

ARTICLE 24
Meeting of the Parties

1. A Meeting of the Parties is hereby established. The Secretariat shall convene the first Meeting of the Parties in conjunction with the first meeting of the Conference of the Parties to the Convention after entry into force of the Protocol.
2. Subsequent ordinary Meetings of the Parties shall be held in conjunction with meetings of the Conference of the Parties to the Convention unless the Meeting of the Parties decides otherwise. Extraordinary Meetings of the Parties shall be held at such other times as may be deemed necessary by a Meeting of the Parties, or at the written request of any Contracting Party, provided that within six months of such a request being communicated to them by the Secretariat, it is supported by at least one third of the Contracting Parties.
3. The Contracting Parties, at their first meeting, shall adopt by consensus rules of procedure for their meetings as well as financial rules.
4. The functions of the Meeting of the Parties shall be:
    (a) To review the implementation of and compliance with the Protocol;
    (b) To provide for reporting and establish guidelines and procedures for such reporting where necessary;
    (c) To consider and adopt, where necessary, proposals for amendment of the Protocol or any annexes and for any new annexes; and
    (d) To consider and undertake any additional action that may be required for the purposes of the Protocol.

ARTICLE 25
Secretariat

1. For the purposes of the Protocol, the Secretariat shall:
    (a) Arrange for and service Meetings of the Parties as provided for in Article 24;
    (b) Prepare reports, including financial data, on its activities carried out in implementation of its functions under the Protocol and present them to the Meeting of the Parties;

(c) Ensure the necessary coordination with relevant international bodies, and in particular enter into such administrative and contractual arrangements as may be required for the effective discharge of its functions;
(d) Compile information concerning the national laws and administrative provisions of Contracting Parties implementing the Protocol;
(e) Cooperate with Contracting Parties and with relevant and competent international organisations and agencies in the provision of experts and equipment for the purpose of rapid assistance to States in the event of an emergency situation;
(f) Encourage non-Parties to attend the Meetings of the Parties as observers and to act in accordance with the provisions of the Protocol; and
(g) Perform such other functions for the achievement of the purposes of this Protocol as may be assigned to it by the Meetings of the Parties.

2. The secretariat functions shall be carried out by the Secretariat of the Basel Convention.

ARTICLE 26
Signature

The Protocol shall be open for signature by States and by regional economic integration organizations Parties to the Basel Convention in Berne at the Federal Department of Foreign Affairs of Switzerland from 6 to 17 March 2000 and at United Nations Headquarters in New York from 1 April to 10 December 2000.

ARTICLE 27
Ratification, acceptance, formal confirmation or approval

1. The Protocol shall be subject to ratification, acceptance or approval by States and to formal confirmation or approval by regional economic integration organizations. Instruments of ratification, acceptance, formal confirmation, or approval shall be deposited with the Depositary.
2. Any organization referred to in paragraph 1 of this Article which becomes a Contracting Party without any of its member States being a Contracting Party shall be bound by all the obligations under the Protocol. In the case of such organizations, one or more of whose member States is a Contracting Party, the organization and its member States shall decide on their respective responsibilities for the performance of their obligations under the Protocol. In such cases, the organization and the member States shall not be entitled to exercise rights under the Protocol concurrently.
3. In their instruments of formal confirmation or approval, the organizations referred to in paragraph 1 of this Article shall declare the extent of their competence with respect to the matters governed by the Protocol. These organizations shall also inform the Depositary, who will inform the Contracting Parties, of any substantial modification in the extent of their competence.

Appendix II: Text of the Basel Protocol

ARTICLE 28
Accession

1. The Protocol shall be open for accession by any States and by any regional economic integration organization Party to the Basel Convention which has not signed the Protocol. The instruments of accession shall be deposited with the Depositary.
2. In their instruments of accession, the organizations referred to in paragraph 1 of this Article shall declare the extent of their competence with respect to the matters governed by the Protocol. These organizations shall also inform the Depositary of any substantial modification in the extent of their competence.
3. The provisions of Article 27, paragraph 2, shall apply to regional economic integration organizations which accede to the Protocol.

ARTICLE 29
Entry into force

1. The Protocol shall enter into force on the ninetieth day after the date of deposit of the twentieth instrument of ratification, acceptance, formal confirmation, approval or accession.
2. For each State or regional economic integration organization which ratifies, accepts, approves or formally confirms the Protocol or accedes thereto after the date of the deposit of the twentieth instrument of ratification, acceptance, approval, formal confirmation or accession, it shall enter into force on the ninetieth day after the date of deposit by such State or regional economic integration organization of its instrument of ratification, acceptance, approval, formal confirmation or accession.
3. For the purpose of paragraphs 1 and 2 of this Article, any instrument deposited by a regional economic integration organization shall not be counted as additional to those deposited by member States of such organization.

ARTICLE 30
Reservations and declarations

1. No reservation or exception may be made to the Protocol. For the purposes of the Protocol, notifications according to Article 3, paragraph 1, Article 3, paragraph 6, or Article 14, paragraph 5, shall not be regarded as reservations or exceptions.
2. Paragraph 1 of this Article does not preclude a State or a regional economic integration organization, when signing, ratifying, accepting, approving, formally confirming or acceding to the Protocol, from making declarations or statements, however phrased or named, with a view, inter alia, to the harmonization of its laws and regulations with the provisions of the Protocol, provided that such declarations or statements do not purport to exclude or to modify the legal effects of the provisions of the Protocol in their application to that State or that organization.

ARTICLE 31
Withdrawal

1. At any time after three years from the date on which the Protocol has entered into force for a Contracting Party, that Contracting Party may withdraw from the Protocol by giving written notification to the Depositary.
2. Withdrawal shall be effective one year from receipt of notification by the Depositary, or on such later date as may be specified in the notification.

ARTICLE 32
Depositary

The Secretary-General of the United Nations shall be the Depositary of the Protocol.

ARTICLE 33
Authentic texts

The original Arabic, Chinese, English, French, Russian and Spanish texts of the Protocol are equally authentic.

ANNEX A
List of States of Transit as referred to in Article 3, subparagraph 3 (d)

1. Antigua and Barbuda
2. Bahamas
3. Bahrain
4. Barbados
5. Cape Verde
6. Comoros
7. Cook Islands
8. Cuba
9. Cyprus
10. Dominica
11. Dominican Republic
12. Fiji
13. Grenada
14. Haiti
15. Jamaica
16. Kiribati
17. Maldives
18. Malta
19. Marshall Islands
20. Mauritius
21. Micronesia (Federated States of)
22. Nauru
23. Netherlands, on behalf of Aruba, and the Netherlands Antilles
24. New Zealand, on behalf of Tokelau
25. Niue

Appendix II: Text of the Basel Protocol   333

26. Palau
27. Papua New Guinea
28. Samoa
29. Sao Tome and Principe
30. Seychelles
31. Singapore
32. Solomon Islands
33. St. Lucia
34. St. Kitts and Nevis
35. St. Vincent and the Grenadines
36. Tonga
37. Trinidad and Tobago
38. Tuvalu
39. Vanuatu

ANNEX B
Financial limits

1. Financial limits for the liability under Article 4 of the Protocol shall be determined by domestic law.
2. The limits of liability shall:
   (a) For the notifier, exporter or importer, for any one incident, be not less than:
      (i) 1 million units of account for shipments up to and including 5 tonnes;
      (ii) 2 million units of account for shipments exceeding 5 tonnes, up to and including 25 tonnes;
      (iii) 4 million units of account for shipments exceeding 25 tonnes, up to and including 50 tonnes;
      (iv) 6 million units of account for shipments exceeding 50 tonnes, up to and including to 1,000 tonnes;
      (v) 10 million units of account for shipments exceeding 1,000 tonnes, up to and including 10,000 tonnes;
      (vi) Plus an additional 1,000 units of account for each additional tonne up to a maximum of 30 million units of account;
   (b) For the disposer, for any one incident, be not less than 2 million units of account for any one incident.
3. The amounts referred to in paragraph 2 shall be reviewed by the Contracting Parties on a regular basis taking into account, inter alia, the potential risks posed to the environment by the movement of hazardous wastes and other wastes and their disposal, recycling, and the nature, quantity and hazardous properties of the wastes.

Bibliography

Part 1: Literature, Press, Encyclopaedias

Abrams, David J., 'Regulating the International Hazardous Waste Trade: A Proposed Global Solution', 28 *Columbia Journal of Transnational Law* (1990), pp. 801-845.

Accioly, Hildebrando, 'Principes Généraux de la Responsibilité Internationale d'après la Doctrine et la Jurisprudence', 96 *Recueil des Cours* (1959), pp. 349-436.

Ago, Roberto, 'Le Délit International', 68 *Recueil des Cours* (1939), pp. 415-554.

Akehurst, Michael B., 'International Liability for Injurious Consequences Arising Out of Acts not Prohibited by International Law', 16 *Netherlands Yearbook of International Law* (1985), pp. 3-16.

Akiwumi, Paul/Melvasalo, Terttu, 'UNEP's Regional Seas Programme: Approach, Experience and Future Plans', 22 *Marine Policy* (1998), pp. 229-234.

Alam, Shawkat, 'Trade Restrictions Pursuant to Multilateral Environmental Agreements: Developmental Implications for Developing Countries', 41 *Journal of World Trade* (2007), pp. 983-1014.

Altfuldisch, Rainer, Haftung und Entschädigung nach Tankerunfällen auf See : Bestandsaufnahme, Rechtsvergleich und Überlegungen de lege ferenda, (Berlin, Heidelberg, New York: Springer, 2007).

Anand, Ruchi, *International Environmental Justice : A North-South Dimension*, (Aldershot, Burlington: Ashgate, 2004).

Anderson, David H., 'Resolution and Agreement Relating to the Implementation of Part XI of the UN Convention on the Law of the Sea: A General Assessment', 55 *Zeitschrift für ausländisches öffentliches Recht und Völkerrecht* (1995), pp. 275-289.

Anzilotti, Dionisio, *Lehrbuch des Völkerrechts, Band 1: Einführung - Allgemeine Lehren*, translated 3rd ed., (Berlin, Leipzig: Walter de Gruyter & Co., 1929).

de Aréchaga, Eduardo Jiménez, 'International Responsibility', in: Sørensen, Max (ed.), *Manual of Public International Law*, (London, Melbourne, Toronto: St Martin's Press, 1968), pp. 531-603.

de Aréchaga, Eduardo Jiménez, 'International Law in the past Third of a Century', 159 *Recueil des Cours* (1978), pp. 1-334.

Avery, Ian, 'Our Rubbish: Someone Else's Problem?', 2 *International Journal of Human Rights* (1998), pp. 19-28.

Barboza, Julio, 'International Liability for the Injurious Consequences of Acts not Prohibited by International Law and Protection of the Environment', 247 *Recueil des Cours* (1994), pp. 291-405.

Barnidge, Robert P., Jr., 'The Due Diligence Principle under International Law', 8 *International Community Law Review* (2006), pp. 81-121.

Barrios, Paula, 'The Rotterdam Convention on Hazardous Chemicals: A Meaningful Step Toward Environmental Protection?', 16 *Georgetown International Environmental Law Review* (2004), pp. 679-762.

Basedow, Jürgen, 'Rome II at Sea - General Aspects of Maritime Torts', 74 *The Rabel Journal of Comparative and International Private Law* (2010), pp. 118-138.

'Basel Convention: No Agreement on Draft Protocol', 29 *Environmental Policy and Law* (1999), p. 154.

'Basel Convention: Compensation and Liability Protocol Adopted', 30 *Environmental Policy and Law* (2000), pp. 43-45.

Bergkamp, Lucas, 'Proposals for International Environmental Liability in Respect of Waste and Biotechnology Products', 8 *European Environmental Law Review* (1999), pp. 324-327.

Bergkamp, Lucas, *Liability and Environment: Private and Public Law Aspects of Civil Liability for Environmental Harm in an International Context*, (The Hague, London, New York: Kluwer Law International, 2001).

Beyer, Peter, 'Eine neue Dimension der Umwelthaftung in Europa? Eine Analyse der europäischen Richtlinie zur Umwelthaftung', 16 *Zeitschrift für Umweltrecht* (2004), pp. 257-266.

Beyerlin, Ulrich, 'Grenzüberschreitender Umweltschutz und allgemeines Völkerrecht', in: Hailbronner, Kay/Ress, Georg/Stein, Torsten (ed.), *Staat und Völkerrechtsordnung : Festschrift für Karl Doehring*, (Berlin, Heidelberg, et. al.: Springer, 1989), pp. 37-61.

Beyerlin, Ulrich/Marauhn, Thilo, *International Environmental Law*, (Oxford: Hart Publishing, 2011).

Birnie, Patricia W./Boyle, Alan E./Redgwell, Catherine, *International Law and the Environment*, 3rd ed., (Oxford: Oxford University Press, 2009).

Black's Law Dictionary, Garner, Bryan A. (ed.), 9th ed., (St. Paul: West, 2009).

Blay, Sam/Green, Julia, 'The Development of a Liability Annex to the Madrid Protocol', 25 *Environmental Policy and Law* (1995), pp. 24-37.

de Boer, Jan Engel, 'The New Draft CRTD: Modernising the International Civil Liability and Compensation Regime for the Inland Transport of Dangerous Goods', 9 *Uniform Law Review* (2004), pp. 51-82.

Boga, Çiğdem, 'How to Promote the Acceptance of the HNS Convention? (II)', 10 *Legal Hukuk Dergisi* (2012), pp. 109-132.

Boyle, Alan E., 'Marine Pollution under the Law of the Sea Convention', 79 *American Journal of International Law* (1985), pp. 347-372.

Boyle, Alan E., 'Nuclear Energy and International Law: An Environmental Perspective', *British Year Book of International Law* (1989), pp. 257-313.

Boyle, Alan E., 'State Responsibility and International Liability for Injurious Consequences of Acts Not Prohibited by International Law: A Necessary Distinction?', 39 *International and Comparative Law Quarterly* (1990), pp. 1-26.

Boyle, Alan E., 'Globalising Environmental Liability: The Interplay of National and International Law', 17 *Journal of Environmental Law* (2005), pp. 3-26.

Brockhaus - Enzyklopädie in 30 Bänden, Vol. 1 : A-Anat, Zwahr, Annette (ed.), 21st ed., (Leipzig, Mannheim: F.A. Brockhaus, 2006).

Brown Weiss, Edith, 'Invoking State Responsibility in the Twenty-First Century', 96 *American Journal of International Law* (2002), pp. 798-816.

Brownlie, Ian, *System of the Law of Nations : State Responsibility Part 1*, (Oxford: Clarendon Press, 1983).

Bruno, Kenny, 'Philly Waste Go Home', 19 *Multinational Monitor* (1998).

Bryde, Brun-Otto, 'Völker- und Europarecht als Alibi für Umweltschutzdefizite?', in: Selmer, Peter/von Münch, Ingo (ed.), *Gedächtnisschrift für Wolfgang Martens*, (Berlin, New York: Walter de Gruyter, 1987), pp. 769-787.

Bryde, Brun-Otto, 'Umweltschutz durch allgemeines Völkerrecht?', 31 *Archiv des Völkerrechts* (1993), pp. 1-12.

Caron, David D., 'The ILC Articles on State Responsibility: The Paradoxical Relationship between Form and Authority', 96 *The American Journal of International Law* (2002), pp. 857-873.

Cassese, Antonio, *International Law*, 2nd ed., (Oxford, New York: Oxford University Press, 2005).

Cassing, James/Kuhn, Thomas, 'Strategic Environmental Policies when Waste Products are Tradable', 11 *Review of International Economics* (2003), pp. 495-511.

Cheng, Bin, *General Principles of Law as applied by International Courts and Tribunals*, (London: Stevens & Sons, 1953).

Choksi, Sejal, 'The Basel Convention on the Control of Transboundary Movements of Hazardous Wastes and Their Disposal: 1999 Protocol on Liability and Compensation', 28 *Ecology Law Quarterly* (2001), pp. 509-539.

Churchill, Robin R., 'Facilitating (Transnational) Civil Liability Litigation for Environmental Damage by Means of Treaties: Progress, Problems, and Prospects', 12 *Yearbook of International Environmental Law* (2001), pp. 3-41.

Churchill, Robin R./Lowe, Alan V., *The Law of the Sea*, 3rd ed., (Manchester: Manchester University Press, 1999).

Clapp, Jennifer, 'Africa, NGOs, and the International Toxic Waste Trade', 3 *Journal of Environment & Development* (1994), pp. 17-46.

Clapp, Jennifer, 'The Toxic Waste Trade with Less-Industrialised Countries: Economic Linkages and Political Alliances', 15 *Third World Quarterly* (1994), pp. 505-518.

Claußen, Simone, *Die Abwrackung von Seeschiffen in Nicht-OECD-Staaten : Zur Anwendbarkeit der Abfallverbringungsverordnung*, (Hamburg: Dr. Kovač, 2009).

Coenen, René, 'Dumping of Wastes at Sea: Adoption of the 1996 Protocol to the London Convention 1972', 6 *Review of European Community and International Environmental Law* (1997), pp. 54-61.

Combacau, Jean/Alland, Denis, '"Primary" and "Secondary" Rules in the Law of State Responsibility: Categorizing International Obligations', 16 *Netherlands Yearbook of International Law* (1985), pp. 81-109.

Combacau, Jean/Sur, Serge, *Droit International Public*, 9th ed., (Paris: Montchrestien, 2010).

Copeland, Brian R., 'International Trade in Waste Products in the Presence of Illegal Disposal', 20 *Journal of Environmental Economics and Management* (1991), pp. 143-162.

Corbett, Adam, 'Implications from 'Probo Koala' ruling', Trade Winds, of 30 July 2010.

Crawford, James, 'The ILC's Articles on Responsibility of States for Internationally Wrongful Acts: A Retrospect', 96 *The American Journal of International Law* (2002), pp. 874-890.

Crawford, James, *Brownlie's Principles of Public International Law*, 8th ed., (Oxford: Oxford University Press, 2013).

Crawford, James/Peel, Jacqueline/Olleson, Simon, 'The ILC's Articles on Responsibility of States for Internationally Wrongful Acts: Completion of the Second Reading', 12 *European Journal of International Law* (2001), pp. 963-991.

Cubel, Pablo, 'Transboundary Movements of Hazardous Wastes in International Law: The Special Case of the Mediterranean Area', 12 *International Journal of Marine and Coastal Law* (1997), pp. 447-487.

Cujo, Eglantine, 'Invocation of Responsibility by International Organizations', in: Crawford, James/Pellet, Alain/Olleson, Simon (ed.), *The Law of International Responsibility*, (Oxford, New York: Oxford University Press, 2010), pp. 969-983.

van Daele, Stijn/Vander Beken, Tom/Dorn, Nicholas, 'Waste Management and Crime: Regulatory, Business and Product Vulnerabilities', 37 *Environmental Policy and Law* (2007), pp. 34-38.

Dahm, Georg/Delbrück, Jost/Wolfrum, Rüdiger, *Völkerrecht, Band I/1 : Die Grundlagen. Die Völkerrechtssubjekte*, 2nd ed., (Berlin, New York: Walter de Gruyter, 1989).

Dahm, Georg/Delbrück, Jost/Wolfrum, Rüdiger, *Völkerrecht, Band I/2 : Der Staat und andere Völkerrechtssubjekte. Räume unter internationaler Verwaltung*, 2nd ed., (Berlin: De Gruyter Recht, 2002).
Dahm, Georg/Delbrück, Jost/Wolfrum, Rüdiger, *Völkerrecht, Band I/3 : Die Formen des völkerrechtlichen Handelns. Die inhaltliche Ordnung der internationalen Gemeinschaft*, 2nd ed., (Berlin: De Gruyter Recht, 2002).
Daniel, Anne, 'Civil Liability Regimes as a Complement to Multilateral Environmental Agreements: Sound International Policy or False Comfort?', 12 *Review of European Community and International Environmental Law* (2003), pp. 225-241.
Dhokalia, Ramaa P., *The Codification of Public International Law*, (Manchester: Manchester University Press, 1970).
Dieckmann, Martin, 'Die neue EG-Abfallverbringungsverordnung', 17 *Zeitschrift für Umweltrecht* (2006), pp. 561-567.
Doehring, Karl, *Völkerrecht : Ein Lehrbuch*, 2nd ed., (Heidelberg: C.F. Müller, 2004).
Donald, J. Wylie, 'The Bamako Convention as a Solution to the Problem of Hazardous Waste Exports to Less Developed Countries', 17 *Columbia Journal of Environmental Law* (1992), pp. 419-458.
Dowell, Katy, 'Trafigura Settlement: A Drop in the Ocean?', The Lawyer, of 28 September 2009.
Dreher, Kelly/Pulver, Simone, 'Environment as "High Politics"? Explaining Divergence in US and EU Hazardous Waste Export Policies', 17 *Review of European Community and International Environmental Law* (2008), pp. 308-320.
Drel, Marina I., 'Liability for Damage Resulting from the Transport of Hazardous Cargoes by Sea', in: Couper, Alastair/Gold, Edgar (ed.), *The Marine Environment and Sustainable Development: Law, Policy, and Science : Proceedings of The Law of the Sea Institute, Twenty-fifth Annual Conference, August 6-9, 1991, Malmö, Sweden*, (Honolulu: The Law of the Sea Institute, University of Hawaii, 1993), pp. 349-376.
Dugard, John, 'Diplomatic Protection', in: Crawford, James/Pellet, Alain/Olleson, Simon (ed.), *The Law of International Responsibility*, (Oxford, New York: Oxford University Press, 2010), pp. 1051-1071.
Eagleton, Clyde, *The Responsibility of States in International Law*, (Washington, New York: The New York University Press, 1928).
Eguh, Edna Chinyere, 'Regulations of Transboundary Movement of Hazardous Wastes: Lessons from Koko', 9 *African Journal of International and Comparative Law* (1997), pp. 130-155.
Engels, Urs Daniel, *European Ship Recycling Regulation : Entry-Into-Force Implications of the Hong Kong Convention*, (Berlin: Springer, 2013).
Epiney, Astrid, 'Das "Verbot erheblicher grenzüberschreitender Umweltbeeinträchtigungen": Relikt oder konkretisierungsfähige Grundnorm?', 33 *Archiv des Völkerrechts* (1995), pp. 309-360.
Epping, Volker/Gloria, Christian, 'Der Staat im Völkerrecht', in: Ipsen, Knut (ed.), *Völkerrecht*, 5th ed., (München: C.H. Beck, 2004), pp. 257-341.
Erichsen, Sven, 'Das Liability-Projekt der ILC: Fortentwicklung des allgemeinen Umweltrechts oder Kodifizierung einer Haftung für besonders gefährliche Aktivitäten?', 51 *Zeitschrift für ausländisches öffentliches Recht und Völkerrecht* (1991), pp. 94-132.
European Environment Agency (ed.), *Hazardous Waste Generation in EEA Member Countries : Comparability of Classification Systems and Qualities*, (Copenhagen: EEA, 2002).
European Environment Agency (ed.), *Transboundary Shipments of Waste in the EU : Developments 1995-2005 and Possible Drivers*, (Copenhagen: EEA, 2008).
Evans, Malcolm D. (ed.), *International Law*, 2nd ed., (Oxford, New York: Oxford University Press, 2006).
Evans, Rob, 'Trafigura fined €1m for exporting toxic waste to Africa', The Guardian, of 23 July 2010.

Eze, Chukwuka N., 'The Bamako Convention on the Ban of the Import Into Africa and the Control of the Transboundary Movement and Management of Hazardous Wastes Within Africa: A Milestone in Environmental Protection?', 15 *African Journal of International and Comparative Law* (2007), pp. 208-229.

Fagbohun, Olanrewaju A., 'The Regulation of Transboundary Shipments of Hazardous Waste: A Case Study of the Dumping of Toxic Waste in Abidjan, Cote d' Ivoire', 37 *Hong Kong Law Journal* (2007), pp. 831-858.

de la Fayette, Louise Angélique, 'The ILC and International Liability: A Commentary', 6 *Review of European Community and International Environmental Law* (1997), pp. 322-333.

de la Fayette, Louise Angélique, 'Compensation for Environmental Damage in Maritime Liability Regimes', in: Kirchner, Andree (ed.), *International Marine Environmental Law : Institutions, Implementation and Innovations*, (The Hague, New York, London: Kluwer Law International, 2003), pp. 231-265.

de la Fayette, Louise Angélique, 'New Approaches for Addressing Damage to the Marine Environment', 20 *The International Journal of Marine and Coastal Law* (2005), pp. 167-224.

Feess, Eberhard, *Umweltökonomie und Umweltpolitik*, 3rd ed., (München: Vahlen, 2007).

Fitzmaurice, Malgosia, 'International Liability for Injurious Consequences of Acts Not Prohibited by International Law (the "Liability Draft")', 24 *Polish Yearbook of International Law* (1999/2000), pp. 47-76.

French, Duncan, 'Developing States and International Environmental Law: The Importance of Differentiated Responsibilities', 49 *International and Comparative Law Quarterly* (2000), pp. 35-60.

Frenk, Kira, 'Was geschah an Bord der "Probo Koala"?', Frankfurter Allgemeine Zeitung, of 27 October 2006.

Frenz, Walter, 'Grenzüberschreitende Abfallverbringungen und gemeinschaftliche Warenverkehrsfreiheit', 20 *Umwelt- und Planungsrecht* (2000), pp. 210-215.

Friedrich-Ebert-Stiftung (ed.), *Zehn Jahre Basler Übereinkommen : Internationaler Handel mit gefährlichen Abfällen; Gutachten im Auftrag der Friedrich-Ebert-Stiftung*, (Bonn: Friedrich-Ebert-Stiftung, 1999).

Fröhler, Ludwig/Zehetner, Franz, *Rechtsschutzprobleme bei grenzüberschreitenden Umweltbeeinträchtigungen, Band I*, (Linz: Rudolf Trauner Verlag, 1979).

von Gadow-Stephani, Inken, *Der Zugang zu Nothäfen und sonstigen Notliegeplätzen für Schiffe in Seenot*, (Berlin, Heidelberg: Springer, 2006).

Gaines, Sanford E., 'International Principles for Transnational Environmental Liability: Can Developments in Municipal Law Help Break the Impasse?', 30 *Harvard International Law Journal* (1989), pp. 311-349.

Ganten, Reinhard H., 'Die Regulierungspraxis des internationalen Ölschadensfonds - Ergänzende Anmerkung zu dem Beitrag von Pfennigstorf VersR 88, 1201 - ', 40 *Versicherungsrecht* (1989), pp. 329-334.

Ganten, Reinhard H., 'HNS and Oil Pollution : Developments in the Field of Compensation for Damage to the Marine Environment', 27 *Environmental Policy and Law* (1997), pp. 310-314.

Geisler, Alexander, *Das Internationale Übereinkommen von 1990 über Vorsorge, Bekämpfung und Zusammenarbeit auf dem Gebiet der Ölverschmutzung (OPRC)*, (Münster: Lit Verlag, 2004).

Giampetro-Meyer, Andrea, 'Captain Planet Takes on Hazard Transfer: Combining the Forces of Market, Legal and Ethical Decisionmaking to Reduce Toxic Exports', 27 *UCLA Journal of Environmental Law & Policy* (2009), pp. 71-92.

Gilmore, Lori, 'The Export of Nonhazardous Waste', 19 *Environmental Law* (1988/1989), pp. 879-907.

Glazewski, Jan I., 'Regulating Transboundary Movement of Hazardous Waste: International Developments and Implications for South Africa', 26 *Comparative and International Law Journal of Southern Africa* (1993), pp. 234-249.

Goldie, Louis F. E., 'International Principles of Responsibility for Pollution', 9 *Columbia Journal of Transnational Law* (1970), pp. 283-330.
Göransson, Magnus, 'The HNS Convention', 2 *Uniform Law Review* (1997), pp. 249-270.
Graaff, Nora, *Staatenverantwortlichkeit : Vom "völkerrechtlichen Verbrechen" zur "schwerwiegenden Verletzung einer zwingenden Völkerrechtsnorm" anhand der ILC-Kodifikationsarbeit*, (Marburg: Frankfurt a.M., 2007).
Graf Vitzthum, Wolfgang (ed.), *Handbuch des Seerechts*, (München: C.H. Beck, 2006).
Graf Vitzthum, Wolfgang, 'Begriff, Geschichte und Rechtsquellen des Seerechts', in: Graf Vitzthum, Wolfgang (ed.), *Handbuch des Seerechts*, (München: C.H. Beck, 2006), pp. 1-62.
Graf Vitzthum, Wolfgang, 'Maritimes Aquitorium und Anschlusszone', in: Graf Vitzthum, Wolfgang (ed.), *Handbuch des Seerechts*, (München: C.H. Beck, 2006), pp. 63-160.
Graf Vitzthum, Wolfgang (ed.), *Völkerrecht*, 5th. ed., (Berlin: De Gruyter, 2010).
Graf Vitzthum, Wolfgang, 'Begriff, Geschichte und Rechtsquellen des Völkerrechts', in: Graf Vitzthum, Wolfgang (ed.), *Völkerrecht*, 5th ed., (Berlin: De Gruyter, 2010), pp. 1-77.
Greenfield, Erin F., 'CERCLA's Applicability Abroad: Examining the Reach of a U.S. Environmental Statute in the Face of a Cross-Border Pollution Dispute', 19 *Emory International Law Review* (2005), pp. 1697-1731.
Griggs, Patrick/Shaw, Richard, 'The IMO Legal Committee Considers the Adoption of a Protocol to the HNS Convention and the Problem of Fair Treatment of Seafarers', 111 *Il Diritto Marittimo* (2009), pp. 277-283.
Griggs, Patrick/Williams, Richard/Farr, Jeremy, *Limitation of Liability for Maritime Claims*, 4th ed., (London, Singapore: LLP, 2005).
Grotius, Hugo, 'The Law of War and Peace, Book III', in: Grotius, Hugo, *De Jure Belli Ac Pacis Libri Tres*, translated reprint, (Oxford, London: Clarendon Press, 1925), pp. 597-862.
Grotius, Hugo, *The Free Sea or a Disputation Concerning the Right which the Hollanders Ought to Have to the Indian Merchandise for Trading*, transl. ed., (Indianapolis: Liberty Fund, 2004).
Gründling, Lothar, 'Verantwortlichkeit der Staaten für grenzüberschreitende Umweltbeeinträchtigungen', 45 *Zeitschrift für ausländisches öffentliches Recht und Völkerrecht* (1985), pp. 265-292.
Gudofsky, Jason L., 'Transboundary Shipments of Hazardous Waste for Recycling and Recovery Operations', 34 *Stanford Journal of International Law* (1998), pp. 219-286.
Guggenheim, Paul, *Traité de Droit International Public, Tome II*, (Geneva: Librairie de L'Université, Georg & Cie, 1954).
Gunasekera, Dan Malika, *Civil Liability for Bunker Oil Pollution Damage*, (Frankfurt a.M., Berlin, et al.: Peter Lang, 2010).
Gwam, Cyril Uchenna, 'Adverse Effects of the Illicit Movement and Dumping of Hazardous, Toxic, and Dangerous Wastes and Products on the Enjoyment of Human Rights', 14 *Florida Journal of International Law* (2001/2002), pp. 427-474.
Gwam, Cyril Uchenna, 'Travaux Preparatoires of the Basel Convention of Transboundary Movements of Hazardous Wastes and Their Disposal', 18 *Journal of Natural Resources & Environmental Law* (2003/2004), pp. 1-78.
Haak, Krijn, 'New Developments in the Field of Transport of Dangerous Goods: Presence and Prospects of the CRTD Convention', in: Basedow, Jürgen/Magnus, Ulrich/Wolfrum, Rüdiger (ed.), *The Hamburg Lectures on Maritime Affairs 2007 & 2008*, (Heidelberg, Dordrecht, et al.: Springer, 2010), pp. 9-20.
Hackett, David P., 'An Assessment of the Basel Convention on the Control of Transboundary Movements of Hazardous Wastes and Their Disposal', 5 *American University Journal of International Law and Policy* (1989/1990), pp. 291-323.
Hackmann, Johannes, 'International Trade in Waste Materials', 29 *Intereconomics* (1994), pp. 292-302.
Hafner, Gerhard, 'Meeresumwelt, Meeresforschung und Technologietransfer', in: Graf Vitzthum, Wolfgang (ed.), *Handbuch des Seerechts*, (München: C.H. Beck, 2006), pp. 347-460.

Handl, Günther, 'Territorial Sovereignty and the Problem of Transnational Pollution', 69 *American Journal of International Law* (1975), pp. 50-76.

Handl, Günther, 'Liability as an Obligation Established by a Primary Rule of International Law', 16 *Netherlands Yearbook of International Law* (1985), pp. 49-79.

Handl, Günther, 'Compliance Control Mechanisms and International Environmental Obligations', 5 *Tulane Journal of International and Comparative Law* (1997), pp. 29-50.

Hansen, Robert G./Thomas, Randall S., 'The Efficiency of Sharing Liability for Hazardous Waste: Effects of Uncertainty Over Damages', 19 *International Review of Law and Economics* (1999), pp. 135-157.

Harndt, Raimund, *Völkerrechtliche Haftung für die schädlichen Folgen nicht verbotenen Verhaltens : Schadensprävention und Wiedergutmachung : Typologische Betrachtungen der völkerrechtlichen Haftungstatbestände mit einem rechtsvergleichenden Überblick über die des innerstaatlichen Zivilrechts in ausgewählten Rechtsordnungen*, (Berlin: Duncker & Humblot 1993).

Hegel, Georg Wilhelm Friedrich, *Grundlinien der Philosophie des Rechts*, edited by Georg Lasson, 3rd ed., (Leipzig: Felix Meiner, 1930).

Heintschel von Heinegg, Wolff, 'Die völkerrechtlichen Verträge als Hauptrechtsquelle des Völkerrechts', in: Ipsen, Knut (ed.), *Völkerrecht*, 5th ed., (München: C.H. Beck, 2004), pp. 112-209.

Heintschel von Heinegg, Wolff, 'Internationales öffentliches Umweltrecht', in: Ipsen, Knut (ed.), *Völkerrecht*, 5th ed., (München: C.H. Beck, 2004), pp. 973-1064.

Helfenstein, Allegra, 'U.S. Controls on International Disposal of Hazardous Waste', 22 *International Lawyer* (1988), pp. 775-790.

Helm, Carsten, 'How Liable Should An Exporter Be? The Case of Trade in Hazardous Goods', 28 *International Review of Law and Economics* (2008), pp. 263-271.

Herber, Rolf, *Seehandelsrecht : Systematische Darstellung*, (Berlin, New York: Walter de Gruyter, 1999).

Hessbruegge, Jan Arno, 'The Historical Development of the Doctrines of Attribution and Due Diligence in International Law', 36 *New York University Journal of International Law & Politics* (2003/2004), pp. 265-306.

'Homeless for 16 Years, Barge of Garbage Returns to Pa.', Los Angeles Times, of 11 August 2002.

van Hoogstraten, David/Lawrence, Peter, 'Protecting the South Pacific from Hazardous and Nuclear Waste Dumping: The Waigani Convention', 7 *Review of European Community and International Environmental Law* (1998), pp. 268-273.

Ipsen, Knut (ed.), *Völkerrecht*, 5th ed., (München: C.H. Beck, 2004).

Ipsen, Knut, 'Regelungsbericht, Geschichte und Funktion des Völkerrechts', in: Ipsen, Knut (ed.), *Völkerrecht*, 5th ed., (München: C.H. Beck, 2004), pp. 1-54.

Ipsen, Knut, 'Völkerrechtliche Verantwortlichkeit und Völkerstrafrecht', in: Ipsen, Knut (ed.), *Völkerrecht*, 5th ed., (München: C.H. Beck, 2004), pp. 615-673.

Jacobsson, Måns, 'Entwicklung des Schadensbegriffs im Recht der Haftung für Ölverschmutzungsschäden', 70 *Schriften des Deutschen Vereins für Internationales Seerecht : Reihe A: Berichte und Vorträge* (1990), pp. 1-18.

Jacobsson, Måns, 'Oil Pollution Liability and Compensation: An International Regime', 1 *Uniform Law Review* (1996), pp. 260-273.

Jacobsson, Måns, 'Internationales Schadensersatzrecht für Ölverschmutzungsschäden beim Seetransport - Entwicklung in den letzten Jahren und Zukunftsperspektiven', 90 *Schriften des Deutschen Vereins für Internationales Seerecht : Reihe A: Berichte und Vorträge* (1998), pp. 1-20.

Jellinek, Georg, *Allgemeine Staatslehre*, 4th reprint of the 3rd ed., (Berlin: Julius Springer, 1914).

Jenks, C. Wilfred 'Liability for Ultra-Hazardous Activities in International Law', 117 *Recueil des Cours* (1966), pp. 99-200.

Jennings, Robert/Watts, Arthur (ed.), *Oppenheim's International Law, Volume I, Peace, Introduction and Part 1*, 9th ed., (Harlow: Longman, 1992).

Johnstone, Nick, 'The Implications of the Basel Convention for Developing Countries: The Case of Trade in Non-ferrous Metal-bearing Waste', 23 *Resources, Conservation and Recycling* (1998), pp. 1-28.

Jones, Wordsworth Filo, 'The Evolution of the Bamako Convention: An African Perspective', 4 *Colorado Journal of International Environmental Law and Policy* (1993), pp. 324-342.

Kaminsky, Howard S., 'Assessment of the Bamako Convention on the Ban of Import into Africa and the Control of Transboundary Movement and Management of Hazardous Wastes within Africa', 5 *Georgetown International Environmental Law Review* (1992/1993), pp. 77-90.

Kant, Immanuel, *Critik der reinen Vernunft*, (Riga: Johann Friedrich Hartknoch, 1781).

Kappet, Liliane C., *Tankerunfälle und der Ersatz ökologischer Schäden : Am Beispiel des Vertragssystems des Internationalen Übereinkommens über die zivilrechtliche Haftung für Ölverschmutzungsschäden in Verbindung mit dem Internationalen Übereinkommen über die Errichtung eines Internationalen Fonds zur Entschädigung für Ölverschmutzungsschäden*, (Hamburg: Lit, 2006).

Kasten, Verena, *Europarechtliche und völkerrechtliche Aspekte der grenzüberschreitenden Abfallverbringung*, (Frankfurt a.M., Berlin, et al.: Peter Lang, 1997).

Kelsen, Hans, *General Theory of Law and State*, (Cambridge/Mass.: Harvard University Press, 1949).

Kelsen, Hans, *Reine Rechtslehre*, 2nd ed., (Wien: Franz Deuticke, 1960).

Kelson, John M., 'State Responsibility and the Abnormally Dangerous Activity', 13 *Harvard International Law Journal* (1972), pp. 197-244.

Kirstein, Roland, 'Internationaler Müllhandel aus Sicht der ökonomischen Analyse des Rechts', in: Eger, Thomas/Bigus, Jochen/Ott, Claus/von Wangenheim, Georg (ed.), *Internationalization of the Law and its Economic Analysis: Festschrift für Hans-Bernd Schäfer zum 65. Geburtstag*, (Wiesbaden: Gabler, 2008), pp. 443-453.

Kiss, Alexandre, 'Strict Liability in International Environmental Law', in: Führ, Martin/Wahl, Rainer/von Wilmowsky, Peter (ed.), *Umweltrecht und Umweltwissenschaft: Festschrift für Eckard Rehbinder*, (Berlin: Erich Schmidt, 2007), pp. 213-221.

Kiss, Alexandre/Shelton, Dinah, 'Strict Liability in International Environmental Law', in: Ndiaye, Tafsir Malick/Wolfrum, Rüdiger (ed.), *Law of the Sea, Environmental Law and Settlement of Disputes : Liber Amicorum Judge Thomas A. Mensah*, (Leiden, Boston: Martinus Nijhoff, 2007), pp. 1131-1151.

Kitt, Jennifer R., 'Waste Exports to the Developing World: A Global Response', 7 *Georgetown International Environmental Law Review* (1994/1995), pp. 485-514.

Knauer, Sebastian / Thielke, Thilo / Traufetter, Gerald, 'Profits for Europe, Industrial Slop for Africa', SPIEGEL ONLINE, of 18 September 2006.

Köck, Heribert Franz, 'Staatenverantwortlichkeit und Staatenhaftung im Völkerrecht', in: Köck, Heribert Franz/Lengauer, Alina/Ress, Georg (ed.), *Europarecht im Zeitalter der Globallisierung : Festschrift für Peter Fischer*, (Wien: Linde, 2004), pp. 191-248.

Krueger, Jonathan, 'Prior Informed Consent and the Basel Convention: The Hazards of What Isn't Known', 7 *Journal of Environment & Development* (1998), pp. 115-137.

Krueger, Jonathan, *International Trade and the Basel Convention*, (London: Earthscan Publications, 1999).

Krueger, Jonathan, 'The Basel Convention and the International Trade in Hazardous Wastes', *Yearbook of International Co-operation on Environment and Development* (2001/2002), pp. 43-51.

Kummer, Katharina, *International Management of Hazardous Wastes: The Basel Convention and Related Legal Rules*, (Oxford: Clarendon Press, 1995).

Kummer, Katharina, 'The Basel Convention: Ten Years On', 7 *Review of European Community and International Environmental Law* (1998), pp. 227-236.

Kummer, Katharina, 'Prior Informed Consent for Chemicals in International Trade: The 1998 Rotterdam Convention', 8 *Review of European Community and International Environmental Law* (1999), pp. 323-330.

Kunig, Philip, 'Reform der Charta der Vereinten Nationen aus völkerrechticher Sicht - Bestandsaufnahme und einige Utopien', in: Albrecht, Ulrich (ed.), *Die Vereinten Nationen am Scheideweg : Von der Staatenorganisation zur internationalen Gemeinschaftswelt?*, (Hamburg: Lit, 1998), pp. 137-156.

Lagoni, Rainer, 'Umweltvölkerrecht : Anmerkungen zur Entwicklung eines Rechtsgebiets', in: Thieme, Werner (ed.), *Umweltschutz im Recht*, (Berlin: Duncker & Humblot, 1988), pp. 233-250.

Lagoni, Rainer, 'Die Abwehr von Gefahren für die marine Umwelt', 32 *Berichte der Deutschen Gesellschaft für Völkerrecht* (1992), pp. 87-152.

Lagoni, Rainer, 'Altöl und Seeschiffahrt', in: Becker, Bernd/Bull, Hans Peter/Seewald, Otfried (ed.), *Festschrift für Werner Thieme zum 70. Geburtstag*, (Köln, Berlin, Bonn, et al. : Carl Heymanns, 1993), pp. 997-1022.

Lagoni, Rainer, 'Monitoring Compliance and Enforcement of Compliance Through the OSPAR Commission', in: Ehlers, Peter/Mann-Borgese/Wolfrum, Rüdiger (ed.), *Marine Issues from a Scientific, Political and Legal Perspective*, (The Hague, London, New York: Kluwer Law International, 2002), pp. 155-163.

Lagoni, Rainer/Albers, Jan, 'Schiffe als Abfall? Zur Anwendung des Basler Übereinkommens und der EG-Abfallverbringungsverordnung auf Seeschiffe', 30 *Natur und Recht* (2008), pp. 220-227.

Lallas, Peter L., 'The Stockholm Convention on Persistent Organic Pollutants', 95 *American Journal of International Law* (2001), pp. 692-707.

Lammers, Johan G., 'International Responsibility and Liability for Damage Caused by Environmental Interferences', 31 *Environmental Policy and Law* (2001), pp. 94-105.

Lammers, Johan G., 'Prevention of Transboundary Harm from Hazardous Activities : The ILC Draft Articles', 14 *Hague Yearbook of International Law* (2001), pp. 3-24.

Lammers, Johan G., 'New Developments Concerning International Responsibility and Liability for Damage Caused by Environmental Interferences', 19 *Hague Yearbook of International Law* (2006), pp. 87-112.

Lang, Winfried, 'The International Waste Regime', in: Lang, Winfried/Neuhold, Hanspeter/Zemanek, Karl (ed.), *Environmental Protection and International Law*, (London, Dordrecht, Boston: Graham & Trotman/Martinus Nijhoff, 1991), pp. 147-161.

Lauterpacht, Hersch, *Private Law Sources and Analogies of International Law*, (London: Longman, Green and Co., 1927).

Lawrence, Peter, 'Negotiation of a Protocol on Liability and Compensation for Damage Resulting from Transboundary Movements of Hazardous Wastes and Their Disposal ', 7 *Review of European Community and International Environmental Law* (1998), pp. 249-255.

Lefeber, René, *Transboundary Environmental Interference and the Origin of State Liability*, (The Hague, London, Boston: Kluwer Law International 1996).

Lefeber, René, 'The Liability Provisions of Regional Sea Conventions: Dead Letters in the Sea?', in: Vidas, Davor/Østreng, Willy (ed.), *Order for the Oceans at the Turn of the Century*, (The Hague, London, Boston: Kluwer Law International, 1999), pp. 507-522.

Leigh, David, 'Trafigura Offers £1,000 Each to Toxic Dumping Victims', The Guardian, of 18 September 2009.

Levis, Lara, 'The European Community's Internal Regime on Trade in Hazardous Wastes: Lessons from the US Regime', 7 *Review of European Community and International Environmental Law* (1998), pp. 283-290.

Linderfalk, Ulf, 'State Responsibility and the Primary-Secondary Rules Terminology - The Role of Language for an Understanding of the International Legal System', 78 *Nordic Journal of International Law* (2009), pp. 53-72.

Lipman, Zada, 'Transboundary Movements of Hazardous Waste: Environmental Justice Issues for Developing Countries', *Acta Juridica* (1999), pp. 266-286.

Liu, Sylvia F., 'The Koko Incident: Developing International Norms for the Transboundary Movement of Hazardous Waste', 8 *Journal of Natural Resources & Environmental Law* (1992/1993), pp. 121-154.

Lohnes, Jud, 'Taiwanese Company Dumps 3000 Tons of Toxic Waste in Cambodia', 11 *Colorado Journal of International Environmental Law and Policy* (2000), pp. 262-277.

Long, Jerrold A., 'Protocol on Liability and Compensation for Damage Resulting from the Transboundary Movements of Hazardous Wastes and Their Disposal', 11 *Colorado Journal of International Environmental Law and Policy* (2000), pp. 253-261.

Louka, Elli, 'Bringing Polluters Before Transnational Courts: Why Industry Should Demand Strict and Unlimited Liability for the Transnational Movements of Hazardous and Radioactive Wastes', 22 *Denver Journal of International Law and Policy* (1993/1994), pp. 63-106.

Louka, Elli, *International Environmental Law : Fairness, Effectiveness, and World Order*, (New York: Cambridge University Press, 2006).

Lowe, Vaughan, 'Precluding Wrongfulness or Responsibility: A Plea for Excuses', 10 *European Journal of International Law* (1999), pp. 405-411.

Magraw, Daniel Barstow, 'Transboundary Harm: The International Law Commission's Study of "International Liability"', 80 *American Journal of International Law* (1986), pp. 305-329.

Marbury, Hugh J., 'Hazardous Waste Exportation: The Global Manifestation of Environmental Racism', 28 *Vanderbilt Journal of Transnational Law* (1995), pp. 251-294.

Markus, Francis, 'Taiwanese Waste Sent to Europe', BBC News, of 2 March 2000.

Matz-Lück, Nele, 'Safe and Sound Scrapping of "Rusty Buckets"? : The 2009 Hong Kong Ship Recycling Convention', 19 *Review of European Community and International Environmental Law* (2010), pp. 95-103.

McDorman, Ted L., 'The Rotterdam Convention on the Prior Informed Consent Procedure for Certain Hazardous Chemicals and Pesticides in International Trade: Some Legal Notes', 13 *Review of European Community and International Environmental Law* (2004), pp. 187-200.

Meeson, Nigel/Kimbell, John A., *Admiralty Jurisdiction and Practice*, 4th ed., (London: Informa, 2011).

Mensah, Thomas A., 'Civil Liability and Compensation for Environmental Damage in the 1982 Convention on the Law of the Sea', in: Basedow, Jürgen/Magnus, Ulrich/Wolfrum, Rüdiger (ed.), *The Hamburg Lectures on Maritime Affairs 2007 & 2008*, (Heidelberg, Dordrecht, et al.: Springer, 2010), pp. 3-8.

van der Mensbrugghe, Yves, 'Commentary to "The International Waste Regime"', in: Lang, Winfried/Neuhold, Hanspeter/Zemanek, Karl (ed.), *Environmental Protection and International Law*, (London, Dordrecht, Boston: Graham & Trotman/Martinus Nijhoff, 1991), pp. 161-166.

Meßerschmidt, Klaus, *Europäisches Umweltrecht : Ein Studienbuch*, (München: C. H. Beck, 2011).

Meyers Enzyklopädisches Lexikon, Vol. 1 : A-Alu, 9th ed., (Mannheim, Wien, Zürich: Lexikonverlag, 1971).

Montgomery, Mark A., 'Reassessing the Waste Trade Crisis: What Do We Really Know?', 4 *Journal of Environment & Development* (1995), pp. 1-28.

Moutier-Lopet, Anaïs, 'Contribution to the Injury', in: Crawford, James/Pellet, Alain/Olleson, Simon (ed.), *The Law of International Responsibility*, (Oxford, New York: Oxford University Press, 2010), pp. 639-645.

von Münch, Ingo, *Das völkerrechtliche Delikt in der modernen Entwicklung der Völkerrechtsgemeinschaft*, (Frankfurt a.M.: P. Keppler, 1963).

Murphy, Sean D., 'Prospective Liability Regimes for the Transboundary Movement of Hazardous Wastes', 88 *American Journal of International Law* (1994), pp. 24-75.

Nadelson, Robert, 'After MOX: The Contemporary Shipment of Radioactive Substances in the Law of the Sea', 15 *International Journal of Marine and Coastal Law* (2000), pp. 193-244.

Nanda, Ved P./Bailey, Bruce C., 'Export of Hazardous Waste and Hazardous Technology: Challenge for International Environmental Law', 17 *Denver Journal of International Law and Policy* (1988/1989), pp. 155-206.

Nordquist, Myron H./Rosenne, Shabtai/Yankov, Alexander/Grandy, Neal R. (ed.), *United Nations Convention on the Law of the Sea, 1982 : A Commentary; Volume IV, Articles 192 to 278, Final Act, Annex VI*, (Dordrecht, Boston, London: Martinus Nijhoff Publishers, 1991).

O'Neill, Kate, 'Out of the Backyard: The Problems of Hazardous Waste Management at a Global Level', 7 *Journal of Environment & Development* (1998), pp. 138-163.

O'Neill, Kate, *Waste Trading Among Rich Nations : Building a New Theory of Environmental Regulation*, (Cambridge/Mass., London: MIT Press, 2000).

Obstler, Peter, 'Toward a Working Solution to Global Pollution: Importing CERCLA to Regulate the Export of Hazardous Waste', 16 *Yale International Journal of International Law* (1991), pp. 73-126.

Oexle, Anno, 'Rechtsfragen des neuen Verbringungsrechts', 18 *Zeitschrift für Umweltrecht* (2007), pp. 460-466.

Ognibene, Lara, 'Dumping of Toxic Waste in Côte d'Ivoire : The International Framework', 37 *Environmental Policy and Law* (2007), pp. 31-33.

Oppenheim, Lassa, *International Law : A Treatise, Vol. I. : Peace*, 2nd ed., (New York, Bombay, Calcutta: Longmans, Green & Co., 1912).

Ovink, B. John, 'Transboundary Shipments of Toxic Waste: The Basel and Bamako Conventions: Do Third World Countries Have a Chance?', 13 *Dickinson Journal of International Law* (1994/1995), pp. 281-295.

The Oxford English Dictionary, Vol. XII : V-Z and Bibliography, Murray, James A. H. / Bradley, Henry (eds.), (Oxford: Clarendon Press, 1978).

Park, Rozelia S., 'An Examination of International Environmental Racism Through the Lens of Transboundary Movement of Hazardous Wastes', 5 *Indiana Journal of Global Legal Studies* (1997/1998), pp. 659-709.

Pellow, David Naguib, *Resisting Global Toxics : Transnational Movements for Environmental Justice*, (Cambridge/Mass., London: MIT Press, 2007).

Pineschi, Laura, 'The Transit of Ships Carrying Hazardous Wastes through Foreign Coastal Zones', in: Francioni, Francesco/Scovazzi, Tullio (ed.), *International Responsibility for Environmental Harm*, (London, Dordrecht, Boston: Graham & Trotman, 1991), pp. 299-316.

Pineschi, Laura, 'Non-Compliance Procedures and the Law of State Responsibility', in: Treves, Tullio/Pineschi, Laura/Tanzi, Attila/Pitea, Cesare/Ragni, Chiara/Jacur, Francesca Romanin (ed.), *Non-Compliance Procedures and Mechanisms and the Effectiveness of International Environmental Agreements*, (The Hague: T.M.C. Asser Press, 2009), pp. 483-497.

Pinto-Dobernig, Ilse R., 'Liability for the Harmful Consequences of Instances of Transfrontier Pollution Not Prohibited by International Law', 38 *Österreichische Zeitschrift für öffentliches Recht und Völkerrecht* (1987), pp. 79-133.

Pinto, Moragodage C. W., 'Reflections on International Liability for Injurious Consequences Arising Out of Facts Not Prohibited by International Law', 16 *Netherlands Yearbook of International Law* (1985), pp. 17-48.

Pisillo Mazzeschi, Riccardo, 'Forms of International Responsibility for Environmental Harm', in: Francioni, Francesco/Scovazzi, Tullio (ed.), *International Responsibility for Environmental Harm*, (London: Graham & Trotman, 1991), pp. 15-35.

Pisillo Mazzeschi, Riccardo, 'The Due Diligence Rule and the Nature of the International Responsibility of States', 35 *German Yearbook of International Law* (1992), pp. 9-51.

Poulakidas, Dean M., 'Waste Trade and Disposal in the Americas: The Need for and Benefits of a Regional Response', 21 *Vermont Law Review* (1996/1997), pp. 873-928.

Pratt, Laura A. W., 'Decreasing Dirty Dumping? A Reevaluation of Toxic Waste Colonialism and the Global Management of Transboundary Hazardous Waste', 35 *William and Mary Environmental Law and Policy Review* (2011), pp. 581-623.

Proelß, Alexander, *Meeresschutz im Völker- und Europarecht : Das Beispiel des Nordostatlantiks*, (Berlin: Duncker & Humblot, 2004).
Rabe, Dieter, *Seehandelsrecht : Fünftes Buch des Handelsgesetzbuches mit Nebenvorschriften und Internationalen Übereinkommen*, 4th ed., (München: C.H. Beck, 2000).
Randelzhofer, Albrecht, 'Probleme der völkerrechtlichen Gefährdungshaftung', 24 *Berichte der Deutschen Gesellschaft für Völkerrecht* (1984), pp. 35-73.
Rao, Pemmaraju Sreenivasa, 'International Liability for Transboundary Harm', 34 *Environmental Policy and Law* (2004), pp. 224-231.
Rattray, Kenneth, 'Resolution and Agreement Relating to the Implementation of Part XI of the UN Convention on the Law of the Sea: A General Assessment - Comment', 55 *Zeitschrift für ausländisches öffentliches Recht und Völkerrecht* (1995), pp. 298-309.
Rauscher, Michael, *International Trade, Factor Movements, and the Environment*, (Oxford, New York: Clarendon Press, 1997).
Rauscher, Michael, 'International Trade in Hazardous Waste', in: Schulze, Günther G./Ursprung, Heinrich W. (ed.), *International Environmental Economics : A Survey of the Issues*, (Oxford, New York: Oxford University Press, 2001), pp. 148-165.
Rauschning, Dietrich, 'Verantwortlichkeit der Staaten für völkerrechtswidriges Verhalten', 24 *Berichte der Deutschen Gesellschaft für Völkerrecht* (1984), pp. 7-34.
Reeder, John (ed.), *Brice on Maritime Law of Salvage*, 4th ed., (London: Sweet and Maxwell, 2003).
Reis, Tarcísio Hardmann, *Compensation for Environmental Damages under International Law : The Role of the International Judge*, (Alphen aan den Rijn: Kluwer Law International, 2011).
Rengifo, Antonio, 'The International Convention on Liability and Compensation for Damage in Connection with the Carriage of Hazardous and Noxious Substances by Sea, 1996', 6 *Review of European Community and International Environmental Law* (1997), pp. 191-197.
Rest, Alfred, 'State Responsibility / Liability : Erga Omnes Obligations and Judicial Control', 40 *Environmental Policy and Law* (2010), pp. 298-330.
Røsæg, Erik, 'Compulsory Maritime Insurance', 258 *Scandinavian Institute of Maritime Law Yearbook* (2000), pp. 179-205.
Røsæg, Erik, 'Non-Collectable Contributions to the Separate LNG Account of the HNS Convention', 13 *Journal of International Maritime Law* (2007), pp. 94-99.
Røsæg, Erik, 'The Rebirth of the HNS Convention', in: Berlingieri, Georgio/Boglione, Angelo/Carbone, Sergio M./Siccardi, Francesco (ed.), *Scritti in Onore di Francesco Berlingieri, Volume II : Numero Speciale di Il Diritto Marittimo*, (Genova: 2010), pp. 852-860.
Rose, Francis D., *Kennedy and Rose: Law of Salvage*, 8th ed., (London: Sweet & Maxwell, 2013).
Rosenthal, Robert M., 'Ratification of the Basel Convention: Why the United States Should Adopt the No Less Environmentally Sound Standard', 11 *Temple Environmental Law & Technology Journal* (1992), pp. 61-78.
Rothwell, Donald R./Stephens, Tim, *The International Law of the Sea*, (Oxford: Hart Publishing, 2010).
Rublack, Susanne, 'Fighting Transboundary Waste Streams: Will the Basel Convention Help?', 22 *Verfassung und Recht in Übersee* (1989), pp. 360-391.
Rudolf, Walter, 'Haftung für rechtmäßiges Verhalten im Völkerrecht', in: Damrau, Jürgen/Kraft, Alfons/Fürst, Walther (ed.), *Festschrift für Otto Mühl zum 70. Geburtstag, 10. Oktober 1981*, (Stuttgart, Berlin, *et al*.: W. Kohlhammer, 1981), pp. 535-552.
de la Rue, Colin M./Anderson, Charles B., *Shipping and the Environment : Law and Practice*, 2nd ed., (London: Informa, 2009).
Rummel-Bulska, Iwona, 'The Basel Convention and the UN Convention on the Law of the Sea', in: Ringbom, Henrik (ed.), *Competing Norms in the Law of Marine Environmental Protection : Focus on Ship Safety and Pollution Prevention*, (London, The Hague, Boston: Kluwer Law International, 1997), pp. 83-108.

Rutinwa, Bonaventure, 'Liability and Compensation for Injurious Consequences of the Transboundary Movement of Hazardous Wastes', 6 *Review of European Community and International Environmental Law* (1997), pp. 7-13.
Salvioli, Gabriele, 'Les Régles Générales de la Paix', 46 *Recueil des Cours* (1933), pp. 1-161.
Scelle, Georges, *Manuel Élémentaire de Droit International Public*, (Paris: Domat-Montchrestien, 1943).
Schachter, Oscar, *International Law in Theory and Practice*, (Dordrecht, Boston, London: Martinus Nijhoff, 1991).
Schneider, Jan, 'Codification and Progressive Development of International Environmental Law at the Third United Nations Conference on the Law of the Sea: The Environmental Aspects of the Treaty Review', 20 *Columbia Journal of Transnational Law* (1981), pp. 243-275.
Schneider, William, 'The Basel Convention Ban on Hazardous Waste Exports: Paradigm of Efficacy or Exercise in Futility?', 20 *Suffolk Transnational Law Review* (1996/1997), pp. 247-288.
Schoenbaum, Thomas J., *Admiralty and Maritime Law, Volume 2, Chapters 11-21*, 5th ed., (St. Paul, Minn.: West, 2011).
Schröder, Meinhard, 'Verantwortlichkeit, Völkerstrafrecht, Streitbeilegung und Sanktionen', in: Graf Vitzthum, Wolfgang (ed.), *Völkerrecht*, 5th ed., (Berlin: De Gruyter, 2010), pp. 585-644.
Schweisfurth, Theodor, *Völkerrecht*, (Tübingen: Mohr Siebeck, 2006).
Scovazzi, Tullio, 'New Ideas as Regards the Passage of Ships Carrying Hazardous Wastes: The 1996 Mediterranean Protocol', 7 *Review of European Community and International Environmental Law* (1998), pp. 264-267.
Scovazzi, Tullio, 'The Transboundary Movement of Hazardous Waste in the Mediterranean Regional Context', 19 *UCLA Journal of Environmental Law and Policy* (2000-2002), pp. 231-245.
Scovazzi, Tullio, 'State Responsibility for Environmental Harm', 12 *Yearbook of International Environmental Law* (2001), pp. 43-67.
Scovazzi, Tullio, 'The Developments within the "Barcelona System" for the Protection of the Mediterranean Sea Against Pollution', 26 *Annuaire de Droit Maritime et Oceanique* (2008), pp. 201-218.
Selden, John, *Of the Dominion, or, Ownership of the Sea*, transl. ed., (New York: Arno Press, 1972).
Sharma, Piyush K., 'The Basel Protocol: International Regime on Civil Liability and Compensation for Transboundary Movement of Hazardous Wastes', 26 *Delhi Law Review* (2004), pp. 181-196.
Shaw, Malcolm N., *International Law*, 5th ed., (Cambridge: Cambridge University Press, 2003).
Shaw, Richard, 'Hazardous and Noxious Substances - Is the End in Sight? Proposed Protocol to the HNS Convention 1996', 3 *Lloyd's Maritime and Commercial Law Quarterly* (2009), pp. 279-284.
Shaw, Richard, 'The 1996 HNS Convention - An Impossible Dream?', in: Berlingieri, Georgio/ Boglione, Angelo/Carbone, Sergio M./Siccardi, Francesco (ed.), *Scritti in Onore di Francesco Berlingieri, Volume II : Numero Speciale di Il Diritto Marittimo*, (Genova: 2010), pp. 906-912.
Shaw, Richard, 'IMO Diplomatic Conference Adopts HNS Protocol on 30 April 2010', January/ March 2010 *CMI Newsletter* (2010), pp. 8-11.
Shearer, C. Russell H., 'Comparative Analysis of the Basel and Bamako Conventions on Hazardous Waste', 23 *Environmental Law* (1993), pp. 141-183.
Shibata, Akiho, 'The Basel Compliance Mechanism', 12 *Review of European Community and International Environmental Law* (2003), pp. 183-198.
Shibata, Akiho, 'Ensuring Compliance with the Basel Convention : Its Unique Features', in: Beyerlin, Ulrich/Stoll, Peter-Tobias/Wolfrum, Rüdiger (ed.), *Ensuring Compliance with Multilateral Environmental Agreements : A Dialogue between Practitioners and Academia*, (Leiden, Boston: Martinus Nijhoff Publishers, 2006), pp. 69-87.

Sinclair, Ian, *The International Law Commission*, (Cambridge: Grotius Publications, 1987).
Smith, Brian D., *State Responsibility and the Marine Environment : The Rules of Decision*, (Oxford: Clarendon Press, 1988).
Soares, Guido Fernando Silva/Vargas, Everton Vieira, 'The Basel Liability Protocol on Liability and Compensation for Damage Resulting from Transboundary Movements of Hazardous Wastes and Their Disposal', 12 *Yearbook of International Environmental Law* (2001), pp. 69-104.
Sucharitkul, Sompong, 'State Responsibility and International Liability Under International Law', 18 *Loyola of Los Angeles International and Comparative Law Journal* (1995), pp. 821-839.
Suttles, John T., Jr., 'Transmigration of Hazardous Industry: The Global Race to the Bottom, Environmental Justice, and the Asbestos Industry', 16 *Tulane Environmental Law Journal* (2002/2003), pp. 1-64.
Tan, Alan Khee-Jin, *Vessel-Source Marine Pollution : The Law and Politics of International Regulation*, (Cambridge: Cambridge University Press, 2006).
Tomuschat, Christian, 'International Liability for Injurious Consequences Arising out of Acts not Prohibited by International Law: The Work of the International Law Commission', in: Francioni, Francesco/Scovazzi, Tullio (ed.), *International Responsibility for Environmental Harm*, (London, Dordrecht, Boston: Graham & Trotman, 1991), pp. 37-68.
Tomuschat, Christian, 'Individuals', in: Crawford, James/Pellet, Alain/Olleson, Simon (ed.), *The Law of International Responsibility*, (Oxford, New York: Oxford University Press, 2010), pp. 985-991.
'Trafigura found guilty of exporting toxic waste', BBC News, of 23 July 2010.
Triepel, Heinrich, *Völkerrecht und Landesrecht*, (Leipzig: C.L. Hirschfeld, 1899).
Tsimplis, Michael, 'Liability and Compensation in the International Transport of Hazardous Wastes by Sea: The 1999 Protocol to the Basel Convention', 16 *International Journal of Marine and Coastal Law* (2001), pp. 295-334.
Tsimplis, Michael, 'Marine Pollution from Shipping Activities', 14 *Journal of International Maritime Law* (2008), pp. 101-152.
Tuerk, Helmut, 'The Idea of the Common Heritage of Mankind', in: Martínez Gutiérrez, Norman A. (ed.), *Serving the Rule of International Maritime Law : Essays in Honour of Professor David Joseph Attard*, (London, New York: Routledge, 2010), pp. 156-175.
United Nations (ed.), *The Work of the International Law Commission*, 7th ed., (New York: United Nations, 2007).
Valin, Donna, 'The Basel Convention on the Control of Transboundary Movements of Hazardous Wastes and Their Disposal: Should the United States Ratify the Accord?', 6 *Indiana International & Comparative Law Review* (1995), pp. 267-288.
Vanden Bilcke, Christian, 'The Stockholm Convention on Persistent Organic Pollutants', 11 *Review of European Community and International Environmental Law* (2002), pp. 328-342.
Verdross, Alfred, 'Die systematische Verknüpfung von Recht und Moral', in: Sauer, Ernst (ed.), *Forum der Rechtsphilosophie*, (Köln: Balduin Pick, 1950), pp. 9-19.
Verdross, Alfred, *Völkerrecht*, 5th ed., (Wien: Springer, 1964).
Verdross, Alfred/Simma, Bruno, *Universelles Völkerrecht : Theorie und Praxis*, 3rd ed., (Berlin: Duncker & Humblot, 1984).
Vir, Arti K., 'Toxic Trade with Africa', 23 *Environmental Science & Technology* (1989), pp. 23-25.
de Visscher, Charles, *Théories et Réalités en Droit International Public*, 4th ed., (Paris: Éditions A. Pedone, 1970).
Wägener, Thomas, 'Dank der Schifffahrtskrise floriert die Abwrackindustrie', 149 *HANSA* (2012), pp. 74-79.
Walsh, Maureen T., 'The Global Trade in Hazardous Wastes: Domestic and International Attempts to Cope With a Growing Cisis in Waste Management', 42 *Catholic University Law Review* (1992/1993), pp. 103-140.

Weems, Philip R./Keenan, Kevin D., 'Is the LNG Industry Ready for Strict Liability?', Nov/Dec 2003 *LNG Journal* (2003), pp. 13-18.
Wendel, Philipp, *State Responsibility for Interferences with the Freedom of Navigation in Public International Law*, (Berlin, Heidelberg: Springer, 2007).
Wetterstein, Peter, 'Carriage of Hazardous Cargous by Sea - The HNS Convention', 26 *Georgia Journal of International and Comparative Law* (1996/1997), pp. 595-614.
Widawsky, Lisa, 'In My Backyard: How Enabling Hazardous Waste Trade to Developing Nations Can Improve the Basel Convention's Ability to Achieve Environmental Justice', 38 *Environmental Law* (2008), pp. 577-625.
Wiedemann, Erich, 'Die schlimmste Fracht meines Lebens', DER SPIEGEL, of 30 May 1988.
Williams, Jeffery D., 'Trashing Developing Nations: The Global Hazardous Waste Trade', 39 *Buffalo Law Review* (1991), pp. 275-312.
Winter, Gerd/Jans, Jan H./Macrory, Richard/Krämer, Ludwig, 'Weighing up the EC Environmental Liability Directive', 20 *Journal of Environmental Law* (2008), pp. 163-191.
Wirth, David A., 'Trade Implications of the Basel Convention Amendment Banning North-South Trade in Hazardous Wastes', 7 *Review of European Community and International Environmental Law* (1998), pp. 237-248.
Wolfrum, Rüdiger, 'The Principle of the Common Heritage of Mankind', 43 *Zeitschrift für ausländisches öffentliches Recht und Völkerrecht* (1983), pp. 312-338.
Wolfrum, Rüdiger, 'Purposes and Principles of International Environmental Law', 33 *German Yearbook of International Law* (1990), pp. 308-330.
Wolfrum, Rüdiger, 'Means of Ensuring Compliance with and Enforcement of International Environmental Law', 272 *Recueil des Cours* (1998), pp. 9-154.
Wolfrum, Rüdiger, 'Obligation of Result Versus Obligation of Conduct: Some Thoughts About the Implementation of International Obligations', in: Arsanjani, Mahnoush H./Cogan, Jacob Katz/Sloane, Robert D./Wiessner, Siegfried (ed.), *Looking to the Future : Essays on International Law in Honor of W. Michael Reisman*, (Leiden, Boston: Martinus Nijhoff, 2011), pp. 363-383.
Wolfrum, Rüdiger/Langenfeld, Christine/Minnerop, Petra, *Environmental Liability in International Law: Towards a Coherent Conception : Research Report 202 18 148*, (Berlin: German Federal Environmental Agency [Umweltbundesamt]; Erich Schmidt Verlag, 2005).
Wolfrum, Rüdiger/Matz, Nele, *Conflicts in International Environmental Law*, (Berlin, Heidelberg, New York, et al.: Springer, 2003).
Wood, Michael C., 'International Seabed Authority: The First Four Years', 3 *Max Planck Yearbook of United Nations Law* (1999), pp. 173-241.
Wynne, Brian, 'The Toxic Waste Trade: International Regulatory Issues and Options', 11 *Third World Quarterly* (1989), pp. 120-146.
Zhu, Ling, *Compulsory Insurance and Compensation for Bunker Oil Pollution Damage*, (Berlin, Heidelberg: Springer, 2007).

Part 2: Documents, Reports, Press Releases

A. United Nations and Subsidiary Bodies

I. *United Nations General Assembly:*
A/44/461 (of 18 September 1989): Protection and preservation of the marine environment, Report of the Secretary-General.

II. *United Nations Environment Programme:*
UNEP, Press Release of 14 December 1999: Compensation and Liability Protocol Adopted by Basel Convention on Hazardous Wastes, available at www.unep.org.

III. *Conference of the Parties to the Basel Convention:*

UNEP/CHW.1/5 (of 7 July 1992): Note of the Secretariat on Draft Articles of a Protocol on Liability and Compensation for Damage Resulting from the Transboundary Movement of Hazardous Wastes and their Disposal.
UNEP/CHW.1/24 (of 5 December 1992): Report of the First Meeting of the Conference of the Parties to the Basel Convention, 3 - 4 December 1992.
UNEP/CHW.2/30 (of 25 March 1994): Report of the Second Meeting of the Conference of the Parties to the Basel Convention, 21 - 25 March 1994.
UNEP/CHW.3/35 (of 28 November 1995): Decisions Adopted by the Third Meeting of the Conference of the Parties to the Basel Convention, 18 - 22 September 1995.
UNEP/CHW.4/35 (of 18 March 1998): Report of the Fourth Meeting of the Conference of the Parties to the Basel Convention, 23 - 27 February 1998.
UNEP/CHW.5/29 (of 10 December 1999): Report of the Fifth Meeting of the Conference of the Parties to the Basel Convention, 6 - 10 December 1999.
UNEP/CHW.6/10 (of 14 August 2002): Enlargement of the Scope of the Technical Cooperation Trust Fund.
UNEP/CHW.6/40 (of 10 February 2003): Report of the Sixth Meeting of the Conference of the Parties to the Basel Convention, 9 - 13 December 2002.
UNEP/CHW.7/INF/11 (of 16 September 2004): Report of the Regional Workshop Aimed at Promoting Ratification of the Basel Protocol held in Addis Ababa, Ethiopia from 30 August to 2 September 2004.
UNEP/CHW.7/INF/11/Add. 1 (of 30 September 2004): Report of the Regional Workshop Aimed at Promoting Ratification of the Basel Protocol held in Buenos Aires, Argentina from 22 to 25 June 2005.
UNEP/CHW.7/INF/11/Add. 2 (of 28 October 2004): Report of the Regional Workshop Aimed at Promoting Ratification of the Basel Protocol held in San Salvador, El Salvador from 25 June to 1 July 2004.
UNEP/CHW.7/33 (of 25 January 2005): Report of the Seventh Meeting of the Conference of the Parties to the Basel Convention, 25 - 29 October 2004.
UNEP/CHW.8/INF/16 (of 6 November 2006): Report of the Regional Workshop Aimed at Promoting Ratification of the Basel Protocol held in Cairo, Egypt from 30 October to 1 November 2006.
UNEP/CHW.8/INF/16/Add.1 (of 16 June 2006): Report of the Regional Workshop Aimed at Promoting Ratification of the Basel Protocol held in Yogyakarta, Indonesia from 16 to 18 May 2006.
UNEP/CHW.8/16 (of 5 January 2007): Report of the Parties to the Basel Convention on its Eight Meeting, 27 November - 1 December 2006.
UNEP/CHW.9/29 (17 April 2008): Protocol on Liability and Compensation: Note by the Secretariat.
UNEP/CHW.9/39 (27 June 2008): Report of the Parties to the Basel Convention on its Ninth Meeting, 23 - 27 June 2008.
UNEP/CHW.10/INF/4 (of 10 August 2011): Transboundary Movements of Hazardous Wastes: Quantities Moved, Reasons for Movements and their Impact on Human Health and the Environment.
UNEP/CHW.10/INF/16 (of 7 October 2011): Legal Analysis of the Application of the Basel Convention to Hazardous Wastes and Other Wastes Generated on board Ships.
UNEP/CHW.10/28 (of 1 November 2011): Report of the Conference of the Parties to the Basel Convention on its Tenth Meeting, 17 – 21 October 2011.

IV. *Legal Working Group of the Basel Convention:*
UNEP/CHW/LWG/4/4 (of 4 December 2001): Legal Aspects of the Full and Partial Dismantling of Ships.

Bibliography 351

V. *Open-ended Working Group of the Basel Convention on the Control of Transboundary Movements of Hazardous Wastes and Their Disposal:*

UNEP/CHW/OEWG/4/8 (of 25 May 2005): Instruction manual for the implementation of the Basel Protocol on Liability and Compensation.

UNEP/CHW/OEWG/4/INF/4 (of 26 May 2005): Summary of the obstacles and difficulties faced by the Parties in the process of ratification or accession to the Protocol.

UNEP/CHW/OEWG/4/INF/20 (of 27 June 2005): Environmentally Sound Management of Ship Dismantling : Note by Greenpeace and the Basel Action Network (BAN).

UNEP/CHW/OEWG/5/INF/6 (of 6 February 2006): Report of the Regional Workshop Aimed at Promoting Ratification of the Basel Protocol held in Warsaw, Poland from 18 to 20 January 2006.

UNEP/CHW/OEWG/5/2/Add. 7 (of 2 March 2006): Basel Protocol on Liability and Compensation: insurance, other financial guarantees and financial limits.

UNEP/CHW/OEWG/6/2 (of 2 July 2007): Decision VIII/1 on Côte d'Ivoire.

UNEP/CHW/OEWG/6/14 (of 29 June 2007): Basel Protocol on Liability and Compensation: insurance, bonds or other financial guarantees.

VI. *Ad Hoc Working Group on Legal and Technical Experts to Consider and Develop a Draft Protocol on Liability and Compensation for Damage Resulting from Transboundary Movements of Hazardous Wastes and their Disposal:*

UNEP/CHW.1/WG.1/1/5 (of 16 September 1993): Report on the Work of the 1st Session, 13 - 17 September 1993.

UNEP/CHW.1/WG.1/2/4 (of 24 October 1994): Report on the Work of the 2nd Session, 10 - 14 October 1994.

UNEP/CHW.1/WG.1/3/2 (of 17 March 1995): Report on the Work of the 3rd Session, 20 - 24 February 1995.

UNEP/CHW.1/WG.1/4/2 (of 3 July 1996): Report on the Work of the 4th Session, 24 - 28 June 1996.

UNEP/CHW.1/WG.1/5/5 (of 23 May 1997): Report on the Work of the 5th Session, 20 - 23 May 1997.

UNEP/CHW.1/WG.1/6/2 (of 26 June 1998): Report on the Work of the 6th Session, 24 - 26 June 1998.

UNEP/CHW.1/WG.1/7/2 (of 9 October 1998): Report on the Work of the 7th Session, 7 - 9 October 1998.

UNEP/CHW.1/WG.1/8/5 (of 15 January 1999): Report on the Work of the 8th Session, 11 - 15 January 1999.

UNEP/CHW.1/WG.1/9/2 (of 28 April 1999): Report on the Work of the 9th Session, 19 - 23 April 1999.

UNEP/CHW.1/WG.1/10/2 (of 20 September 1999): Report on the Work of the 10th Session, 30 August - 3 September 1999.

VII. *Expanded Bureau of the Conference of the Parties to the Basel Convention:*

UNEP/SBC/BUREAU/8/1/INF/2 (of 16 April 2007): Report of the mission undertaken in Côte d'Ivoire following the incident of the Probo Koala, in the context of decision V/32.

VIII. *Secretariat of the Basel Convention:*

Bocken, H., de Kezel, E., Bernauw, K., *Limitations of Liability and Compulsory Insurance under the Protocol on Liability for Transboundary Movements of Hazardous Waste and Other Waste : Report Prepared at the Request of the Secretariat of the Basel Convention in Connection with the Preparation of the Protocol on Liability and Compensation for Damage Resulting from the Transboundary Movement of Hazardous Wastes and their Disposal*, available at www.basel.int.

Secretariat of the Basel Convention (ed.), *Global Trends in Generation and Transboundary Movement of Hazardous Wastes and Other Wastes : Analysis of the Data Provided by Parties to the Secretariat of the Basel Convention*, prepared by Wielenga, Kees (Châtelaine: Secretariat of the Basel Convention, 2002).

Secretariat of the Basel Convention (ed.), *Instruction Manual for the Implementation of the Basel Protocol on Liability and Compensation for Damage Resulting from Transboundary Movements of Hazardous Wastes and their Disposal*, as adopted by OEWG Decision IV/7 and as published in OEWG Doc. UNEP/CHW/OEWG/4/8.

Secretariat of the Basel Convention (ed.), *Technical Guidelines for the Environmentally Sound Management of the Full and Partial Dismantling of Ships*, as adopted by OEWG Decision VI/24 (Châtelaine: Secretariat of the Basel Convention, 2003).

Secretariat of the Basel Convention (ed.), *Waste without Frontiers : Global Trends in Generation and Transboundary Movements of Hazardous Wastes and Other Wastes : Analysis of the Data from National Reporting to the Secretariat of the Basel Convention for the Years 2004-2006*, prepared by Wielenga, Kees (Geneva: Secretariat of the Basel Convention, 2010).

IX. *International Law Commission:*

Report to the General Assembly, *Yearbook of the International Law Commission* (1949), pp. 277-290.

A/7610/Rev. 1: Report of the International Law Commission on the Work of its Twenty-First Session, 2 June - 8 August 1969, *Yearbook of the International Law Commission* (1969, Volume II), pp. 203-237.

A/CN.4/246 and Add. 1-3 (of 5 March, 7 April, 28 April and 18 May 1971): Third Report on State Responsibility, by Mr. Roberto Ago, Special Rapporteur: The Internationally Wrongful Act of the State, Source of International Responsibility, *Yearbook of the International Law Commission* (1971, Volume II, Part One), pp. 199-274.

A/9010/Rev. 1: Report of the International Law Commission on the Work of its Twenty-Fifth Session, 7 May - 13 July 1973, *Yearbook of the International Law Commission* (1973, Volume II), pp. 161-235.

A/32/10: Report of the International Law Commission on the Work of its Twenty-Ninth Session, 9 May - 29 July 1977, *Yearbook of the International Law Commission* (1977, Volume II, Part 2), pp. 1-135.

A/33/10: Report of the International Law Commission on the Work of its Thirtieth Session, 8 May - 28 July 1978, *Yearbook of the International Law Commission* (1978, Volume II, Part 2), pp. 1-189.

A/CN.4/325 (of 23 July 1979): Report of the Working Group on Review of the Multilateral Treaty-Making Process, *Yearbook of the International Law Commission* (1979, Volume II, Part 1), pp. 183-212.

A/CN.4/334 and Add. 1 and 2 (of 24 and 27 June and 4 July 1980): Preliminary Report on International Liability for Injurious Consequences Arising out of Acts not Prohibited by International Law, by Mr. Robert Q. Quentin Baxter, Special Rapporteur, *Yearbook of the International Law Commission* (1980, Volume II, Part One), pp. 247-266.

A/35/10: Report of the International Law Commission on the Work of its Thirty-Second Session, 5 May - 25 July 1980, *Yearbook of the International Law Commission* (1980, Volume II, Part 2), pp. 1-174.

A/CN.4/360 (of 23 June 1982): Third Report on International Liability for Injurious Consequences Arising out of Acts not Prohibited by International Law, by Mr. Robert Q. Quentin-Baxter, Special Rapporteur, *Yearbook of the International Law Commission* (1982, Volume II, Part 1), pp. 51-64.

A/37/10: Report of the International Law Commission on the Work of its Thirty-Fourth Session, 3 May - 23 July 1982, *Yearbook of the International Law Commission* (1982, Volume II, Part 2), pp. 1-146.

Bibliography 353

A/CN.4/402 (of 13 May 1986): Second Report on International Liability for Injurious Consequences Arising out of Acts not Prohibited by International Law, by Mr. Julio Barboza, Special Rapporteur, *Yearbook of the International Law Commission* (1986, Volume II, Part One), pp. 145-161.

A/CN.4/428 and Add. 1 (of 15 March 1990): Sixth Report on International Liability for Injurious Consequences Arising out of Acts not Prohibited by International Law, by Mr. Julio Barboza, Special Rapporteur, *Yearbook of the International Law Commission* (1990, Volume II, Part 1), pp. 83-109.

A/47/10: Report of the International Law Commission on the Work of its Forty-Fourth Session, 4 May - 24 July 1992, *Yearbook of the International Law Commission* (1992, Volume II, Part 2), pp. 1-80.

A/51/10: Report of the International Law Commission on the Work of its Forty-Eighth Session, 6 May - 26 July 1996, *Yearbook of the International Law Commission* (1996, Volume II, Part 2), pp. 1-143.

A/52/10: Report of the International Law Commission on the Work of its Forty-Ninth Session, 12 May - 18 July 1997, *Yearbook of the International Law Commission* (1997, Volume II, Part 2), pp. 1-74.

A/CN.4/498 and Add. 1-4 (of 17 March, 1 and 30 April, 19 July 1999): Second Report on State Responsibility, by Mr. James Crawford, Special Rapporteur, *Yearbook of the International Law Commission* (1999, Volume II, Part 2), pp. 3-96.

A/53/10: Report of the International Law Commission on the Work of its Fiftieth Session, 20 April - 12 June and 27 July - 14 August 1998, *Yearbook of the International Law Commission* (1998, Volume 2, Part 2), pp. 1-113.

A/CN.4/510 (of 9 June 2000): Third Report on International Liability for Injurious Consequences Arising out of Acts not Prohibited by International Law (Prevention of Transboundary Damage from Hazardous Activities), by Mr. Pemmaraju Sreenivasa Rao, Special Rapporteur, *Yearbook of the International Law Commission* (2000, Volume II, Part 1), pp. 113-125.

A/56/10: Report of the International Law Commission on the Work of its Fifty-Third Session (23 April - 1 June and 2 July - 10 August 2001), *Yearbook of the International Law Commission* (2001, Volume II, Part 2), pp. 1-208.

A/59/10: Report of the International Law Commission on the Work of its Fifty-Sixth Session (3 May - 4 June and 5 July - 6 August 2004).

A/61/10: Report of the International Law Commission on the Work of its Fifty-Eighth Session (1 May - 9 June and 3 July - 11 August 2006).

International Law Commission, *Commentaries on the Draft Articles on Responsibility of States for Internationally Wrongful Acts*, as adopted in 2001 (ILC Doc. A/56/10).

International Law Commission, *Commentaries on the Draft Articles on Prevention of Transboundary Harm from Hazardous Activities*, as adopted in 2001 (ILC Doc. A/56/10).

X. ***International Conference on Hazardous and Noxious Substances and Limitation of Liability, 1996:***

LEG/CONF.10/8/3 (of 9 May 1996): Adoption of the Final Act and any Instruments, Recommendations and Resolutions Resulting from the Work of the Conference.

XI. ***IMCO/IMO:***

C/ES.III/5 (of 8 May 1967): Conclusions of the IMCO Council on the Action to be Taken on the Problems Brought to Light by the Los of the 'Torrey Canyon'.

LEG 71/3/4 (of 5 August 1994): Consideration of a Draft HNS Convention.

LEG 80/10/2 (of 5 July 1999): Special Consultative Meeting to Discuss the Hazardous and Noxious Substances Convention, Friday, 16 April 1999.

LEG 80/10/3 (6 August 1999): Draft Terms of Reference for a HNS Correspondence Group.

LEG 80/11 (of 25 October 1999): Report of the Legal Committee on the Work of its Eightieth Session.

LEG 87/11 (of 6 August 2003): Report of the Special Consultative Meeting of the HNS Correspondence Group in Ottawa, 3-5 June 2003.
LEG 87/11/1 (of 8 August 2003): Papers Discussed at the Special Consultative Meeting of the HNS Correspondence Group in Ottawa, 3-5 June 2003.
LEG 93/13 (of 2 November 2007): Report of the Legal Committee on the Work of its Ninety-Third Session.
LEG 94/12 (of 31 October 2008): Report of the Legal Committee on the Work of its Ninety-Fourth Session.
LEG 95/10 (of 31 October 2008): Report of the Legal Committee on the Work of its Ninety-Fifth Session.

XII. *IOPC Fund:*
71FUND/EXC.52/9 (20 January 1997): Admissibility of Claims Relating to Salvage Operations and similar Activities.
92FUND/A.1/34 (of 28 June 1996): Record of Decisions of the 1st Session of the Assembly.
92FUND/A.12/25/1 (of 20 September 2007): Report of the Correspondence Group on Annual Contributions to the LNG Account.
92FUND/A.12/25/2 (of 20 September 2007): Implementation of the Definition of 'Receiver' in Article 1.4(A) of the HNS Convention.
92FUND/A.12/25/3 (of 20 September 2007): Depositing Instruments of Ratification without Accompanying Contributing Cargo Reports and Common Ratification of the HNS Convention.
92FUND/A.12/28 (of 19 October 2007): Records of the Decisions of the Twelfth Session of the Assembly.
92FUND/WGR.5/2 (of 18 January 2008): Facilitating the Entry into Force of the HNS Convention. Consideration of a Draft Text of a Protocol to the HNS Convention.
92FUND/WGR.5/3 (of 18 January 2008): Facilitating the Entry into Force of the HNS Convention. Consideration of a Draft Text of a Protocol to the HNS Convention.
92FUND/WGR.5/4 (of 18 January 2008): Facilitating the Entry into Force of the HNS Convention. Consideration of a Draft Text of a Protocol to the HNS Convention.
92FUND/A/ES.13/5, 92FUND/WGR.5/8 (of 16 May 2008): Report on the First Meeting of the Fifth Intersessional Working Group ('HNS Focus Group').
92FUND/A/ES.13/5/2, 92 FUND/WGR.5/10/1 (of 5 June 2008): Draft Protocol - Consolidated Text
92FUND/A.13/22, 92FUND/WGR.5/13 (of 17 September 2008): Report on the Second Meeting of the Fifth Intersessional Working Group ('HNS Focus Group').
IOPC Fund Annual Report 1988.
IOPC Fund Annual Report 1991.
IOPC Fund Annual Report 1992.
IOPC Fund Annual Report 1999.

XIII. *International Conference on the Revision of the HNS Convention:*
LEG/CONF.17/10 (of 4 May 2010): Text adopted by the Conference
LEG/CONF.17/11 (of 4 May 2010): Conference Resolutions.
LEG/CONF.17/12 (of 4 May 2010): Final Act of the International Conference on the Revision of the HNS Convention.

XIV. *Human Rights Council:*
A/HRC/12/26/Add.2 (of 3 September 2009): Report of the Special Rapporteur on the Adverse Effects of the Movement and Dumping of Toxic and Dangerous Products and Wastes on the Enjoyment of Human Rights.

XV. *Office for the Coordination of Humanitarian Affairs:*
OCHA/GVA/2006/0184 (of 7 September 2006): OCHA Situation Report No. 1, Toxic Waste Pollution Crisis – Côte d'Ivoire.

Bibliography

OCHA/GVA/2006/0190 (of 14 September 2006): OCHA Situation Report No. 5, Toxic Waste Pollution Crisis – Côte d'Ivoire.

XVI. *UN Compensation Commission:*

S/AC.26/1991/7/Rev.1 (of 17 March 1992): Decision Taken by the Governing Council of the United Nations Compensation Commission.

B. **OECD**

OECD (ed.), *OECD Environmental Data : Compendium 2006-2008 – Waste* (Paris: OECD, 2008).

C. **European Union**

COM(2006) 430 final (of 1 August 2006): Report on the Implementation of Council Regulation (EEC) No 259/93 of 1 February 1993 on the Supervision and Control of Shipments of Waste within, into and out of the European Community: Generation, Treatment and Transboundary Shipment of Hazardous Waste and Other Waste in the Member States of the European Union, 1997-2000.

SEC(2006) 1053 (of 1 August 2006): Commission Staff Working Document annexed to EC Doc. COM(2006) 430 final.

COM(2007) 269 final (of 22 May 2007): Green Paper on Better Ship Dismantling.

COM(2009) 282 final (of 24 June 2009): Report on the Implementation of Council Regulation (EEC) No 259/93 of 1 February 1993 on the Supervision and Control of Shipments of Waste within, into and out of the European Community: Generation, Treatment and Transboundary Shipment of Hazardous Waste and Other Waste in the Member States of the European Union, 2001-2006.

SEC(2009) 811 final (of 24 June 2009): Commission Staff Working Document accompanying EC Doc. COM(2009) 282 final.

COM(2011) 131 final (of 17 March 2011): Report on Statistics Compiled Pursuant to Regulation (EC) No 2150/2002 on Waste Statistics and their Quality.

European Union (ed.), *Europe in Figures : Eurostat Yearbook 2011* (Luxembourg: European Union, 2011).

D. **National Governmental Institutions**

Haitian Government, Press Release of 22 April 2000: The Haitian People Achieve Environmental Justice for Earth Day, available at www.essentialaction.org/return/government.txt.

E. **Other Institutions**

Basel Action Network, Press Release of 10 December 1999: Hazardous Waste Agreement on Liability Protocol Reached at Basel Conference of Parties, available at http://ban.org/ban_news/hazardous3.html.

Basel Action Network and Greenpeace, Press Release of 10 January 2002: Shipbreaking and the Legal Obligations under the Basel Convention, available at: http://ban.org/library/ShipbreakingLegal%20Final.pdf

Greenpeace, Press Release of 29 October 1998: Philadelphia Incinerator Ash to Return from Haiti to the U.S., available at www.essentialaction.org/return/Clean_Up.html.

Table of Cases

I. Permanent Court of International Justice

The S.S. Wimbledon Case, Judgment of 17 August 1923, PCIJ Series A No. 1 (1923), pp. 14–34.

German Settlers in Poland, Advisory Opinion of 10 September 1923, PCIJ Series B No. 6 (1923), pp. 5–43.

The S.S. Lotus Case (France v. Turkey), Judgment of 7 September 1927, PCIJ Series A No. 10 (1927), pp. 3–33.

Factory at Chorzów Case (Germany v. Poland), Claim for Indemnity, Merits, Judgment of 13 September 1928, PCIJ Series A No. 17 (1928), pp. 3–65.

Case Relating to the Territorial Jurisdiction of the International Commission of the River Oder, Judgment of 10 September 1929, PCIJ Series A No. 23 (1929), pp. 4–32.

The Diversion of Water from the Meuse (Netherlands v. Belgium), Judgment of 28 June 1937, PCIJ Series A/B No. 70, pp. 3–33.

Phosphates in Morocco Case (Italy v. France), Preliminary Objections, Judgment of 14 June 1938, PCIJ Series A/B No. 74 (1938), pp. 9–30.

II. International Court of Justice

The Corfu Channel Case (United Kingdom v. Albania), Merits, Judgment of 9 April 1949, ICJ Reports 1949, pp. 2–169.

South West Africa Cases (Ethiopia v. South Africa; Liberia v. South Africa), Preliminary Objections, Judgment of 21 December 1962, ICJ Reports 1962, pp. 319–348.

Case Concerning the Barcelona Traction, Light and Power Company, Limited, (Belgium v. Spain), Second Phase, Judgment of 5 February 1970, ICJ Reports 1970, pp. 3–53.

Case Concerning United States Diplomatic and Consular Staff in Tehran (United States v. Iran), Judgment of 24 May 1980, ICJ Reports 1980, pp. 3–46.

Legality of the Threat or Use of Nuclear Weapons, Advisory Opinion of 8 July 1996, ICJ Reports 1996, pp. 226–267.

The Gabčíkovo-Nagymaros Project Case (Hungary v. Slovakia), Judgment of 25 September 1997, ICJ Reports 1997, pp. 4–84.

Difference Relating to Immunity from Legal Process of a Special Rapporteur of the Commission on Human Rights, Advisory Opinion of 29 April 1999, ICJ Reports 1999, pp. 62–91.

Case Concerning Pulp Mills on the River Uruguay (Argentina v. Uruguay), Judgment of 20 April 2010, ICJ Reports 2010, pp. 14–107.

III. International Tribunal for the Law of the Sea

The M/V "Saiga" (No. 2) Case (St. Vincent and the Grenadines v. Guinea), Judgment of 1 July 1999, ITLOS Reports 1999, pp. 10 *et seq.*

IV. Iran-United States Claims Tribunal

Cases No. A15(IV) and A24 (Islamic Republic of Iran v. United States of America), Partial Award of 28 December 1998, 11 World Trade and Arbitration Materials (1999), pp. 47–164.

V. Permanent Court of Arbitration

The MOX Plant Case (Ireland v. United Kingdom), Order No. 3 (Suspension of Proceedings on Jurisdiction and Merits and Request for Further Provisional Measures) of 24 June 2003.

VI. International Arbitral Tribunals

Davis Case, Merits, Arbitral Award of 1903, 9 RIAA (1959), pp. 460–464.

Salas Case, Arbitral Award of 1903, 10 RIAA (1960), pp. 720–721.

Affaire de Casablanca (Germany v. France), Arbitral Award of 22 May 1909, 11 RIAA (1961), pp. 119–131.

Home Frontier and Foreign Missionary Society of the United Brethren in Christ (United States) v. Great Britain, Arbitral Award of 18 December 1920, 6 RIAA (1955), pp. 42–44.

Owners of the Jessie, the Thomas F. Bayard and the Pescawha (Great Britain) v. United States, Arbitral Award of 2 December 1921, 6 RIAA (1955), pp. 57–60.

China Navigation Co., Ltd. (Great Britain) v. United States (Newchwang Case), Arbitral Award of 9 December 1921, 6 RIAA (1955), pp. 64–68.

British Claims in the Spanish Zone of Morocco (Spain v. United Kingdom), Arbitral Award of 1 May 1925, 2 RIAA (1949), pp. 615–742.

D. Earnshaw and Others (Great Britain) v. United States (Zafiro Case), Arbitral Award of 30 November 1925, 6 RIAA (1955), pp. 160–165.

L. F. H. Neer and Pauline Neer (U.S.A.) v. United Mexican States, Arbitral Award of 15 October 1926, 4 RIAA (1951), pp. 60–66.

H. G. Venable (U.S.A.) v. United Mexican States, Arbitral Award of 8 July 1927, 4 RIAA (1951), pp. 219–261.

Responsabilité de l'Allemagne à raison des dommages causés dans les colonies portugaises du sud de l'Afrique (sentence sur le principe de la responsabilité) (Portugal v. Germany), Arbitral Award of 31 July 1928, 2 RIAA (1949), pp. 1011–1033.

Estate of Jean-Baptiste Caire (France) v. United Mexican States, Arbitral Award of 7 June 1929, 5 RIAA (1952), pp. 516–534.

Walter A. Noyes (United States) v. Panama, Arbitral Award of 22 May 1933, 6 RIAA (1955), pp. 308–312.

Trail Smelter Arbitration Award (United States v. Canada), Arbitral Award of 16 April 1938 and 11 March 1941, 3 RIAA (1949), pp. 1905–1982.

Case Concerning the Difference between New Zealand and France Concerning the Interpretation or Application of two Agreements, Concluded on 9 July 1986 between the two States and which Related to the Problems Arising from the Rainbow Warrior Affair, Decision of 30 April 1990, 20 RIAA (1990), pp. 215–284.

Table of International Conventions and Agreements, OECD, EU and Other Legal Instruments

I. *International Conventions and Agreements:*

Convention for the Unification of Certain Rules of Law Respecting Assistance and Salvage at Sea, Brusses, 23 September 1910, entered into force 1 March 1913; as amended by the Protocol of 27 May 1967 (entered into force 15 August 1977).

International Convention for the Unification of Certain Rules of Law related to Bills of Lading (Hague Rules), Brussels, 25 August 1924, entered into force 2 June 1931; as amended by the Protocol of 23 February 1968 (Visby Protocol) (entered into force 23 June 1977) and by the Protocol of 21 December 1979 (entered into force 14 February 1984).

Charter of the United Nations, San Francisco, 26 June 1945, entered into force 24 October 1945.

General Agreement on Tariffs and Trade (GATT), 30 October 1947, entered into force 1 January 1948.

International Convention for the Prevention of Pollution of the Sea by Oil (OILPOL), London, 12 May 1954, entered into force, 26 July 1958.

European Agreement Concerning the International Carriage of Dangerous Goods by Road (ADR), Geneva, 30 September 1957, entered into force 29 January 1968; as amended by the Protocol of 21 August 1975 (entered into force 19 April 1985).

Convention on the Territorial Sea and the Contiguous Zone, Geneva, 29 April 1958, entered into force 10 September 1964.

Convention on the Continental Shelf, Geneva, 29 April 1958, entered into force 10 June 1964.

Convention on the High Seas, Geneva, 29 April 1958, entered into force 30 September 1962.

Convention on Fishing and Conversation of Living Resources of the High Seas, Geneva, 29 April 1958, entered into force 20 March 1966.

Antarctic Treaty, Washington, 1 December 1959, entered into force 23 June 1961.

Paris Convention on Third Party Liability in the Field of Nuclear Energy, Paris, 29 July 1960, entered into force 1 April 1968; as amended by the Protocol of 28 January 1964 (entered into force 1 April 1968), by the Protocol of 16 November

1982 (entered into force 7 October 1988) and by the Protocol of 12 February 2004 (not yet in force).

Convention on the Organisation for Economic Co-operation and Development (OECD Convention), Paris, 14 December 1960, entered into force 30 September 1961.

Convention on the Liability of Operators of Nuclear Ships, Brussels, 25 May 1962, not yet in force.

Brussels Convention Supplementary to the Paris Convention, Brussels, 31 January 1963, entered into force 4 December 1974; as amended by the Protocol of 28 January 1964 (entered into force 4 December 1974), the Protocol of 16 November 1982 (entered into force 1 August 1991) and the Protocol of 12 February 2004 (not yet in force).

Vienna Convention on Civil Liability for Nuclear Damage, Vienna, 21 May 1963, entered into force 12 November 1977; as amended by the Protocol of 29 September 1997 (entered into force on 4 October 2003).

Treaty on Principles Governing the Activities of States in the Exploration and Use of Outer Space, including the Moon and Other Celestial Bodies (Outer Space Treaty), London, Moscow, Washington, 27 January 1967, entered into force 10 October 1967.

Vienna Convention of the Law of Treaties, Vienna, 23 May 1969, entered into force 27 January 1980.

International Convention on Civil Liability for Oil Pollution Damage (CLC), Brussels, 29 November 1969, entered into force 19 June 1975; as amended by the Protocol of 19 November 1976 (entered into force 8 April 1981) and by the Protocol of 27 November 1992 (entered into force 30 May 1996).

Convention Relating to Civil Liability in the Field of Maritime Carriage of Nuclear Material (NUCLEAR Convention), Brussels, 17 December 1971, entered into force 15 July 1975.

International Convention on the Establishment of an International Fund for Compensation for Oil Pollution Damage (Fund Convention), Brussels, 18 December 1971, entered into force 16 October 1978, ceased to be in force from 24 May 2002; as amended by the Protocol of 19 December 1976 (entered into force 22 November 1994) and of the Protocol of 27 September 2000 (entered into force 27 June 2001).

Convention on International Liability for Damage Caused by Space Objects (Space Liability Convention), London Moscow, Washington, 29 March 1972, entered into force 1 September 1972.

London Convention on the Prevention of Marine Pollution by Dumping of Wastes and Other Matter (London Dumping Convention), London, 29 December 1972, entered into force 24 March 2006; as amended by the Protocol of 7 November 1996 (entered into force 24 March 2006).

International Convention for the Prevention of Pollution from Ships (73/78 MARPOL Convention), London, 2 November 1973, as amended by the Protocol of 17 February 1978, entered into force 2 October 1983; as supplemented by Annex III (entered into force 1 July 1992), Annex IV (entered into force 27

September 2003), Annex V (entered into force 31 December 1988) and Annex VI (entered into force 19 May 2005).

International Convention for the Safety of Life at Sea (SOLAS), London, 1 November 1974, entered into force 25 May 1980.

Athens Convention relating to the Carriage of Passengers and their Luggage by Sea, Athens, 13 December 1974, entered into force 28 April 1987.

Barcelona Convention for the Protection of the Marine Environment and the Coastal Region of the Mediterranean, Barcelona, 16 February 1976, entered into force 12 February 1978; as amended by the Protocol of 10 June 1995 (entered into force 9 July 2004).

Convention on Limitation of Liability for Maritime Claims (LLMC Convention), London, 19 November 1976, entered into force 1 December 1986; as amended by the Protocol of 2 May 1996 (entered into force 13 May 2004) and Amendments to the Protocol of 19 April 2010 (entered into force 8 June 2015).

Convention on Civil Liability for Oil Pollution Damage Resulting from Exploration and Exploitation of Seabed Mineral Resources, London, 1 May 1977, not yet in force.

United Nations Convention on the Carriage of Goods by Sea (Hamburg Rules), Hamburg, 31 March 1978, entered into force 1 November 1992.

Kuwait Regional Convention for Co-operation on the Protection of the Marine Environment from Pollution, Kuwait, 24 April 1978, entered into force 1 July 1979.

Convention the Physical Protection of Nuclear Material, Vienna, 26 October 1979, entered into force 8 February 1987.

Agreement Governing the Activities of States on the Moon and Other Celestial Bodies (Moon Treaty), New York, 5 December 1979, entered into force 11 July 1984.

Convention Concerning International Carriage by Rail (COTIF), Berne, 9 May 1980, as amended by the Protocol of 3 June 1999, entered into force 1 July 2006.

Abidjan Convention for Co-operation in the protection and Development of the Marine and Coastal Environment of the West and Central African Region, Abidjan, 1 October 1981, entered into force 5 August 1984.

Lima Convention for the Protection of the Marine Environment and Coastal Areas of the South-East Pacific, Lima, 12 November 1981, entered into force 19 May 1986.

Jeddah Regional Convention for the Conservation of the Red Sea and Gulf of Aden Environment, Jeddah, 14 February 1982, entered into force 20 August 1985.

United Nations Convention for the Law of the Sea (UNCLOS), Montego Bay, 10 December 1982, entered into force 16 November 1994;

Agreement Relating to the Implementation of Part XI of the United Nations Convention on the Law of the Sea of 10 December 1982, 28 July 1994, entered into force 28 July 1996.

Cartagena Convention for the Protection and Development of the Marine Environment of the Wider Caribbean Region, Cartagena, 24 March 1983, entered into force 11 October 1986.

Nairobi Convention for the Protection, Management and Development of the Marine and Coastal Environment of the Eastern African Region, Nairobi, 21 June 1985, entered into force 30 May 1996; as amended on 31 March 2010, not yet in force.

Noumea Convention for the Protection of Natural Resources and Environment of the South Pacific Region, Noumea, 24 November 1986, entered into force 22 August 1990.

Joint Protocol Relating to the Application of the Vienna Convention and the Paris Convention, Vienna, 21 September 1988, entered into force 27 April 1992.

Basel Convention on the Control of Transboundary Movements of Hazardous Wastes and Their Disposal, Basel, 22 March 1989, entered into force 5 May 1992; Ban Amendment to the Basel Convention, Geneva, 22 September 1995, not yet in force.

International Convention on Salvage, London, 28 April 1989, entered into force 14 July 1996.

Convention on Civil Liability for Damage Caused During Carriage of Dangerous Goods by Road, Rail and Inland Navigation Vessels (CRTD Convention), Geneva, 10 October 1989, not yet in force.

The Fourth African, Caribbean and Pacific States – European Economic Community Convention of Lomé (Lomé IV Convention), Lomé, 15 December 1989, entered into force 1 March 1990.

International Convention on Oil Pollution Preparedness, Response and Cooperation (OPRC Convention), Paris, 30 November 1990, entered into force 13 May 1995.

Bamako Convention on the Ban on the Import into Africa and the Control of Transboundary Movement and Management of Hazardous Wastes within Africa, Bamako, 30 January 1991, entered into force 22 April 1998.

Espoo Convention on Transboundary Environmental Impact Assessment, Espoo, 25 February 1991, entered into force 10 September 1997.

Madrid Protocol on Environmental Protection to the 1959 Antarctic Treaty, Madrid, 4 October 1991, entered into force, 14 January 1998.

Helsinki Convention on the Transboundary Effects of Industrial Accidents, Helsinki, 17 March 1992, entered into force 19 April 2000.

Helsinki Water Convention on the Protection and Use of Transboundary Watercourses and International Lakes, Helsinki, 17 March 1992, entered into force 6 October 1996.

Helsinki Convention on the Protection of the Marine Environment of the Baltic Sea Area, Helsinki, 9 April 1992, entered into force 17 January 2000.

Bucharest Convention on the Protection of the Black Sea Against Pollution, Bucharest, 21 April 1992, entered into force 15 January 1994.

OSPAR Convention for the Protection of the Marine Environment of the North-East Atlantic, Paris, 22 September 1992, entered into force 25 March 1998.

International Maritime Organization Protocol of 1992 to amend the International Convention on the Establishment of an International Fund for Compensation for Oil Pollution Damage of 18 December 1971 (1992 Fund

Convention), London, 27 November 1992, entered into force 30 May 1996; as amended by the 2003 Protocol on the Establishment of a Supplementary Fund for Oil Pollution Damage, 16 May 2003, entered into force on 3 March 2005.

Convention on Civil Liability for Damage Resulting from Activities Dangerous to the Environment (Lugano Convention), Lugano, 21 June 1993, not yet in force.

Waigani Convention to Ban the Importation into Forum Island Countries of Hazardous and Radioactive Wastes and to Control the Transboundary Movement and Management of Hazardous Wastes within the South Pacific Region, Waigani, 16 September 1995, entered into force 12 December 2008.

International Convention on Liability and Compensation for Damage in Connection with the Carriage of Hazardous and Noxious Substances by Sea (HNS Convention), London, 3 May 1996, not yet in force; as amended by the Protocol of 30 April 2010 (not yet in force).

Izmir Protocol on the Prevention of Pollution of the Mediterranean Sea by Transboundary Movements of Hazardous Wastes and Their Disposal (to the Barcelona Convention), Izmir, 1 October 1996, entered into force 19 January 2008.

Protocol to the Convention on the Prevention of Marine Pollution by Dumping of Wastes and Other Matter (to the London Convention), London, 7 November 1996, entered into force 24 March 2006.

Vienna Convention on Supplementary Compensation for Nuclear Damage, Vienna, 12 September 1997, not yet in force.

Kyoto Protocol to the United Nations Framework Convention on Climate Change, Kyoto, 11 December 1997, entered into force 16 February 2005.

Tehran Protocol on the Control of Marine Transboundary Movements and Disposal of Hazardous Wastes to the Kuwait Regional Convention for Cooperation on the Protection of the Marine Environment from Pollution (to the Kuwait Convention), Tehran, 17 March 1998, entered into force 6 September 2005.

Rotterdam PIC Convention on the Prior Informed Consent Procedure for Certain Hazardous Chemicals and Pesticides in International Trade, Rotterdam, 10 September 1998, entered into force 24 February 2004.

Basel Protocol on Liability and Compensation for Damage Resulting from Transboundary Movements of Hazardous Wastes and Their Disposal (to the Basel Convention), Basel, 10 December 1999, not yet in force.

Protocol on Preparedness, Response and Co-operation to Pollution Incidents by Hazardous and Noxious Substances (OPRC-HNS Protocol), London, 15 March 2000, entered into force 14 June 2007.

European Agreement concerning the International Carriage of Dangerous Goods by Inland Waterways (ADN), Geneva, 26 May 2000, entered into force 29 February 2008.

Partnership Agreement between the Members of the African, Caribbean and Pacific Group of States, of the One Part, and the European Community and its Member States, of the Other Part (Cotonou Agreement), Cotonou, 23 June 2000, entered into force 1 April 2003.

International Convention on Civil Liability for Bunker Oil Pollution Damage (Bunker Oil Convention), London, 23 March 2001, entered into force 21 November 2008.

Stockholm Convention on Persistent Organic Pollutants (Stockholm POPs Convention), Stockholm, 22 May 2001, entered into force 17 May 2004.

Antigua Convention for Cooperation in the Protection and Sustainable Development of the Marine and Coastal Environment of the Northeast Pacific, Antigua, 18 February 2002, not yet in force.

Kiev Protocol on Civil Liability and Compensation for Damage Caused by the Transboundary Effects of Industrial Accidents on Transboundary Waters to the Convention on the Protection and Use of Transboundary Watercourses and International Lakes and to the 1992 Convention on the Transboundary Effects of Industrial Accidents, Kiev, 21 May 2003, not yet in force.

Tehran Framework Convention for the Protection of the Marine Environment of the Caspian Sea, Tehran, 4 November 2003, not yet in force.

Hong Kong International Convention for the Safe and Environmentally Sound Recycling of Ships, Hong Kong, 15 May 2009, not yet in force.

United Nations Convention on Contracts for the International Carriage of Goods Wholly or Partly by Sea (Rotterdam Rules), Rotterdam and New York, 23 September 2009, not yet in force.

II. *OECD Instruments:*

C(83)180/FINAL (of 1 February 1984): Decision-Recommendation of the Council on Transfrontier Movements of Hazardous Waste.

C(86)64/FINAL (of 5 June 1986): Decision-Recommendation of the Council on Exports of Hazardous Wastes from the OECD area.

C(89)1/FINAL (of 30 January 1989): Resolution of the Council on the Control of Transfrontier Movements of Hazardous Wastes.

C(89)112/FINAL (of 18-20 July 1989): Resolution of the Council on the Control of Transfrontier Movements of Hazardous Wastes.

C(92)39/FINAL (of 30 March 1992): Decision of the Council Concerning the Control of Transfrontier Movements of Wastes Destined for Recovery Operations.

C(2001)107/FINAL (of 14 June 2001): Decision of the Council concerning the Control of Transboundary Movements of Wastes Destined for Recovery Operations, as amended by C(2001)107/ADD1, C(2004)20, C(2005)141 and C(2008)156.

III. *EU Legislation:*

Council Directive 75/439/EEC of 16 June 1975 on the Disposal of Waste Oils.

Council Directive 78/176/EEC of 20 February 1978 on Waste from the Titanium Dioxide Industry.

Council Directive 82/883/EEC of 3 December 1982 on Procedures for the Surveillance and Monitoring of Environments Concerned by Waste from the Titanium Dioxide Industry.

Council Directive 91/689/EEC of 12 December 1991 on Hazardous Wastes.

Council Directive 92/112/EEC of 15 December 1992 on Procedures for Harmonizing the Programmes for the Reduction and Eventual Elimination of Pollution Caused by Waste from the Titanium Dioxide Industry.

European Parliament and Council Directive 94/62/EC of 20 December 1994 on Packaging and Packaging Waste.

Council Directive 96/59/EC of 16 September 1996 on the Disposal of Polychlorinated Biphenyls and Polychlorinated Terphenyls.

Council Directive 1999/31/EC of 26 April 1999 on the Landfill of Waste.

Directive 2000/53/EC of the European Parliament and of the Council of 18 September 2000 on End-of-life Vehicles.

Directive 2000/76/EC of the European Parliament and of the Council of 4 December 2000 on the Incineration of Waste.

Council Regulation (EC) No 44/2001 of 22 December 2000 on Jurisdiction and the Recognition and Enforcement of Judgments in Civil and Commercial Matters.

Regulation (EC) No 2150/2002 of the European Parliament and of the Council of 25 November 2002 on Waste Statistics.

Directive 2002/96/EC of the European Parliament and of the Council of 27 January 2003 on Waste Electrical and Electronic Equipment.

Directive 2004/35/CE of the European Parliament and of the Council of 21 April 2004 on Environmental Liability with Regard to the Prevention and Remedying of Environmental Damage.

Regulation (EC) No 850/2004 of the European Parliament and of the Council of 29 April 2004 on Persistent Organic Pollutants and Amending Directive 79/11/EEC.

Directive 2006/12/EC of the European Parliament and of the Council of 5 April 2006 on Waste.

Regulation (EC) No 1013/2006 of the European Parliament and of the Council of 14 June 2006 on Shipments of Waste.

Directive 2006/66/EC of the European Parliament and of the Council of 6 September 2006 on Batteries and Accumulators and Waste Batteries and Accumulators and Repealing Directive 91/157/EEC.

Regulation (EC) No 864/2007 of the European Parliament and of the Council of 11 July 2007 on the Law Applicable to Non-contractual Obligations (Rome II).

Directive 2008/68/EC of the European Parliament and of the Council of 24 September 2008 on the Inland Transport of Dangerous Goods.

Directive 2008/98/EC of the European Parliament and of the Council of 19 November 2008 on Waste and Repealing Certain Directives.

IV. *Non-binding Instruments:*

Stockholm Declaration on Human Environment, adopted by the United Nations Conference on the Human Environment, Stockholm, 16 June 1972.

Cairo-Guidelines and Principles of the Environmentally Sound Management of Hazardous Wastes, of 17 June 1987.

IAEA Code of Practice on the International Transboundary Movement of Radioactive Waste, IAEA, Vienna 1990.

Rio Declaration on Environment and Development, adopted by the United Nations Conference on Environment and Development, Rio de Janeiro, 14 June 1992.

2001 ILC Draft Articles on Responsibility of States for Internationally Wrongful Acts, adopted at the 53rd Session of the ILC (23 April - 1 June and 2 July - 10 August 2001).

2001 ILC Draft Articles on the Prevention of Transboundary Harm from Hazardous Activities, adopted at the 53rd Session of the ILC (23 April - 1 June and 2 July - 10 August 2001).

2006 ILC Draft Principles on the Allocation of Loss in the Case of Transboundary Harm Arising out of Hazardous Activities, adopted at the 58th Session of the ILC (1 May - 9 June and 3 July - 11 August 2006).

IAEA Regulations for the Safe Transport of Radioactive Material, IAEA, Vienna 2009.

About the International Max Planck Research School for Maritime Affairs at the University of Hamburg

The International Max Planck Research School for Maritime Affairs at the University of Hamburg was established by the Max Planck Society for the Advancement of Science, in co-operation with the Max Planck Institute for Foreign Private Law and Private International Law (Hamburg), the Max Planck Institute for Comparative Foreign Public Law and International Law (Heidelberg), the Max Planck Institute for Meteorology (Hamburg) and the University of Hamburg. The School's research is focused on the legal, economic, and geophysical aspects of the use, protection, and organization of the oceans. Its researchers work in the fields of law, economics, and natural sciences. The School provides extensive research capacities as well as its own teaching curriculum. Currently, the School has 22 Directors who determine the general work of the School, act as supervisors for dissertations, elect applicants for the School's PhD-grants, and are the editors of this book series:

Prof. Dr. Dr. h.c. mult. Jürgen Basedow is Director of the Max Planck Institute for Foreign Private Law and Private International Law; *President and Professor Monika Breuch-Moritz* is the President of the German Federal Maritime and Hydrographic Agency; *Prof. Dr. Dr. h.c. Peter Ehlers* is the Director ret. of the German Federal Maritime and Hydrographic Agency; *Prof. Dr. Dr. h.c. Hartmut Graßl* is Director emeritus of the Max Planck Institute for Meteorology; *Dr. Tatiana Ilyina* is the Leader of the Research Group "Ocean Biogeochemistry" at the Max Planck Institute for Meteorology in Hamburg; *Prof. Dr. Florian Jeßberger* is Head of the International and Comparative Criminal Law Division at the University of Hamburg; *Prof. Dr. Lars Kaleschke* is Junior Professor at the Institute of Oceanography of the University of Hamburg; *Prof. Dr. Hans-Joachim Koch* is Director emeritus of the Seminar of Environmental Law at the University of Hamburg; *Prof. Dr. Robert Koch* is Director of the Institute of Insurance Law at the University of Hamburg; *Prof. Dr. Doris König* is the President of the Bucerius Law School; *Prof. Dr. Rainer Lagoni* is Director emeritus of the Institute of Maritime Law and the Law of the Sea at the University of Hamburg; *Prof. Dr. Gerhard Lammel* is Senior Scientist and Lecturer at the Max Planck Institute for Chemistry, Mainz; *Prof. Dr. Ulrich Magnus* is Managing Director of the Seminar of Foreign Law and Private International Law at the University of Hamburg; *Prof. Dr. Peter Mankowski* is Director of the Seminar of Foreign and Private

International Law at the University of Hamburg; *Prof. Stefan Oeter* is Managing Director of the Institute for International Affairs at the University of Hamburg; *Prof. Dr. Marian Paschke* is Managing Director of the Institute of Maritime Law and the Law of the Sea at the University of Hamburg; *PD Dr. Thomas Pohlmann* is Senior Scientist at the Centre for Marine and Climate Research and Member of the Institute of Oceanography at the University of Hamburg; *Dr. Uwe A. Schneider* is Assistant Professor at the Research Unit Sustainability and Global Change of the University of Hamburg; *Prof. Dr. Detlef Stammer* is Professor in Physical Oceanography and Remote Sensing at the Institute of Oceanography of the University of Hamburg; *Prof. Dr. Jürgen Sündermann* is Director emeritus of the Centre for Marine and Climate Research at the University of Hamburg; *Prof. Dr. Rüdiger Wolfrum* is Director emeritus at the Max Planck Institute for Comparative Foreign Public Law and International Law and a judge at the International Tribunal for the Law of the Sea; *Prof. Dr. Wilfried Zahel* is Professor emeritus at the Centre for Marine and Climate Research of the University of Hamburg.

At present, *Prof. Dr. Dr. h.c. Jürgen Basedow* and *Prof. Dr. Ulrich Magnus* serve as speakers of the Research School.